TRENDS DUE TO CLIMATE CHANGE

THE EFFECTS OF CLIMATE, WEATHER, AND ENVIRONMENTAL CHANGES ON THE ANNUAL CYCLE OF LIFE ON THE FLORA AND FAUNA IN A SUBURBAN BACKYARD

FREDERIC BUSE

Author's Tranquility Press
MARIETTA, GEORGIA

Copyright © 2025 by Frederic Buse

All rights reserved. No part of this publication may be reproduced, distributed or transmitted in any form or by any means, including photocopying, recording, or other electronic or mechanical methods, without the prior written permission of the publisher, except in the case of brief quotations embodied in critical reviews and certain other noncommercial uses permitted by copyright law. For permission requests, write to the publisher, addressed "Attention: Permissions Coordinator," at the address below.

Frederic Buse/Author's Tranquility Press
531 Roselane Street NW Suite 400-175
Marietta, GA 30060
www.authorstranquilitypress.com

Ordering Information:
Quantity sales. Special discounts are available on quantity purchases by corporations, associations, and others. For details, contact the "Special Sales Department" at the address above.

Trends Due to Climate Change: The Effects of Climate, Weather, and Environmental Changes on the Annual Cycle of Life on the Flora and Fauna in a Suburban Backyard/Frederic Buse
Paperback: 978-1-966088-05-9
eBook: 978-1-965463-76-5

Contents

Foreword 1
Evolution 5
Weather 17
March 33
April 61
May 100
June 132
July 164
August 203
September 225
October 256
November 294
December 338
January 365
February 404
References 431

In memory of my wife Dorothy Buse

Thanks and gratitude

To Penn State Master Gardener Diane Dorn for proofreading and editing my drafts with her knowledge of the flora and fauna and English language. To the staff of the Lehigh County Penn State Extension and other Master Gardeners for their guidance and comments and Marie Mayer and Bob Braem.

Foreword

You do not have to own a garden, but just looking at one from a window or walking or sitting in one, have you ever wondered when will the bulbs rise in the spring, the geese go north, robins return, the azaleas bloom, the yearly canopy of foliage be complete, and the house wrens return? You can't wait to smell the perfume of a freshly cut lawn, see the lightning bugs rise, see the kaleidoscope of summer perennials, pick that first tomato, see the leaves change color, see the flocks of geese return, and feel the chill of the first frost. It is all "anticipation."

We all have a general idea when things happen, but I wanted a record of the occurrences. In the 1980's I started to jot down when I observed some of these occurrences, mainly what species of birds were in my yard and when they migrated. Then I added information about the flora and fauna. Soon I was doing that weekly. In 2004, I started a document in a laptop not only as notes, but also tabular. Now I could compare the yearly cycles and come up with anticipation for each week and month. Weekly may seem too finite, but there can be significant changes from the beginning of a month to the last week. I did not know until a later date that the study and observation of the yearly cycle of the occurrences is called phenology, a term which most of us are not familiar with. To me it is the gratification of the anticipation of the annual events relative to the actual happening. For over 50 years I have kept records of things that should be happening in my backyard. I have over 1,000 handwritten pages of a log. Some are about incidents and tales of events, but most are the daily observations of the flora and fauna. This book is the phenology of my backyard. It is a specific location in a suburban development in the west side of Allentown, Pennsylvania. This is not a novel; it is more of a reference book that I hope will inspire the reader to write notes or logs of their own environment so they can anticipate the happenings of their cycle. The book is in two sections. The first portion is about years of the evolution of the garden from bare acreage to a mature garden. Why? So, the reader can obtain a flavor of the "lay of the land" for the next section. The following picture is what the environment looked like in 1969. Then there are pictures of spring years later when basic maturity of the garden had stabilized, but not stopped.

1969 – the beginning

FOREWORD

Statistically, there are over 35 years of data about the birds, animals, insects, perennials, and weather. It is hoped that the reader will pursue them for their own interest.

2015 – 46 years later, mature, and still changing

Spring

The second section is the monthly-weekly anticipation of the fauna and flora starting in the month of March, which is the reawakening of life. It is the summary of over 40 years of the observations and data in following format.

General Occurrences for the Month

Each month is arranged the same way.
Starting with the week numbers of the month.
Amount of daylight. High and low record temperatures.
Overall description of the month.
Weather of the month.
Occurrence during each of the four weeks.
Review of the data.
A short overall summary of the month.

Information that is Reviewed Each Week

The information is arranged alphabetically. First listed by fauna, then flora. Not all the subjects may have an entry.

<u>Fauna</u>
Animals
Birds
Insects
Pond

<u>Flora</u>
Weather
Annuals
Fruits
Perennials
Pond
Shrubs
Trees
Vegetables
Review of Data
Fauna
Flora

At first, I thought that the yearly occurrences would be a cyclic repetition by date, but then at closer inspection of the data there is a change in occurrence dates. I first noticed that there was a change in the dates when the peak color of the leaves occurred. From 1985 to 2024, the date of peak color is over two weeks later. Then I reviewed when migrating birds arrived or passed through and again it is at a later date. The same thing is occurring as to the date when I picked apples. There is not always a forward movement of dates every year because weather is the big factor. But from the big picture, dates of general occurrences are later. This all adds up to climate change. How much of this change is cyclic, and how much is due to man's contribution is still up for discussion. It makes for an interesting prediction for anticipation.

Number of Weeks in A Year and How Numbered

The traditional calendar has 52 weeks, because most months have more than 28 days. However, these extra days of the 29th, 30th, and 31st when using excel to make graphs or charts did not work out when an occurrence of flora or fauna spilled over from one month to another. I have used a four-week, 28-day month, for every month, which gives me a 48-week year. My last week can have seven to 10 days. The other difference is my year starts when the earth is reborn with life, which is March. My first week is the first week in March. As the chapters proceed, I continue the numbers to the last week of February, week 48. This will be used for reference to the period for the cycle of birds, plants blooming, and length of the canopy.

Numbers for the Weeks of the Month				
Week of Month	First	Second	Third	Fourth
Month	Week Number			
March	1	2	3	4
April	5	6	7	8
May	9	10	11	12
June	13	14	15	16
July	17	18	19	20
August	21	22	23	24
September	25	26	27	28
October	29	30	31	32
November	33	34	35	36
December	37	38	39	40
January	41	42	43	44
February	45	46	47	48

How monthly information is arranged.

Evolution

This section of the book describes how my change over evolution of the flora of the garden has changed the type and activity of the fauna.

The photos are ones I have taken. Birds and animals do not pose for a first photo, no less a second shot. The incidents and tales that happen along the way of the evolution can be complex or long.

The location of the garden could be general for a state, a county, a park, or a town. It is not. It takes place in my backyard. A lot has happened in this small environment, from a bare yard with a few trees to many trees that are now 60 to 70 feet tall; from no squirrels, chipmunks or hawks, to a wish there were less of them; an annual bird population of a dozen to 53 to 30 then back to 40 plus; planting trees in the wrong spot to cutting them down because they were too close to the house. The main objective of the book is for the reader to become aware of the nature activity they may observe by just looking out a window or stepping outdoors.

To comprehend the chapters, the reader needs to understand the "lay of the land" of the backyard and the terms I have labeled for the location of flora and fauna.

The company I worked for transferred me in 1969 to Allentown, Pennsylvania. We bought a home in a two-year-old development that was five miles west of the city. The property is three quarters of an acre that is on a partial hill. Being a new development from a corn field there were only a few trees planted by the developer.

In 1982 I started to keep records of the vegetable gardens, but also of the flower gardens, shrubs and birds. As time went by, the wildlife records not only included the birds, but also the animals and insects. I extended the database to the weather and the daily temperatures. I recorded when the perennials and trees came to life in spring, their rate of growth, when they bloom, fade and close for winter. What color do the leaves of the trees and shrubs change? When is the peak color and when do the leaves fall? I recorded when the lightning bugs first appear, when the bats emerge as the sun sets, where the sun rises and sets. Who are the early risers and when is their regular feeding time; how many types of birds are present in the spring, summer, fall and winter; and which species pass through in the spring and fall? I call them the bed and breakfast species.

This writing is neither a bird book nor a flower book, it is the gathering of what can and does happen within the many disciplines of the annual cycle of life in the environment in a yard of a suburban development. An environment that has been continuously changing not only by my hands, but also from Mother Nature.

From 1969 to 2003, I recorded the development and observations in a hand-written notebook. From 2004 to 2020, I kept more precise records in my computer file. The diary includes the morning temperature and weather condition, and daily animal, bird, insect, and plant activity. I wrote where and when the sun rose and set (I only had to do this once). Besides the sun, I concluded that the weather is the **major** factor that rules the changes in the seasons. The falling of the leaves will vary by the conditions of the weather in September and October. From the data, I show how the life cycle of the flora has changed with climate change. The events within the cycle are basically constant, but not always close to the same date.

There are the four distinct seasons which in our minds are always the same. From all data, I conclude that there are no normal seasons. Within a season the outcome changes with the weather and environment. We know that spring gives life, summers are hot, fall gives us bounty of the growing season, and winter is cold. Within cycles of the seasons there is no consistency, but a close similarity. As I changed the flora, how it grows and dies, so does the fauna change, be it birds or animals. Most of our home plants are not wild, we imported them, resulting in the evolution of our own desired environment.

The weather can and will be as calm or as wild as nature wants it to be. The environment of the "circle" was developed for the encouragement of visitation and habitat of the wildlife and does not represent a typical backyard in a development.

For the reader to understand the terms and references in the "chapters," I have written an abbreviated history of the development of the basic environment or "lay of the land" of the circle and then present a month by month description of the Anticipation of the Cycle.

This property is on the north side of a cul-de-sac circle, hence the term the circle. A cul-de-sac is an area of a development which is a dead end that has a circle at its end and in this case four houses on its periphery. My land is located on a hill that rises about 24 feet from the base of the circle.

The circle

An aerial view of the property shows that it shaped like a triangle that has been pushed to the west with its base on the circle. The long side of the triangle has five neighbors attached to it. The other side of the triangle has one neighbor.

There was not much of a workable backyard. The trees the developer planted were a couple of maples in the front, a 15-foot birch and a flowering crabapple in the back. The previous owner had put in a concrete block patio and had a nice canvas awing over it. For privacy around the patio, the owner planted arborvitae on two sides and had a six-foot-high wooden fence on the other.

Excavated hill. Note the braced weeping willow, arborvitaes, spindly birch, and moved wooden fence.

I decided to expand the yard back another 25 feet. It would be shaped like an amphitheater from one end of the house to the other. I hired a backhoe contractor. I showed him what I wanted. He came while I was at work; halfway through the morning I got a call that he was having problems. His backhoe could not dig into the pan-hard rock-clay soil as far back as I wanted. He was only able to excavate back 17 feet. I told him to make three terraces for the rest of the distance. This was a blessing in disguise. If he had gone back the 25 feet, I would have to put up a four- to five-foot retaining wall that would have needed supporting. Second, I would have a bad drainage situation. This was my first encounter with clay and small built-in rock soil.

This is the beginning of the evolvement. Now that I had this raw terraced hill, the surfaces still needed to be faced. I thought of using railroad ties and then brick. Then I found that there was a limestone quarry five miles away. Using a half-ton pick-up on the weekends I literally picked up 10 tons of rocks to make retaining walls. Now-a-days, for insurance reasons, they do not allow individuals into the quarry.

Tossing the rocks

Beginning of 10 tons of limestone

1969 – round one of wall building *1971*

We were beginning to enjoy the back patio, but we realized that in the evening mosquitoes were enjoying us. The mosquitoes used the arborvitaes next to patio as their haven. Moved 10 arborvitaes and fencing to the east borderline.

I now had a three-tier terraced "area." I planted Canadian hemlocks on the periphery and a weeping willow in the center of the third tier. Dug out the clay pan and replaced it with topsoil so I could plant grass on the newly-leveled area.

The back of the house faces north. The back garden is now an amphitheater. Think of it as a theater with a reversed stage. We are sitting on the stage and the garden is where the chairs would be. Except these chairs are small trees and shrubs we hoped would be filled with wildlife. The semi-circular extremity of the amphitheater is a hundred feet long that terminates with gates at each end of the house. The theater is defined by a four-foot-high hedge of Canadian hemlocks and arborvitae. The front row seats are the lawn that goes for the length of the house. From there, the gardens raise up three levels (tiers), like orchestra seating.

1971 – note how the birch, hemlock, and willow have grown

In 1970 there were no hawks, squirrels, chipmunks or skunks, but there were mockingbirds, pheasants and bobwhite. It was a sight to have a harem of ring-necked pheasants cautiously stroll across the lawn on a spring evening. The male would always strut in last. Unfortunately, pheasants across the nation got an intestinal disease and they died off in about five years.

The birch in the back was now 30 feet high. The willow on the third tier was 20 feet high and bushy. The hemlocks that were four feet are filling out.

I planted rhododendrons by the back of the house. In five years, they grew to be six-foot high. They not only provide beautiful flowers, but also offer protection for the birds.

The spruce trees on the hill that were four feet when planted are now 60 to 70 feet in 2016. The hedge of Canadian hemlocks grew to over seven feet; I keep them pruned to five feet.

1971

June 1974

July 1974 – the wall is not as straight as when it was first installed. Now have added steps going to the upper garden. The birch and willow are growing fast.

We wanted more flowers along the first wall; however, I knew that the soil in that area was clay and small rock. I dug out three feet of lawn along the wall and made a raised garden using railroad ties.

1972 – note the willow and lower basin for the water works

1973 – the lower tier is now in; hundreds of daffodils fill the upper garden. The rock walls are in bad shape.

The arborvitaes that I planted on the east boundary were now eight feet tall. I got an inexpensive motley-looking weeping willow tree and put it up at the far end of the upper hill.

In 1975 I decided that I wanted a substantial shade tree on the hill, one that would grow tall and shade the amphitheater in the summer. I chose a pin oak. I was advised that a pin oak may not grow in clay soil because the taproot wants to go straight down. I decided to go ahead. Well, 50 years later, it is 75 feet tall and still growing. And every November it drops thousands of red leaves on the garden.

The willow on the hill is now 25 feet tall; I tell you this bad-looking soil has the right growing stuff. The twig-of-a-maple is almost 30 feet. The diameter of the willow in the third tier is over a foot in diameter. Bought a dogwood tree. The sales lady said that I should put a rock under the root bundle to force the roots to spread. I complied; I put it on the first tier just west of the water basin. In 2022, it died.

In the back, the weeping willow is now 40 feet. The long weeping, swinging yellow branches seem to glow in the March sun. In early June, the Baltimore oriole lands there on its migration. The maples on the second tier and the one on the lawn are 25 feet. The arborvitaes on the east side are now 10 to 12 feet tall.

1977 – the willow tree in the winter. Note how tall the spruces are to the right of the willow. That is a maple in the middle of the lawn. By April of 1979, the weeping willow on the hill has now gotten to the size that I built a treehouse in it.

The two maples on the third tier are 40 feet tall. By 1980, the spruce trees on the hill had grown to 12 to 16 feet. If I only knew how high they were going to be when I planted them.

It was 1985 the first time I put the second level wall up. I piled the rocks on top of each other and filled behind them with our local clay soil. When this stuff settled in, it packed like concrete. As a result, after it was frozen and thawed out, the fill kept pushing and knocking down the wall. Even when I went from a wall that was straight up to one that was slanted back, Mother Nature kept knocking it down. I switched from rocks to wood planks. This was better, but after a while Mother Nature pushed the wood over. Finally, I dug out behind the planks and put the them on a slight backward angle, using lengths of steel pipe to hold them in place, then I used topsoil with compost as back fill. It took 15 years to get that far, however, it is still holding in place 23 years later. In 2000, I tore out most of the rock walls and rebuilt them; they have stayed in place. In 2024, the wall is still okay.

May 1982 – japonica in bloom. Basin surrounded by a creeping cedar.

1982 – The birch replaced the willow. Willow on the hill is 50 feet.

June 1983 – the garden is lush. Note the dogwood, cattail, lupine, coral bells, and tall hemlock hedge.

The Pond

I had put in a small basin as a fishpond but then I decided to put in a larger pond next to the existing basin. I was going to dig down three feet. After two feet, the ground was just too hard, so I raised the pond above ground by over a foot. This was another Godsend because it would be safer for visiting children. I lined the hole with layers of newspaper and peat before putting down a protective tarp. This was done to prevent unseen sharp stones from puncturing the liner as the weight of the 600 gallons of water pressed on the bottom.

July 1988

What have I gotten myself into?

.06-inch thick polypropylene liner

Finished! Plants and circulating water.

*July 1989 – a year later, 20 years from the beginning.
Note the size of the dogwoods and birch.*

Now the pond is the focal point of the garden. It attracts adults and children like a magnet. Especially since I added fish and frogs.

"If you build it, they will come." In 1995 a pair of ducks decided to have a nest next to the pond. They came in May and ducklings were in the yard by June. This mother duck was nasty. She would hiss and come after us if we put out food or got close to her brood.

May 1995 – the string is to prevent birds from landing on the pole.

June 1995 – up and running because when they did, they would relieve themselves.

August 1995 – a mourning dove is literally feeding out of my hand

October 1996 – silver maple

October 1996 – pin oak in back planted 1982

To the left is the maroon leaves of a dogwood. In the foreground is a holly that replaced a white pine that got too big and was too close to the house. The holly is now 40 feet, but not too close to the house.

From the hot-cold cycle weather over the years the concrete basin was leaking. It got to the stage to replace it. In April 1997, purchased a molded basin.

Replacing the concrete basin with a molded one

May 1997 – azaleas transplanted in 1969. Still blooming. They are least 70 years old.

After many years, the railroad tie border was rotting out. By April 1999, anything with creosote was a no-no. I paid someone $75 to haul the old ties away. I replaced them with two tons of slab rocks.

1999 – stone to replace the railroad ties

Then to the right

2004 – tree growth after 35 years

2005 – mountain pink and PJ are out

EVOLUTION

May 3, 2007 – now the change can really be seen

July 2019 – lush

That gives a description of the basic evolution of the circle.
The next chapters are the **anticipation** of the 12 months from 2005 to 2024.

Weather

Weather is caused by unequal heating of the Earth's surface by the sun.

After reviewing the weather data, I realize how much we, the plants, the birds, and animals are dependent on the weather.

The weather regulates how we dress, eat, drive, shop, vacation, heating and cooling of our home, when to use pesticides on the plants, when we plant, when we harvest, what kind of recreation we do, when we shop, when to buy salt to put on the driveway. The business section of the local newspaper in February read, "Warm weather in January made retail sales a gem." The article attributed the strong sales to the mildest January in more than a century.

On an early windless spring day, when the temperature is above 50F, I spray dormant or horticultural oil on fruit trees and shrubs. Put down fungicide for grass mold hopefully before the humidity sets in July.

When the summer is too hot and dry, there are times when there are not enough insects for the parent birds to feed their fledglings. I have heard that the birds can sense this before it happens and only lay the number of eggs according to what they think they will be able to feed! Likewise, the number of insects for those evening bats to feed on will be less. The summer of 2005 was hot and dry. To substitute the insects to feed the fledglings, the amount of peanut butter that the birds consumed was high and continuous.

The movement and location of the jet stream is what affects when we have cold or hot weather. The fronts coming across the country from west to east and how they clash with the fronts coming from the south spawn the tornadoes and produce most of our winter storms and blizzards.

The birth of the storms off the west coast of Africa that go east to our coast or into the Gulf of Mexico from June to November can grow into hurricanes.

The size and location of El Niño and La Niña that form in the south Pacific contribute to the birth, path and strength of our winter storms and nor'easters.

Cause of the Seasons

Standing in the same locations in my house for sunrise and sunset, I observe that the sun rises in the southeast on December 21st, the shortest day of the year. The sunrise moves north each day until Mach 21st (equinox). Equinox is the day of equal light and dark. On the first day of spring the sun is coming up about due east and then it continues north every day until it comes up in the northeast on June 21st, the longest day of the year, at which point it reverses its progress and starts south. On September 21st (equinox), the first day of fall, it sets almost due west. It then progresses south back to the southeast until December 21st, then it reverses course. It sets in the southwest, it then continues to the northwest until June 21st.

What happens that causes the change in seasons in the upper and lower hemispheres?

We all know that the northern and southern hemispheres have seasonal changes, but do you know how it is done? The earth orbits around the sun approximately every 365 days and rotates around its axis in 24 hours. Knowing how and why the seasons change is another question. Get a dinner plate or a drawing compass, a 12- x 12-inch piece of plain paper or cardboard, a straight edge; ruler; an orange, lime or small soft ball; a skew or chopstick; a marker or pen or whiteout tape.

First, let's make the "orbit." Lay the paper on a flat surface, place the dinner plate upside down in the middle of the paper or carboard or use the compass. Trace around the edge of the plate to make a circle. Remove the plate or draw a six-inch radius circle with a compass. Mark where 12:00, 3:00, 6:00 and 9:00 o'clock would be. Write "winter" at 12:00, "spring equinox" at 9:00, "summer" at 6:00, and "fall equinox" at 3:00. Draw a straight line from 12:00 to 6:00. Set this aside.

Second, we make "Earth." Take the fruit or soft material ball and pierce the skew or chopstick through it from top to bottom so the skew protrudes three inches. The skew represents the Earth's axis that it spins around. Holding the ball straight up on its axis, mark a horizontal line around the center as you spin of the Earth around its axis; this is the equator. Mark the top of the ball "NP," north pole and the bottom "SP" south pole. Draw a vertical line from the NP to SP. Now at a distance a third of the length from the NP to the equator, put a big dot and mark it "NH," northern hemisphere. Repeat this for the SP along the same vertical line but mark it "SH," southern hemisphere.

Place the orbit on a flat surface, put the fruit or small ball, the sun, in the middle. Now place the vertical Earth at four o'clock, taking note that there is no change in the equator position, it is always horizontal. Likewise, the northern and southern hemisphere are always in the same horizontal positions.

Now comes the hard part to grasp that our Creator had the wisdom to put in place. Why the following stays in place, I do not know, but it does, and I accept it. Put the vertical Earth at 12:00 and push or tilt the axis away from the sun 23 degrees, about a quarter to the distance if it was pushed flat to the paper. That is the position of our Earth. The equator is no longer horizontal to the paper orbit. This is NH winter and the SH summer. When you spin the Earth around its axis note how short the day (nine hours) is for the NH. Looking from outer space, we would see that with the tilt of the planet that South America and Australia are having summer when we have winter and vice versa.

Now move the Earth along the orbit toward 3:00 but always keeping the tilt in the same direction (parallel to the line you drew from 12:00 to 6:00) as it was at 12:00. When you arrive at 3:00, it is spring equinox, equal day and night. As you spin the Earth, the sun always stays on the equator. Remember the tilt is still parallel to the 12:00-6:00 line, pointing in the same direction as it was at 12:00. The tilting also results in position and length of shadows changing as the seasons change. The exposure of the ground to the sun affects how the soil warms up in the spring. Keep moving along the orbit, keeping the tilt in the same direction. When you arrive at 6:00, the tilt is now pointing toward the sun. This is NH summer and SH winter. It is the NH longest days (15 hours). In the winter when the sun sets, it is almost immediately dark, but in the summer in late July and beginning of August, we still have light for almost an hour after official sunset, this is due to the tilt of the planet that we get the late sun rays. Now move along the orbit from 6:00 to 9:00, keeping the tilt in the same direction and parallel to the 12:00-6:00 line. At 9:00 it is now the fall equinox. Now go back to 12:00 for winter to complete the cycle. Without the tilt which is always held in the same direction we would not have the four seasons in the NH and SH.

Many people think that the coldest part of the night may be 2:00 to 4:00 a.m., but in the winter it is just before dawn, maybe 6:00 to 6:30 a.m. As the sun comes up to warm the Earth, there is a semi-quick rise in temperature. This difference in temperature will produce the morning chilling breezes.

In the summer, you probably observed a rainbow after a thunderstorm. No rainbow can be produced at high noon because the sun is directly overhead, and refraction cannot happen. During the summer of 2006 around 4:00 p.m. the sun was low in the sky, and I happened to look directly overhead to see an upside-down rainbow; it is called a circumzenithal arc and has all the colors of a regular rainbow. It is caused by ice crystal clouds. When there is an exceptionally bright moon after a storm, it is possible to see a rainbow at night.

The moon also has an effect on the weather and how we behave. Have you heard that during a full moon in May, some farmers believe that it is best to plant at night? Or farmers using the light of an October full moon to harvest, or as it is called the "Harvest Moon?" Have you ever noticed a halo around the moon on a clear winter night? If so, get ready for a storm within two days. The halo is formed from ice crystals.

Birds and squirrels come to the feeders in droves just before a rainstorm. They will slow down or not be at the feeders during the storm. As the rain lets up the birds will be back at the feeders. However, during a blizzard they feed in droves of three times the normal population. Very few are to be seen the day after the storm because they had pigged out during the storm. The second day after the storm they are back to normal population, feeding, and activities.

Insects too are more active before a storm. I have not observed this but have heard that ants build dikes in front of their nests. An old saying is, "A bee's wing never gets wet." They will be very active just before a storm and will disappear as the raindrops start to fall.

On a clear morning most grass will be wet with dew, but if the grass is dry it is an indication that we are under a low pressure and rain is coming. If you notice that the flower scents are vivid, it is a sign that rain is on the way. High humidity before a storm allows the scents to travel better. When a storm is coming the tulips close in the spring and morning glory close in the summer. When you see the bottom of a red or silver maple, or silver (Lombardy) poplar leaves it means that a storm is coming. The leaves are normally hanging to the prevailing wind and when there is wind change, they show their bottoms. The wind change is caused by the reduction in the barometric pressure.

How about this one? A sign of rain for farmers was when their pigs picked up sticks and began walking around with them in their mouths.

I strongly go by old sailor's adage:
"Red sun in the morning, sailors take warning.
Red sun at night is a sailor's delight."

For me, this has proven to be more correct than not.

Temperature

The newspaper gives three temperatures; the actual high and low temperature of the preceding day, the normal high and low temperature of the preceding day, and the expected extreme high and low records of the present day. The average difference between the normal high and low for a day is 16F in the summer and fall. In the winter it increases to 24F, in late spring it is about 20F.

The high temperature for a day increases from month to month, increasing rapidly in the spring, it levels off in the summer, and then decreases rapidly in the fall into winter.

Month	Jan	Feb	Mar	Apr	May	Jun	July	Aug	Sep	Oct	Nov	Dec
Temperature change from beginning to end of month	35-36 1F	36-42 6F	42-55 13F	52-65 13F	66-76 10F	78-81 3F	82-84 2F	84-79 5F	78-70 8F	70-58 12F	58-45 13F	45-36 9F
Average high monthly temperature	37	37	49	54	65	72	78	76	70	63	51	45

Yearly average temperatures from 1992 to 2019

To review a bigger picture of what is happening, I determined what is the average temperature of the whole year. The average of the 12 months ranged from 50 to 55F. The trendline from 1992 to 2023, shows a three-degree rise, which is a lot, resulting in future environmental changes.

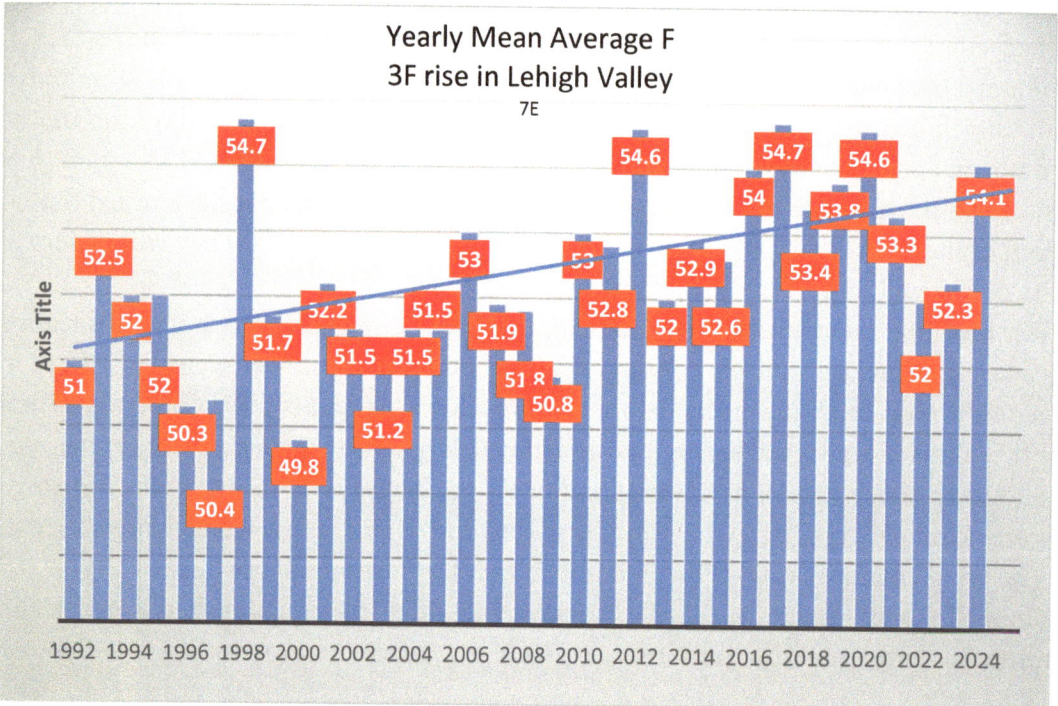

Yearly average temperatures from 1992 to 2024

March is the beginning of the cycle. It is when things come to life. The heads of the bulbs force their way above ground. Shrub and tree buds start to swell, which means the sap is rising from the roots to the tip of the upper twigs. Sap in the sugar maples is flowing. The daytime temperatures are usually above freezing. Yet, this is the month of the heaviest snows. This is the month of the quickest snowmelts, resulting in major flooding, especially if the ground is still frozen.

The quick rise in temperatures in April and May leads to spring showers, thunderstorms, and possible tornadoes. The low temperature for a day increases rapidly in the spring (20F in January to 62F in July), levels off in the summer, and then decreases rapidly in the fall into winter.

The mean temperature is arrived at by taking the average for the high and low for every day of the month and then calculating the average for the month. Plotting the data for a 32-year period shows the trend of monthly temperature. An example is March.

TRENDS DUE TO CLIMATE CHANGE

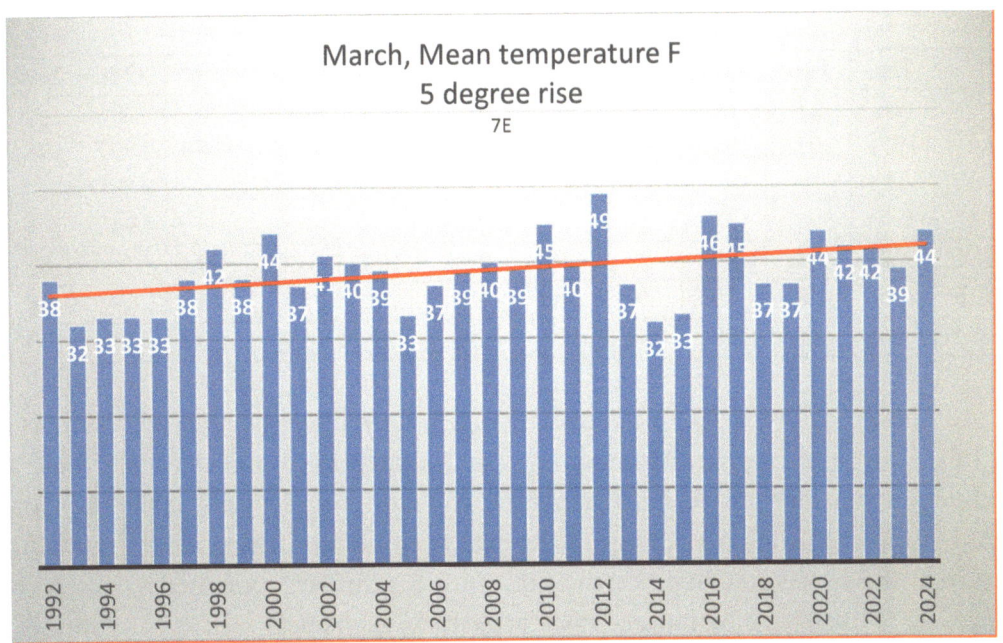

March mean temperature

The trendlines show hotter springs, summers, falls, and colder Decembers and Januarys.

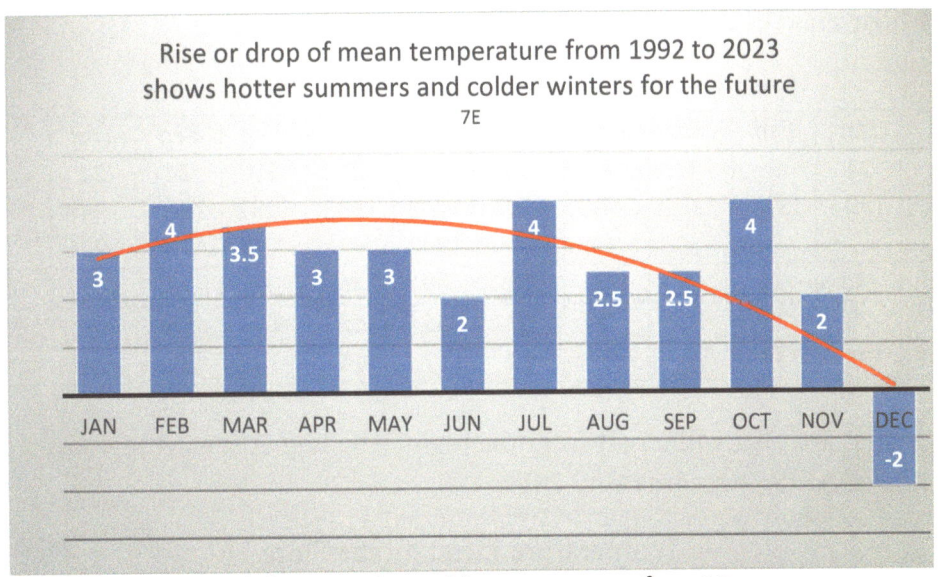

Increase and decrease of monthly temperatures for a 30-year span

December has a four-degree drop in temperature. No wonder the heating fuel is disappearing. July with almost a five-degree increase does not fare well for the air conditioner.

The average low March temperature at the beginning of the month is below freezing, but near the end of the month the night lows are above freezing, so plant your peas and onions on Saint Patrick's Day.

Then there are the extreme high and low temperatures. These are record temperatures and year of occurrence that have been reached on a given day from the 1920's to 2018. These extremes are another factor that make predicting weather more of an art than a science. Depending on the time of the year, the extreme value may be 20 to over 40 degrees different from the normal.

Extreme High Temperatures F

Month	Jan	Feb	Mar	Apr	May	Jun	July	Aug	Sep	Oct	Nov	Dec
Day	26	21	18	19	19	6	104	2-5	2	2	1	1
Year	1972	2018	1926	1926	1962	1925 20 1923	2001	1955	1980	1927 5 1942	1951	2006 4 1990
Extreme F	72	81	95	93	97	100	104	100	99	9	81	72
Normal F	36	40	49	59	63	76	83	81	74	69	58	46
Difference F	36	41	38	34	34	34	21	19	25	23	23	26

The extremes may only last a day or two. Many times, they are associated with storms or unsettled weather. In June, July, and August they may represent a portion of a heat wave. These are the pure numbers and do not consider the "feel like" temperatures from frigid winds or the 3H's (hazy, hot, and humid) of mid-summer. A high percentage of the extremes, hot and cold, occurred during the 1920's-1949. As a boy in the 1940's I remember the hot, humid summers and then heavy snows and cold winters. In the 1930's-1960's the high percentage of extreme highs occurred in the summer, then in the 1970's-1980's, except for September, the percentages were pretty much even through the months, but in the 1990's the higher percentages are in the winter. In the 2000's-2010's the percentage are much higher in the winter. In most cases the new values in the 2000's-2010's replaced a former high of the 1930's-1940's which could be due to climate change.

Extreme Low Temperatures F

Month	Jan	Feb	Mar	Apr	May	Jun	July	Aug	Sep	Oct	Nov	Dec
Day	21	9	1	1	10	1	5	29	24	22	26	28
Year	1994	1934	1934	1923	1943	1995 1932	1933	1982	1963	1940 1936	1938	1950
Extreme F	-15	-12	-5	12	28	39	46	43	31	21	3	-8
Normal F	20	22	22	38	47	56	62	62	53	47	37	29
Difference F	35	34	25	50	19	17	16	19	22	26	34	37

The low extremes are totally different from the highs. The highest percentages were in the 1925 to 1940's range and still hold the records as of 2019.

Temperature Effect on the Flora

The blooming and leafing of the various species of flora through the cycle depends on the temperature and precipitation. The temperature seems to be the more prominent element.

In the winter, many times, we have abnormally high temperatures that will melt snow cover, causing flooding. Warmed perennial bulbs will force foliage through the frozen ground. The cycle of high to low temperatures will kick perennials and bulbs out of the ground. Hot to warm air against a cold ground will give a false sense that spring has arrived. Sap will rise and freeze as temperatures drop back to normal levels.

With heat waves and drought, plants will wither or not blossom. Fall's hot air brings fire alerts. Fall frost and freeze will end the growing season.

Rain

We look at monthly and annual rainfall. The average rainfall in the Lehigh Valley is 45.4 inches. That is nice but can be very misleading from a flora standpoint. I prefer snow to rain in the winter because the snow is an insulator for the flora. Then comes spring warming the snow; it gradually melts and soaks into the ground. If the ground is bare and frozen, a heavy rain is just a runoff causing river ice to break up, ice jams and massive flooding. When we have no rain, the ground gets like concrete and cracks. Then when it does rain, it runs off. In late 2017-early 2018 we had lots of rain in spring and none in the summer or fall. Yet at the end of the year records show that we have normal or even excess rainfall.

Review of the records have shown:

Excess rain in March of over two inches will delay bloom by one week.

Excess rain in April of five inches will delay bloom by two weeks.

Excess rain in May of five inches will delay bloom by three weeks.

Excess rain in July of five inches will delay bloom by two weeks.

Excess rain in August, October or September of three inches will not cause change to activity.

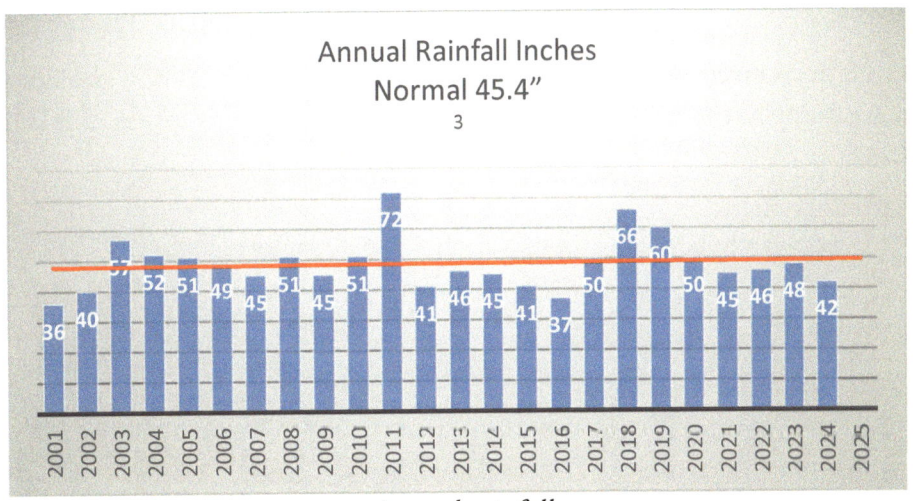

Annual rainfall

A monthly summary may be meaningful as to what may happen. Reality is another story.

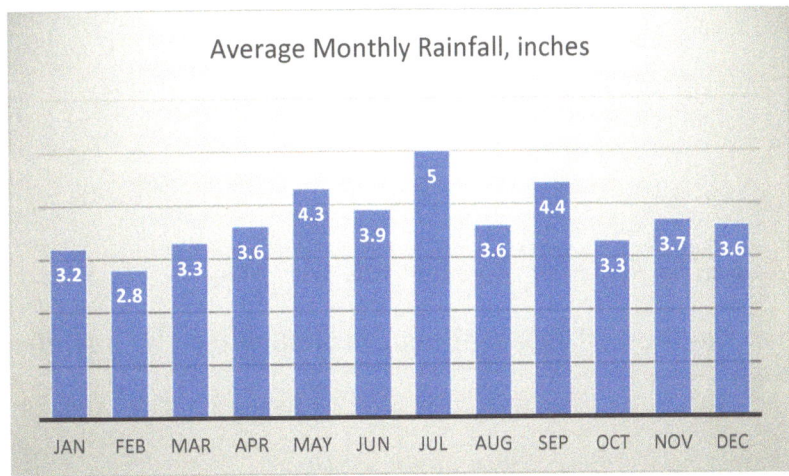

Average monthly rainfall

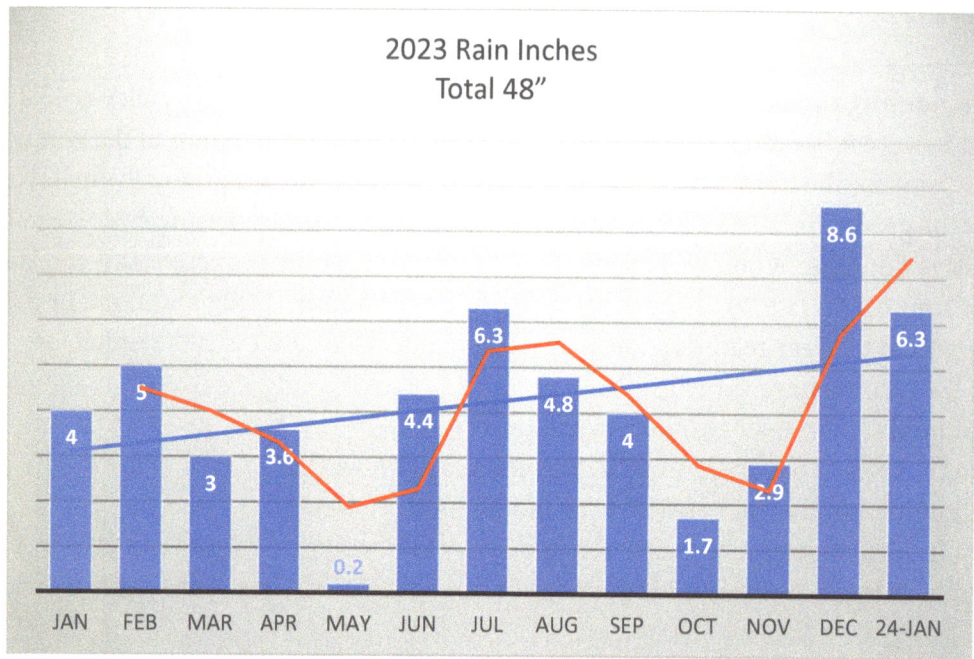

The amount of rain received in January, February, and March dictates how the flora, especially the deciduous trees, are going to perform for the rest of the year. Going into spring, this is the summary of the three months.

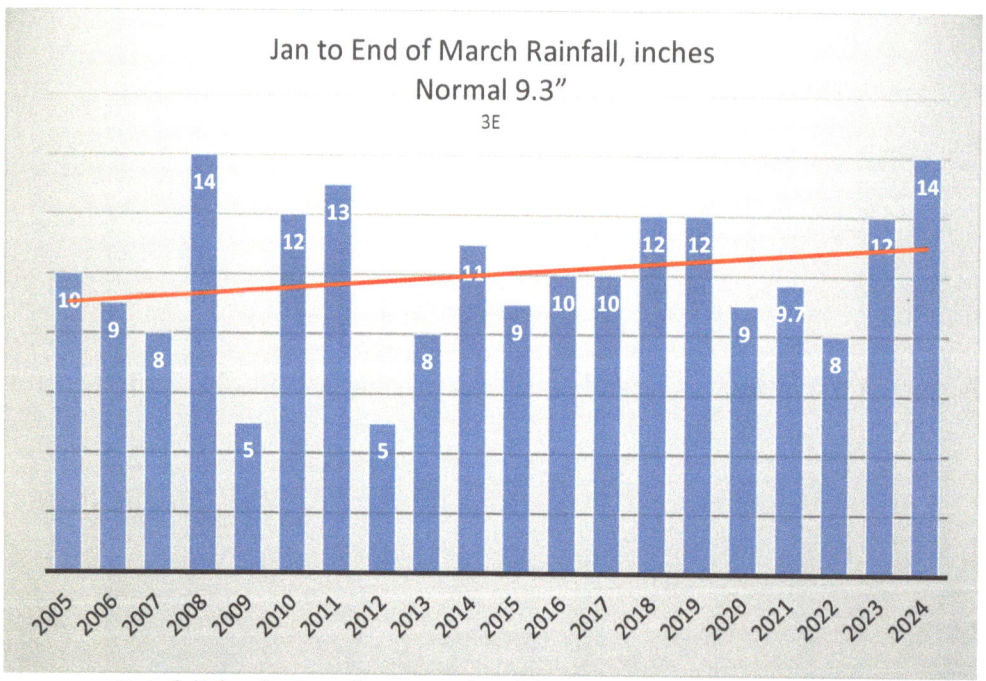

Rainfall for the first three months of the year. Many years look lean.

A more dramatic way to show what is happening is plotting the plus and minus amounts of rain from the start of the year.

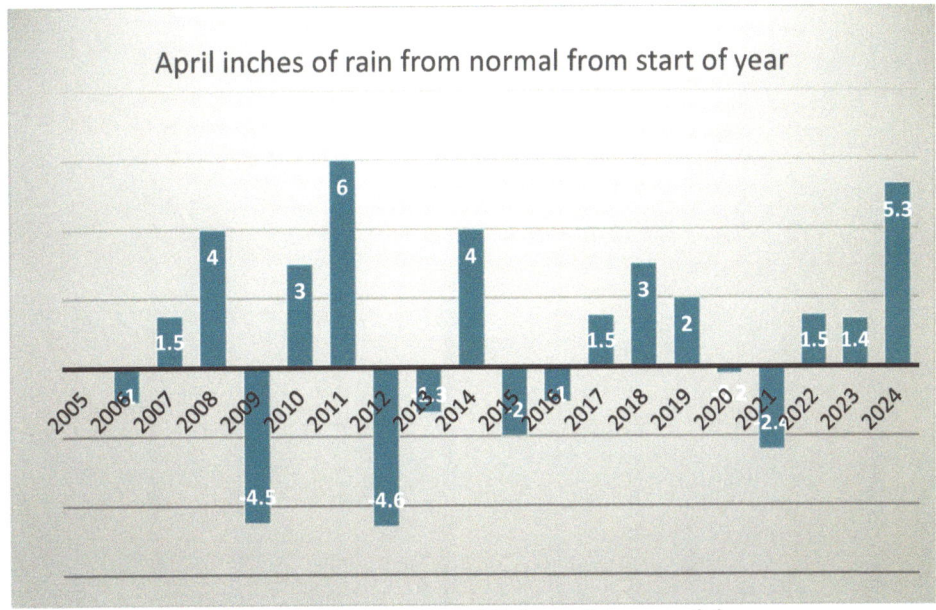

April plus and minus rainfall from the start of the year

Of course, this is all hindsight. The art or science on how to anticipate what is going to happen is the real challenge.

Snow

Snow season starts in September and goes into April. Usually, we don't receive snow until the last week of December. In the past, most of the snow fell in January and advanced to February, but by 2015 the greatest amount fell in March.

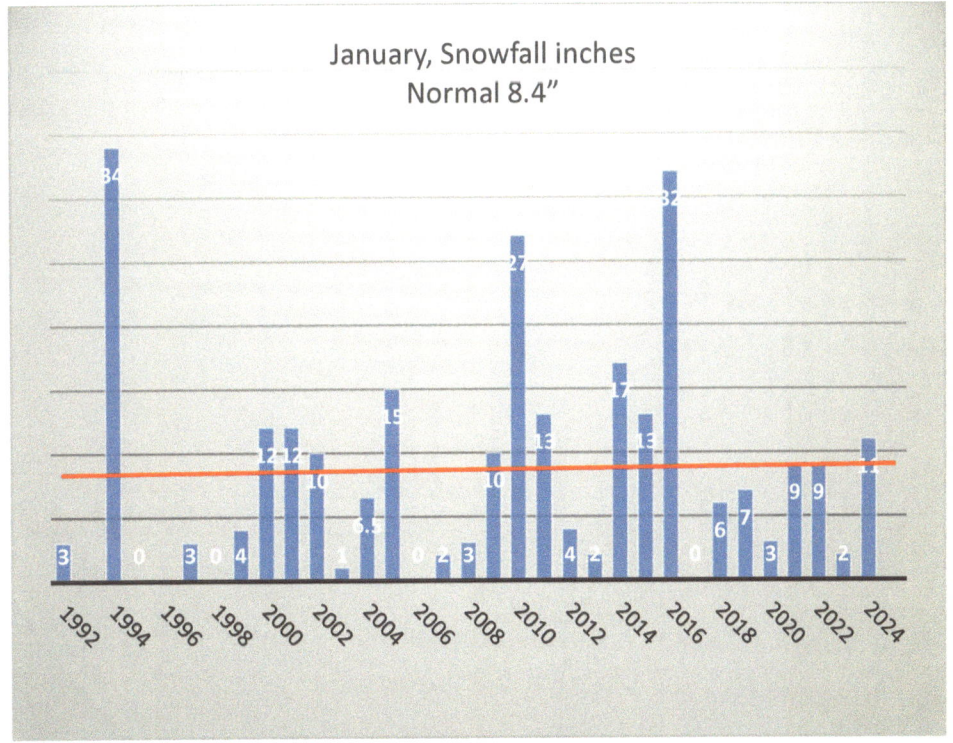

January snowfall

WEATHER

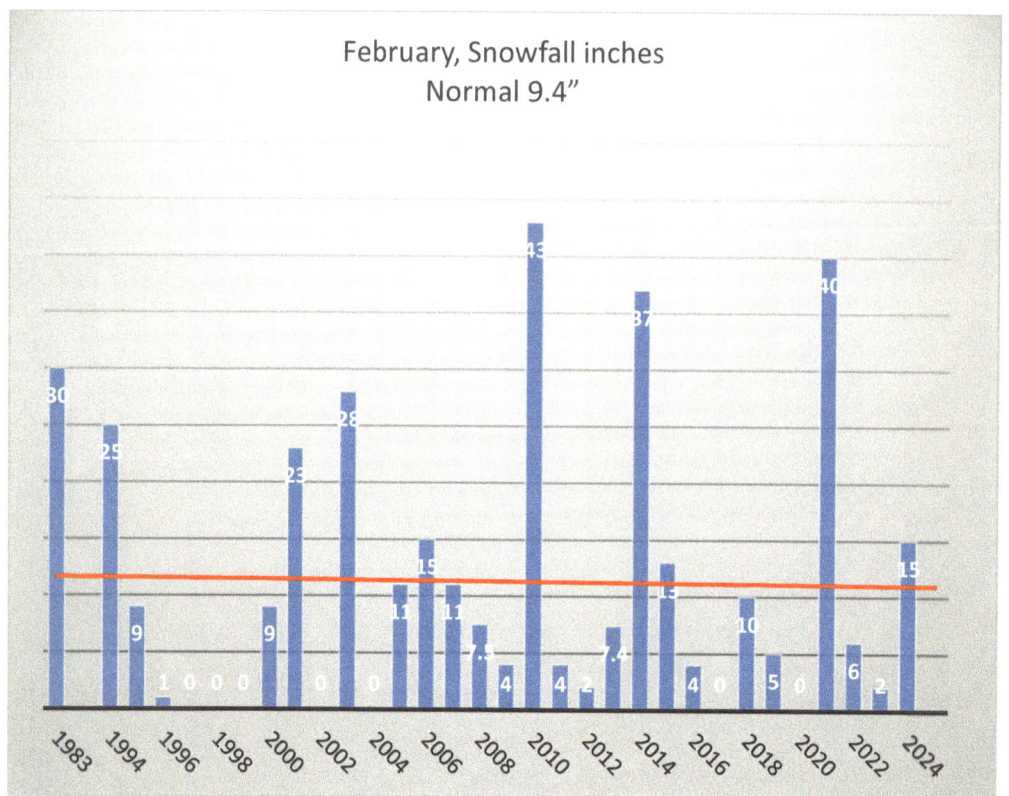

February snowfall

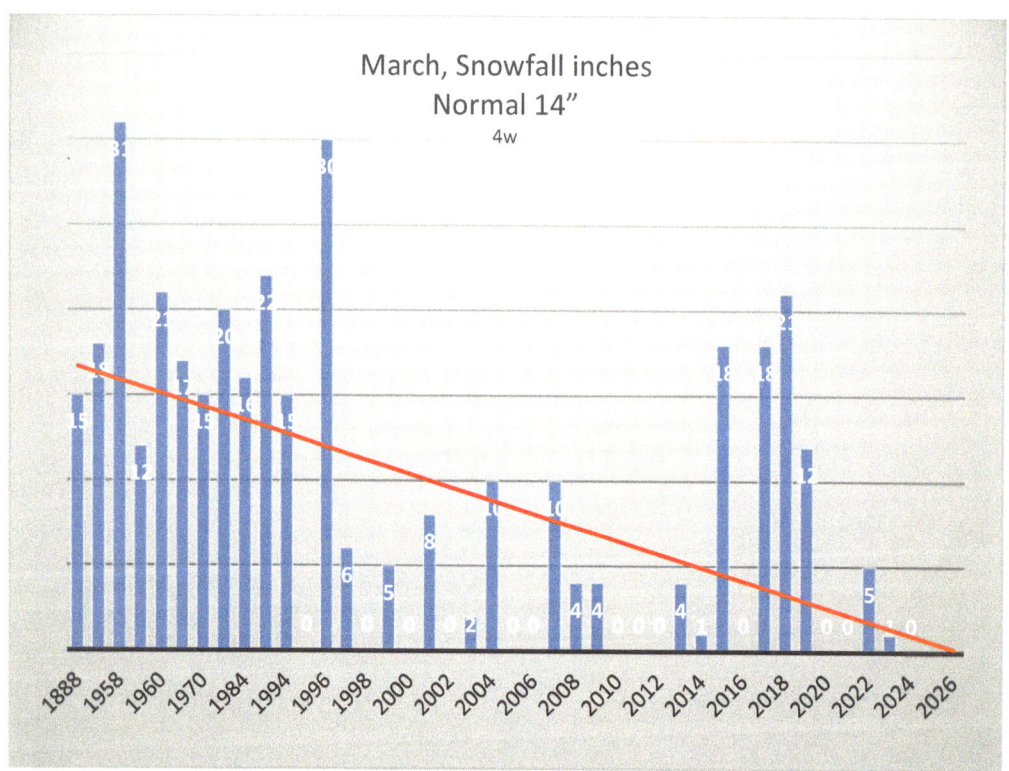

March snowfall

TRENDS DUE TO CLIMATE CHANGE

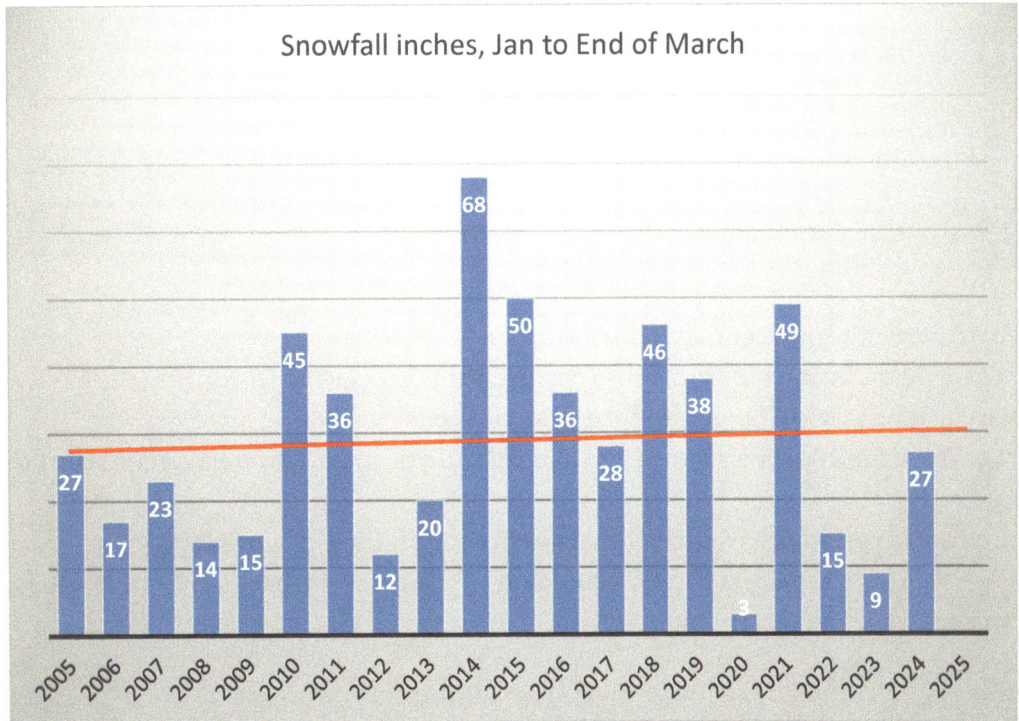

Snowfall from January to the end of March

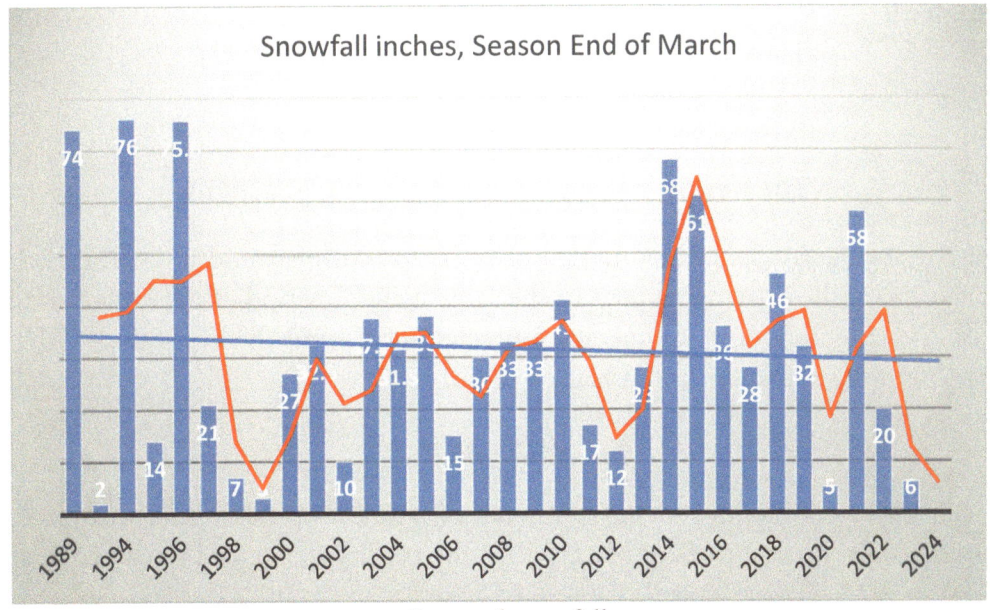

Seasonal snowfall

My observation is many years when we have a large amount of snow, the following year there is a low amount!

I had read that in the Lehigh Valley area one inch of rain would equal 10 inches of snow. I had accepted this without question. However, we receive a lot of heavy snow that quickly turns into a round cylinder of ice in the snow blower shoot. I thought something was not right. Dividing the amount of snow by the amount of rain from January to March, I found that one inch of rain equals approximately three inches of snow!

From My Collection of Trivia

Wind

On average the afternoon is the windiest time of day. A 30-mph wind is needed to make a flag stand straight out.

Hurricanes

The strength scale for hurricanes is Saffir-Simpson.
The word hurricane is derived from Huracan, a Caribbean God of Evil.
The ocean must be at least 79F to produce a hurricane.
On average there are 10 tropical storms in the Atlantic per year and on average six of these storms form hurricanes.
Over a three-year period, an average of five hurricanes hit the U.S. coastline.
The most active hurricane season was 2005 with 27 tropical storms and 15 hurricanes; the 1940's had 10 hurricanes.

Wind Chill

The rule of thumb to determine wind chill is to subtract one degree of temperature for every one mph of wind speed.

How Does Weather Affect the Flora and Birds?

Flora

The weekly "spread" of the beginning of activity

Reviewing the spring activity of blooming, leafing or breaking ground of perennials from 2005 to 2019, in a given specie the activity happens in the same general period of the cycle from year to year. Depending on the combination of temperature-precipitation-wind the beginning of the activity can have a "spread" of three to five weeks (basically a month).

Perennials

Daffodils: 2005-2007 peak bloom was the first week of April; 2008-2009 peak was the third week of April. In 2014 peak was the last week of March; 2024 the peak was the third week of March.

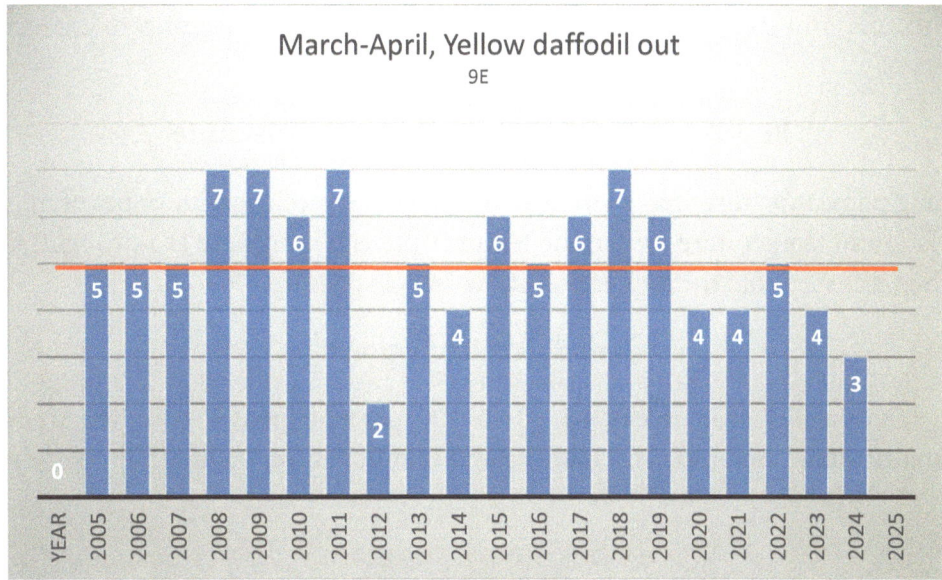

Week daffodils bloom

As the activity changes in one species most of the other species follow suit.

For example, if daffodils change blooming from week 4 (the last week of March) to week 5 (the first week of April), then the summer pink phlox that blooms week 18 (the third week of July) should now bloom week 19 of July. Now that is stretching Anticipation.

The exception was 2012. The temperatures and precipitation were normal for the first three months, but the activity started the second week of March instead of the fourth or fifth. This early activity continued until the middle of July. The cause was an abnormally warm November and December in 2011.

Shrubs

The shrubs have a four to six week (month-and-a-half) spread in activity. The 2012 early activity shown in the perennials shows in the shrubs, but only through May.

Trees

The trees have a spread of four to seven weeks (almost two months).
Be it perennials, shrubs or trees, the spread of when activity starts are dependent on the weather.

Overall, of Flora Relative to Temperature

Whenever nature pushes the button to start the cycle there is no stopping the progress for the rest of the year.

Except for the earlier completion in the canopy in May, the data of the past 15 years does not show an increase to firmly show an overall earlier start in activity. On the other end of the cycle, the completion of the release of the leaves is occurring at least two weeks later. Both the earlier completion of the canopy and later dropping of the leaves are probably due to climate change, which means earlier spring and later fall and shorter winters. That does not necessarily mean a warmer winter, just a shorter time to get winter things done, like later cold and stronger storms.

Perennials, shrubs, and trees blooming activity has shown no relationship to change in the mean temperature.

Bird Population Relative to Mean Temperature

Review of gang population through the months of individual species and number of migrating species show a firm relation to temperature. When the temperature is hot or there is a drought, the population decreases or when it is very cold the population at the feeders goes up.

Bird Spike Populations

Unusual high population values can be 20 to 100 percent greater than the trend value. Low populations usually follow the preceding spike year with much lower values than the trend.

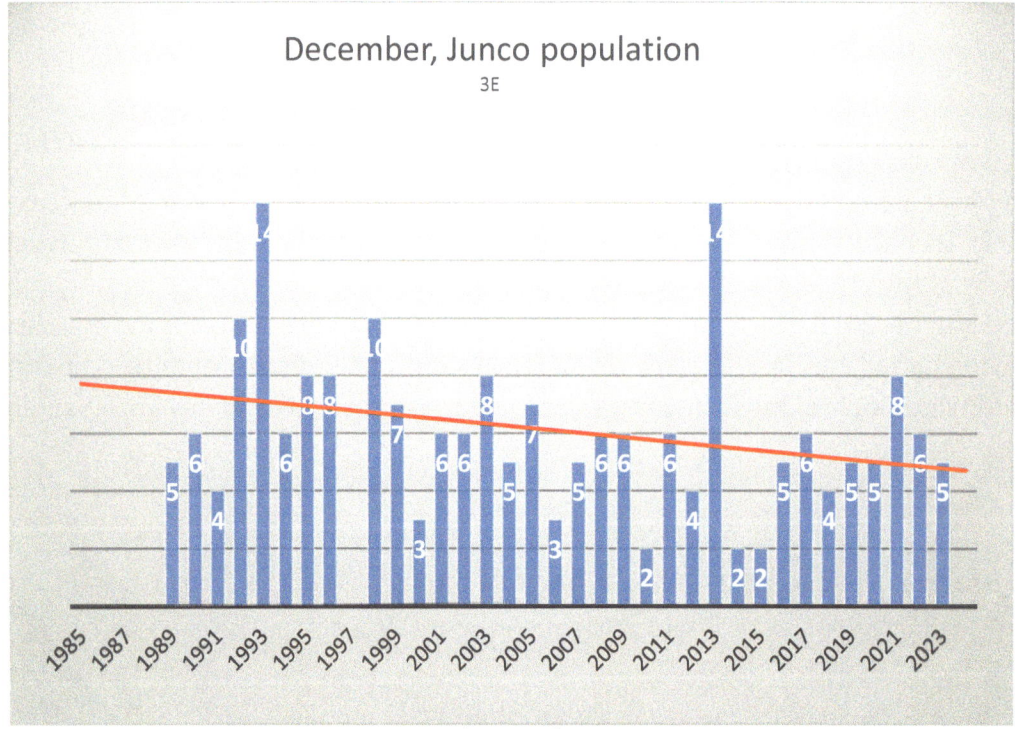

Population of junco

In some species the spikes occurred at 10-year intervals, then changed to six, and lowered to four intervals. The spike used to start in November, then backed up into August, then June, now they start in January and February. Spikes often occur with the following combination of temperature and precipitation.

High bird population		Low bird population	
Winter	Cold and snow	Winter	Warm and wet
Winter	Cold and wet	Summer	Hot and wet
Summer	Cool and wet		

That has been a review of the weather, its cycles, and its effects on the fauna and flora. My conclusions show that temperature affects the fauna and flora more than the precipitation.

It would be interesting to investigate our births to see if there are similar spikes. It is known that after a blackout in New York City that there was a significant spike in births nine months later. That blackout may have been due to hot weather!

Now to the beginning of the journey of the anticipation cycle which is the month of March.

Index of March

- March Overall 33
- Weather 34
- First Week 37
- Second Week 41
- Third Week 43
- Fourth Week 46
- Review of Data 51
  - Animals 51
  - Birds 51
  - Number of Species 52
  - Gang 52
  - Table of the Gang 54
  - Population of Species 54
  - Fledglings 56
  - Migration 56
  - Migrating Table 57
  - Insects 59
  - Flora 59
  - Lawn 59
  - Perennials 59
  - Vegetables 59
- So, ends March 59

March

March, not January, is the beginning of the new year of my phenology. It is the month that flora life in the yard is awaking in earnest. So, March is the beginning chapter. I use four weeks in a month, even when there are 30 to 31 days in the month. The last week may have seven to 10 days. It makes it much easier to make graphs. The first week of March is week 1 and the last week of February is week 48.

	Week number for the year of 48 weeks				
Relative Month	Month/Week	First	Second	Third	Fourth
Previous	February	45	46	47	48
Present	**March**	1	2	3	4
Following	April	5	6	7	8

The amount of light on the 1st is 11 hours, 24 minutes; on the 31st 12 hours, 49 minutes.
The average high temperature on the 1st is 42F, the low is 24F.
The average high temperature on the 31st is 55F and the low is 34F.
Extreme daytime high for the month was 88F on the 30th of 1998.
Extreme nighttime low for the month was -5F on the 1st of 1934.
The full moon in March is called the Worm or Crow Moon because that is when the earthworms come to the surface for the robins to feed and when the crows call for a mate.

Overall Summary of March

March is the month that the annual cycle really starts. It is the month that life is reawakened. Bulbs have unfathomable strength to erupt through the frozen ground. Can you imagine what force it takes to push up through the frozen ground that even an ice pick cannot penetrate? Yet once the plant gets above the ground, the stems are so soft that a late snow will bend and kink the stem over so it cannot recover to bloom. But, watch out, March is the month of the heavy, deep snows; we can get 15 to 20 inches overnight. You walk around the yard and the sun may be warm, but underfoot it is still cold, damp, and muddy.

The male birds are wooing for a mate. The squirrels and chipmunks are on the chase behind a potential mate or fighting another male for territorial rights. Pairs of juncos are doing unbelievable acrobatics over, under and around the shrubbery. Even arrogant blue jays are giving sweet songs. The Carolina wren loudly sings out one of the 30 songs in its repertoire.

The temperatures go up and down like a yo-yo. So does the growth and blooming of the perennials, by one or two weeks. March can come in like a lion and out like a lamb or vice versa. The temperatures can be freezing or up to the mid-70's until the end of the month. Winds can be 30 to 70 mph. The fluctuating temperatures produce heavy fog, frost or dew. This is the beginning of the season that we may get tornado warnings. The ground is usually frozen to the middle of the month, we can get two inches of rain in one storm during that same period. With the frozen ground, the rains just run off, resulting in bad floods. The ice in the rivers breaks up causing ice dams and over-flowing banks. We can have tornado warnings. Sometimes the weather is nice and I am cutting the grass by the fourth week. Other times the month is very dry, causing April to have only a few or smaller flowers. Also, with a dry month, robins cannot find enough water to build their nests, no less feed the chicks. With the unsettled weather, not only the emergence of the perennials can vary by weeks, but the arrival and departure of the migrating birds will change as well as when animals come out of hibernation. The rebirth is all a wonderful gift to observe.

Weather

The chief observation of data from 1988 to present is that the makeup of the weather controls the phenology of the wildlife, especially when the plants emerge, flower and wane.

Temperature Trend

The day-time temperatures during March can vacillate from the teens to the 70's. In 24 hours, the temperature can change 50 degrees; with 40 mph winds it feels like 3F. A plot of average temperatures from 1992 to 2024 shows a five-degree rise. The rise has changed when the perennials bloom and birds breed and migrate.

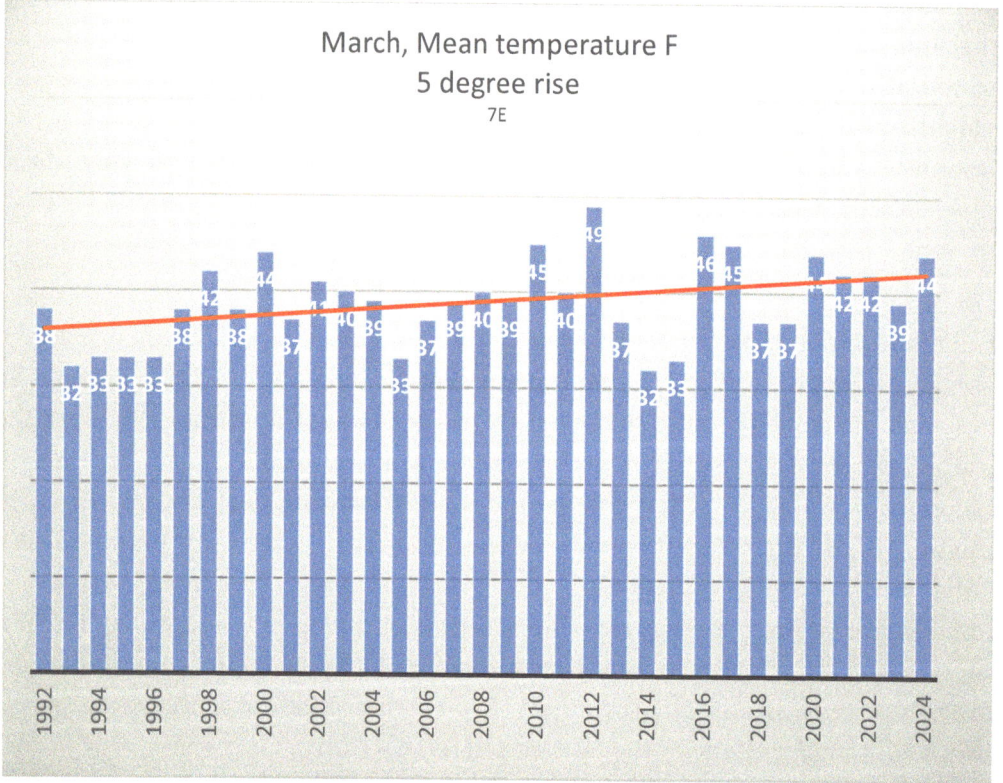

March mean temperature F 1992-2024

<u>Snowfall</u> in March represents the end of the snow season. We have heard about the blizzards and heavy snows that occurred in March. Most of heavy storms occurred pre-1997. From 1998 to 2016, there was very little to no snow. Looking at the latest years it appears that a revival of storms is occurring. Years that March had a total of 18 to 21 inches, it occurred as a series of storms of three, six, eight or ten

inches. These storms occurred from the first to fourth week. The accumulation of snows is cyclic; the next heavy seasonal accumulation should be 2021-2022. No, it was 2015-2018

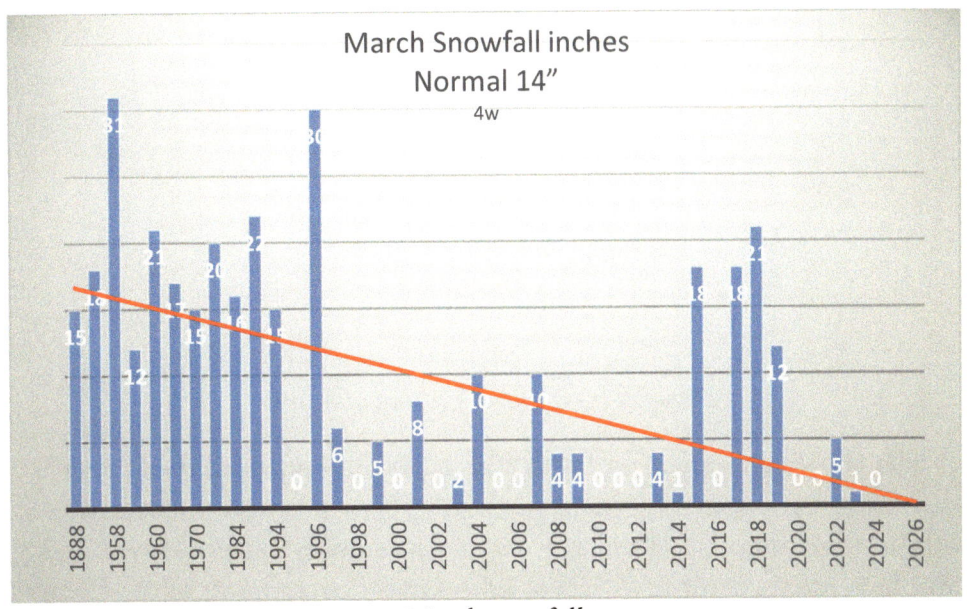

March snowfall

A blizzard is a snowstorm with winds over 35 mph and lasting for a prolonged period of time.

Data from The Morning Call newspaper of Allentown, Pennsylvania:

Largest March Blizzards
19th-21st, 1958: 20.3"
13th-14th, 1993: 17.6"
11th-12th, 1888: 15"
4th, 1960: 14.2"

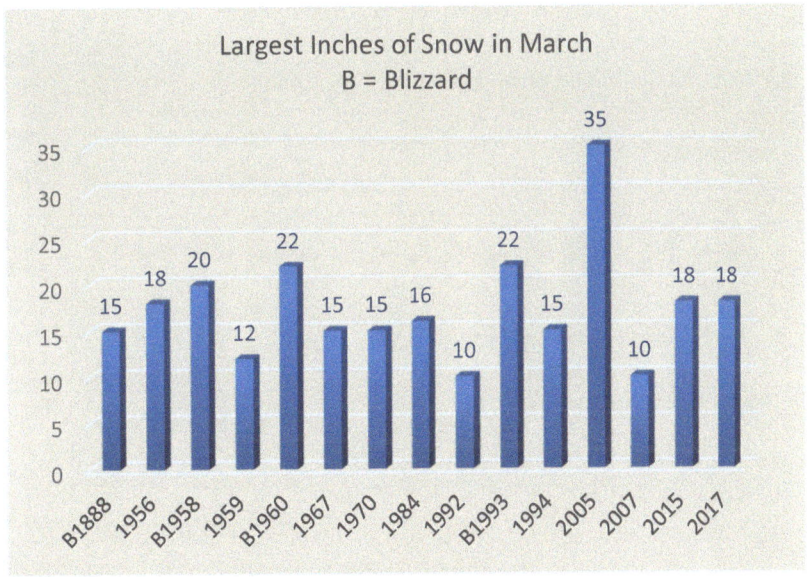

Largest inches of snow in March

MARCH

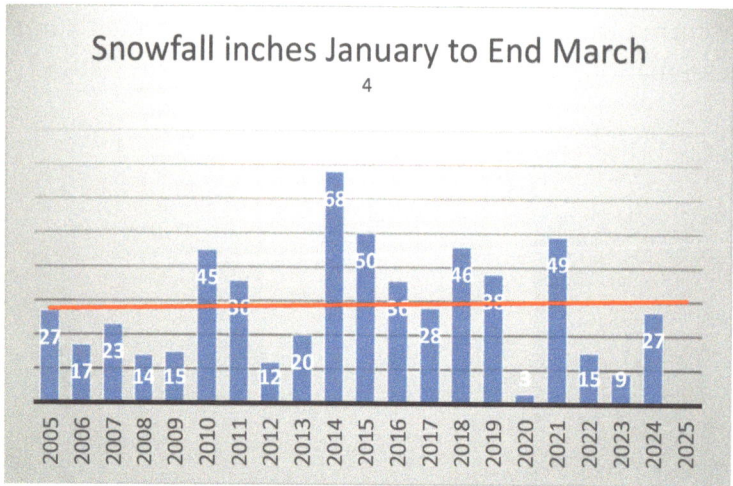

Snowfall from January to end of March

The snow season goes from September 1st to March 31st, but on April Fool's Day in 1997 we got a blizzard of five to 30 inches depending where you lived. The winds were 40 mph. Interstate highways were closed with stranded motorists. They were rescued by snowmobiles over two days.

It appears that large snows occur in a cycle of 10 to 12 years in a cluster of three years like 1992-93-94. Blizzards have occurred from a period of two years to a period of 30 to 70 years, otherwise not too often. When they do, it happens during the third week of March. So, from all of that there should be a large snowstorm in the third week of 2018 and there was. The next blizzard is anticipated in 2023.

When the winds come on strong, little snow devils (mini-tornadoes) formed on the still snow-covered lawn.

<u>Rain</u>: The amount of rain also goes up and down from year to year. March 2006 was the tenth driest on record; 2016 was the fourth warmest March on record, 7.2F above normal (2012 was the third warmest). Consequently, lots of perennials broke through the ground two to three weeks ahead of time. March 2020 rainfall was one inch below the average 9.3 inches. Yet, the rainfall from January through March shows an increase.

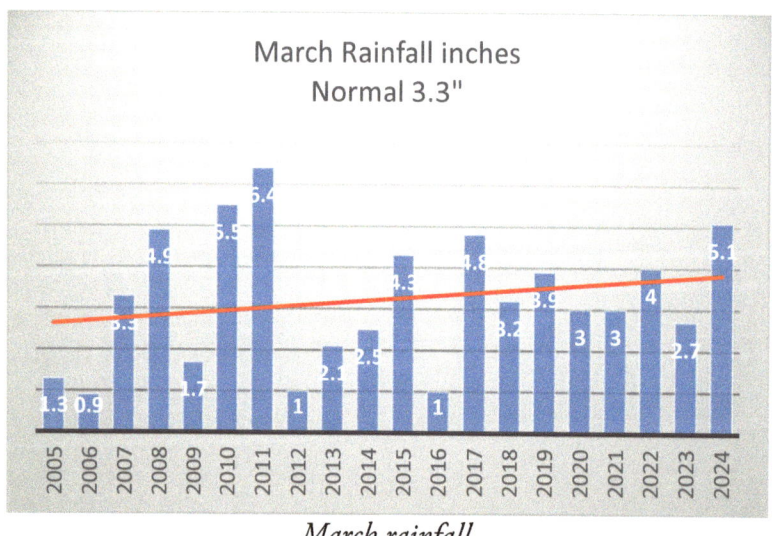

March rainfall

TRENDS DUE TO CLIMATE CHANGE

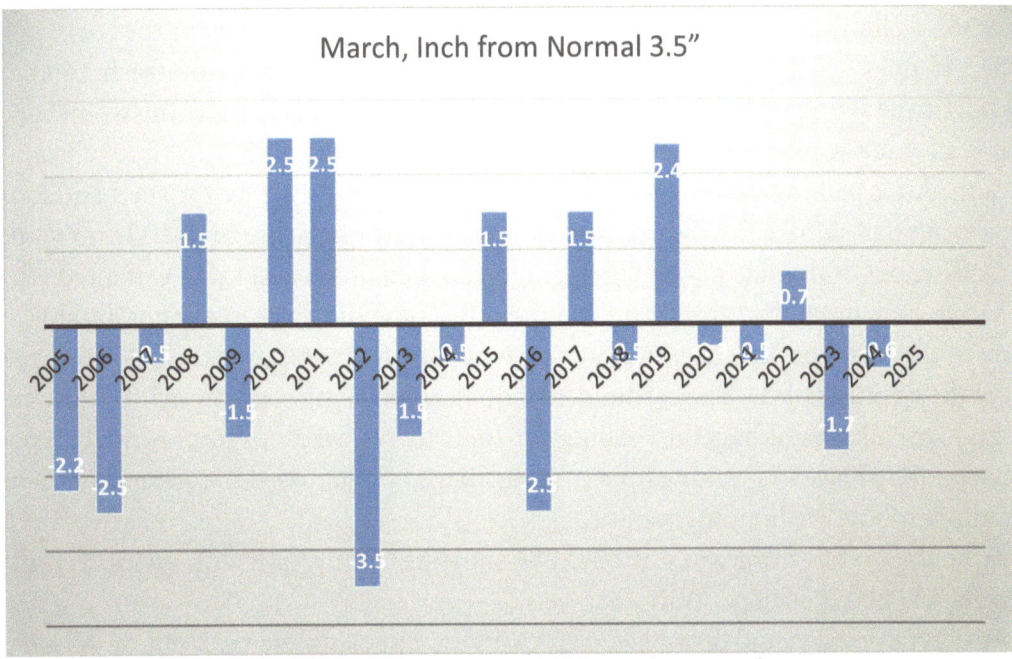

March amount of rainfall from the normal

First Week

<u>Weather</u>: January of 2007 was unseasonably warm for two weeks and then it got cold again. As a result, the sap rose in the trunks of the shrubs and then froze; at least three rhododendrons died from winterkill.

<u>Animals</u>: Two rabbits are running around. Squirrels are taking leaves and cattail punk up an evergreen for nesting. Two squirrels chase each other and then wrestle and roll over each other. The chase is the mating game. When the female slowed down or the male ran faster, they coupled up, did their thing, rested a few minutes and then continued the chase.

On the chase

Hungry chipmunk

When it is 36F a chipmunk comes out of its semi-hibernation from beneath the rock wall to get sun and safflower. It goes scurrying to and around the feeders searching for fresh food. The semi-hibernating nocturnal skunks also come to feed. I smell their fragrance in morning when I get the paper. Unfortunately, a lot of skunks, opossums, and raccoons are killed at night this time of year.

It is 6:00 a.m. After picking up the paper at the bottom of the driveway I turned and looked up the hill and saw a couple of skunks slowly crossing the lawn. They had a late night! Then two days later at 11:00 p.m., coming out of the side door of the garage, I looked down and there was a red fox in front of me. We looked at each other, he turned and ran the other way; didn't even say goodnight.

Birds: One of the joyful occurrences in March is when the temperature gets to 40F; it opens the keys to the songs of the courting birds. The cardinal, titmouse, robin, song sparrow, white-throated sparrow, and house finch are all singing together calling for their mates; what a symphony! At the same time, the young sharp-shinned hawk is looking for a meal from the off-guard lovers. After the sun has warmed up the lawn, three robins are hopping around.

Robins arrive the first to third week. It could be snowing lightly, 28F with 20 mph winds. A lonely robin in a crabapple tree is asking, "Why am I here?"

Early morning, red-bellied and downy woodpeckers are knocking the heck out of the telephone pole with their mating call. At 7:00 a.m., 2007, warm, the gang is here and so are eight blue jays; 8:00 a.m., a group of crows fly west. There are 11 species at the feeder, seven as couples. The pecking order is in full force. A male house finch chases a female off the feeder. A red-breasted nuthatch arrives, a cardinal is singing. Carolina wren sings loudly. When there is one, usually another will be close by. A Cooper's hawk is flying over but continues on. This time of the year, Cooper's hawks are on migration to the north.

Cooper's hawk

Thirty to 50 Canada geese are flying to get exercise; it does not matter if it is raining, snowing or foggy. The average temperature is 32F. The number of geese seems to depend on the temperature. When the temperature is above 42F there may be a local flock of 100 to 200.

A sharpie flits over the yard; rabbits are too big a prey, it wants a dove or a cardinal. A dove comes to the feeder, a flicker gives out its mating call. This is too early for a flicker. The call of the flicker is from a young sharpie; the dove quickly flies off. This sharpie tries three times during the day to get a dove but misses each time. It needs some guidance.

Every year the house sparrows chase the newly arrived starlings who are trying to take over their house; sparrows always win. In March, the hormones may be triggered by the amount of daily sunlight and temperature. The male white-throated sparrow's colors are starting the change; the dull yellow streak on the head is now bright yellow and grayish feathers are now cotton ball white. The male's feathers of some species change from drab to bright colors during the change of seasons from winter into spring. The reverse happens as summer goes to fall. The head of the house finch turns red.

White-throated sparrow with winter and spring colors

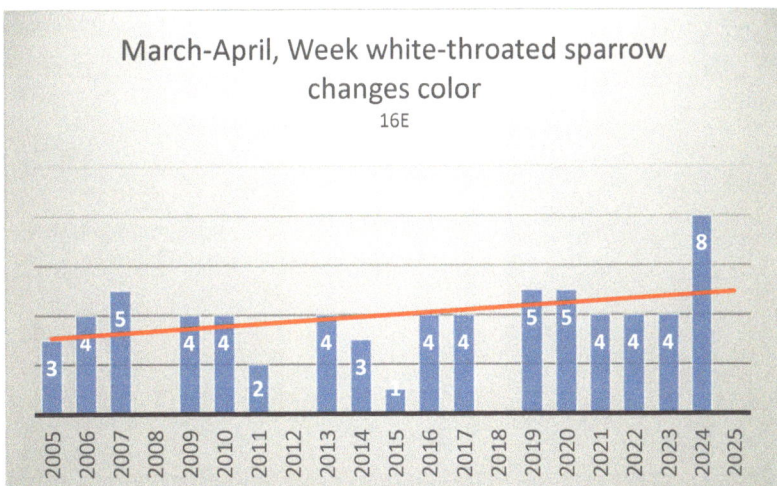

Week white-throated sparrow change color; 2012 and 2020 had very warm winters

A pine siskin is passing through, it is on the feeder, a house finch lands next to it. The siskin chases the finch from the feeder. Fox sparrows may stay a couple of weeks before going north; they like to scratch the fallen leaves for insects.

Pine siskin

Fox sparrow blends into the leaves

Noon and sunny, flocks of 10 to 20 grackles and red-winged blackbirds arrive. Two soaring specks are a pair of migrating broad-winged hawks circling around and around, each circle taking them further east with the drift of the wind. A long shadow quivering over the ground is from a turkey vulture trying to catch a thermal to rise higher. Dusk, 6:00 p.m., the cardinals and song sparrows are still at the feeders. At 2:30 a.m. I am awoken by a deep calling of a great horned owl looking for a mate from the tall spruces in the backyard. This mystical song penetrates the cold air for half an hour, until an echo is heard not too far away. I gaze out the window of a dark room to make out the silhouette of the great bird.

<u>Perennials</u>: In 2007 the daffodils were up two to three inches in January even through the frozen ground. What strength they have! As long as buds did not form, the flowers should still bloom in late March. The purple crocuses are partially blooming. On the third week they are wide open in the sun. We had a record warm February in 2020, resulting in a wide-open crocus the first week of March. A dozen white crocuses are blooming. The first bulbs to bloom are the snowdrops, which may be late February.

Crocus

Snowdrop

Yellow overwintering pansies bloom. Maples on the hill have buds. When the soil is warm enough, I rake off the remaining of last year's leaves. The green tips of the Hosta, iris, and chives are peeking through. The daffodils could be up four inches with dark green sprouts. Last year's cattail punks are blowing around; new cattail sprouts are coming up. There are buds forming on the lilacs, roses, maple trees, and blueberries.

Pond: The pond is usually still covered with ice during the first week. When the ice is melted off the pond, the water is crystal clear. When the air temperature goes to the high 40's it brings out a frog that sits on the edge of the pond.

Frog in winter color

Shrubs: In 2007 the first weeks were warm, the japonicas are in full cluster.

Trees: Sap is running out of the trimmed maple limbs forming icicles. A chickadee lands on the icicle and bends over to get a drop at the bottom. The buds on the silver maple are swelling

Second Week

Animals: A vole scurries across the garden steps. During a winter's day, one sticks its head out of a snow tunnel by the base of the feeder. Squirrels are eating the buds of the silver maple. There was a gathering of three clans of squirrels in the yard, one from the north, west and south; there were a total of 11. There were some conflicts during the gathering. Glad this appears to be only one-of-a-kind gathering. Late in the afternoon I see a red fox in a field, poised and tense over a clump of grass waiting for a mouse to make a move; he wants his dinner.

Birds: The second and third week is when the majority of the snow geese migrate to the north in grand flocks. Right next to one side of the "V" of a flock is an unorganized flock of 60 small birds flying north with the geese! A flock of 40 grackles arrive at the feeder, devour what is there, and then flies on.

A downy is tapping for a mate. The call of a newly arrived flicker is in the air. After a rainy night, the sky becomes a sunny 50F. A pair of turkey vultures is on top of a tall oak with their six-foot wings spread to dry with cocked heads toward the sun. On top of the hill a young broad-winged hawk consumes a preyed rabbit.

Broad winged with rabbit

A male cardinal burst forth with a loud song; within minutes two females arrive. It must be quite a song! Some cardinal couples bring fledglings to the feeders this week. Now and then a cardinal with mange will appear at the feeder, which is different from molting that they do in the summer.

Cardinal with mange

In 2014, walking up the hill next to the hemlock hedge, out of corner of my eye something brown moved. To my astonishment I saw a woodcock! This is the first one I saw in 50 years. It is squat, round, like a big softball, with a five-inch-long needle-like beak to get worms. The next year, two neighbors reported seeing one.

Time to clean out the birdhouses. The houses should not be cleaned in the house or cellar, because uninvited guests such as bees or spiders may emerge.

House sparrow's cluttered house

<u>Insects</u>: When the temperature rises above 50F there are bumble and honeybees.

<u>Perennials</u>: Lots of perennials poking up: columbine, daylily, Montauk daisy, iris, fall sedum. Put 14-14-14 near the iris rhizomes. It is time to cut back the fall clematis vines, cattails, and raspberries plants.

<u>Shrubs</u>: Japonica are out in full blooming clusters.

<u>Trees</u>: The big single trunk birch is like a shedding snake this time of year; its internals swell with the oncoming spring causing the outer bark to stretch and split. Large, thin pieces of bark fly off. When the temperatures stay cold, this action waits until the fourth week.

Third Week

The third week of 2016, maples have full buds, grass is greening up, put copper fungicide powder under the apple trees, spray horticultural oil on the apple trees, bleeding heart is up, daffodils are up eight to 10 inches, skunk cabbage is up 3.5 inches, hyacinth behind the pond are out, hawthorn tree has new leaves, periwinkle are out, summer phlox are up. Lungwort has pink-blue flowers, fall sedum is up, primrose in the east garden is up, yellow-red tulip on the third tier is up, little yellow daffodils out front are out, fall clematis east has new leaves, columbine is up, and miniature iris has new leaves. The magnolia and flowering crabapple trees are out. Dog-tooth violets and the yellow bell are out.

MARCH

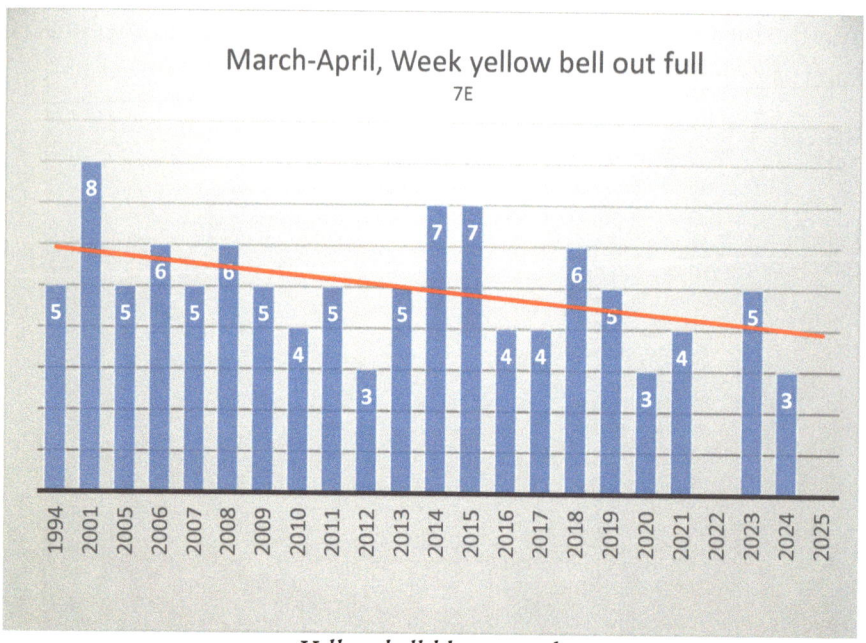
Yellow bell bloom week

Weather: Expect dense fog in the morning. Even without fog there may be heavy dew on the lawn. A most refreshing experience is the sweet fragrance of the air and ground of an early spring morning.

Animals: Now and then a nocturnal opossum is spotted when coming home at night. Next day there was a dead opossum next to the road. How sad for an animal to have wintered hibernation only to be killed a brief time after. Depending on the weather, the male squirrels start to chase for a mate the third week. A week later, the squirrels are carrying leaves up the larger trees to make a nest. The lawn is greening up and the rabbits are looking forward to the new greenery. They always short crop the top of my mountain pink, which always recovers to come out in full bloom.

Making a home

Birds: Third through fourth week the plumage of the overwintering male goldfinch is changing from olive green to bright yellow (gold).

American goldfinch

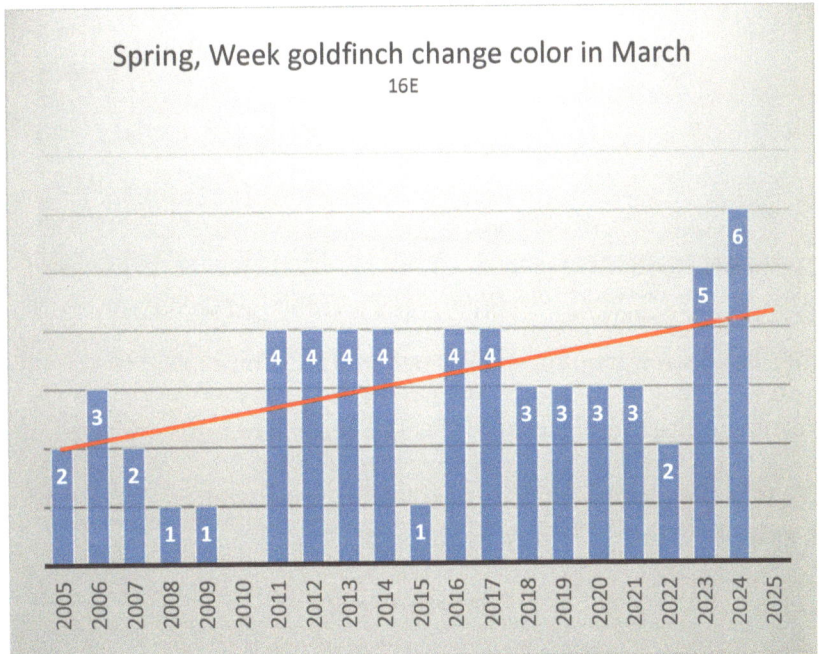
Week goldfinch changes color

    The stripes on the song sparrow's breast get darker on the third week. Many of the male birds are trying to make dates with the girls. The downy woodpecker prefers to use the top of a telephone pole to send his S.O.S love message. Two different woodpecker species attacking the same pole at the same place.

    A tale about female crows is when she flies around with a twig in her beak is that she wants to mate. On the hill three majestic black crows walking under the pine tree. One is collecting a beak full of needles and oak leaves. She is apparently past the twig in her beak stage. After her beak is full, the three fly off.

Fruit: Leaves are forming on the blueberries.

Lawn: The first appraisal of the lawn, it looks matted from the snow and has snow fungus on it. The squirrels have pock-marked the surface of the lawn while trying to bury nuts in the frozen ground.

Perennials: The 15th is the Ides of March; like Caesar, this is the time that winterkill starts to show up on the perennials. Leaves that seemed to look green and healthy suddenly get brown and fall off.

Pond: The 19th is the day to end the winter fasting for the fish. I don't know why? Again, tradition? The fish rise to the surface popping their lips asking for food. Count how many fish are lying belly up. It always seems that the older, larger fish, are the ones that surface belly-up. The non-pregnant fat white fish is 14 years old, but on the fifteenth year a great blue, that arrives the second week, had the fish for breakfast. Have to put the reinforcing rods back across the pond as a defense against the great blue.

The white one in the upper left is not pregnant

Shrubs: If the weather has been warm, the yellow bell and PJM buds may be swelling.

Trees: Leaves are forming on the serviceberry tree. The birch tree buds are swelling to a light yellow.

Vegetables: The 17th is St. Patrick's Day; if the ground is not frozen, it is the traditional time to plant onions. Chives and grape hyacinths are poking through.

Fourth Week

On the equinox, usually the 21st, a couple of things happen. For one, when I step out early in morning, if there is no snow on the ground, I can smell the earth coming to life; second, the male birds are tuned up for the real thing, their symphony is in full volume. The cardinals and robins that have mated up are getting nesting material. Every year at the equinox, I try to balance an egg on its end. It always falls over. I wonder who started this myth. Over the years I have recorded when earthquakes and volcanoes erupt. They mostly seem to happen a month before, during the month of, and the month after equinox, both in the spring and fall! Eight juncos are still flying around aerodynamically chasing each other, doing tight turns and twists over, around and through the bushes.

Weather: In the morning, there still may be ice in the birdbath.

Animals: The squirrels are eating the buds that are out on the silver maples and dogwoods. One chipmunk scurries across the lawn, suddenly a set of large wings swoop down from nowhere but misses. The chipmunk was giving an alarming chip when it got to safety. If there are two or more chipmunks in the area, they start challenging each other. I will hear chip-chip-chip-chip for hours, hence the "chip" of chipmunk. A groundhog ambles along the garden; it keeps on going, good. A frisky squirrel jumps from the ground onto the trunk of a dogwood, pushes off, rolls over on the ground and repeats this ritual six times, and then runs away.

At the bottom of the main feeder on the lawn is a four-foot diameter lawn edging ring. This is used to retain the seed droppings. A chipmunk was inside the ring near the edge eating away. Just outside of the ring was a squirrel eating the peripheral droppings. The two were right next to each other. I think the chipmunk was oblivious to the squirrel, it just kept munching away without realizing it was getting closer and closer to the squirrel. The squirrel was getting annoyed; finally, it lifted its paw and put it on the back of the chipmunk. In horror, the chipmunk came back to reality; it chirped and ran off at a speed. The squirrel just kept on eating.

Squirrels do odd things. One on the hill grabs an eight-inch-long spruce cone, it hugs it against its breast then tumbles and rolls down the hill with the clutched cone. It drops it and scurries away! I have witnessed these many times, but do not know the significance. Yes, of course, spring is in the air.

Birds: The birds do not care for peanut butter in cold weather, none has been touched for three weeks. It is 43F; the coldness is coming out of the ground. Six purple-headed cowbirds take over the feeder for breakfast before moving on. A high shrill chirp comes from an arriving chipping sparrow that landed on the feeder.

Chipping sparrow

Coo-coo clock house

Robins are gathering nesting material. When there is an extra warm beginning of the month, they will start to gather material the third week. Two male robins are going at each other, they fly up and down in a close column, ten feet high; who will own the territory? A female goes to last year's nest that is in pretty good condition; she starts to refurbish it but abandons it. Woodpeckers are still drumming on trees and houses.

A red-bellied and downy woodpecker are here. The downy has taken over the coo-coo clock house for the fourth spring. It will raise its fledgling in it. Nature has timed it so when the downy is fledged the house wren will take residence.

Titmice have chosen a house to reside in. A pair of white-crowned sparrows, fox sparrows, Connecticut warblers, pine siskin and eastern phoebes drop by as they are going north. A male catbird and chipping sparrow have arrived for the summer; the female catbird will come later. Like the grackles, flocks of starling will stop to devour the feed. Savannah sparrows will be leaving soon. White-throated sparrows are going north.

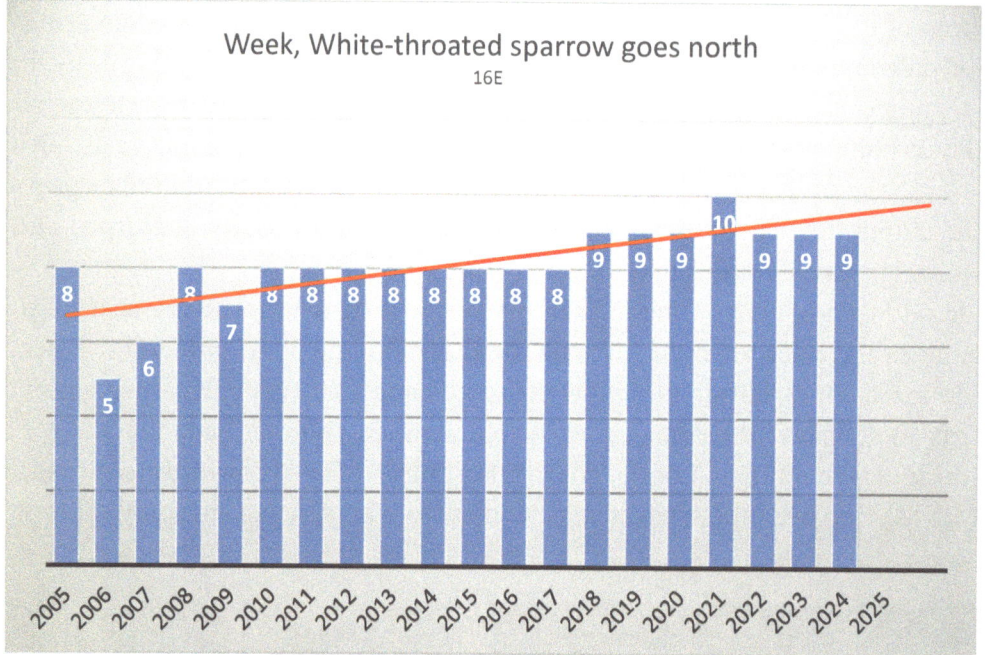

When white-throated sparrows go north

Savannah sparrow

The robin couple is collecting nesting material. As the sun sets the male sings his evening songs, as only he can do. Late in the evening, the hoot of an owl can still be heard.

<u>Insects</u>: On warm days stink bugs and boxelder bugs will come out to sun. A white-sulphur butterfly goes through the garden.

White Sulphur

<u>Lawn</u>: Grass is starting to green up. Put corn-gluten and lime on the lawn.

<u>Perennials</u>: At the end of the month tree leaves are removed from the perennials. There are a few light purple flowers on the periwinkle (vinca or myrtle). The little daffodils in the front yard are blooming. Clematis on the maple shows new leaves, some bleeding heart may be out. The backyard hyacinths are up an inch. The fern fiddleheads are starting to unravel. On the 30th, Mayapples are breaking through the upper crust.

Fiddleheads unraveling

Black-eyed Susan and hosta are showing life. Mountain pink that was chewed by the rabbits in February are blooming. The grape hyacinth leaves are fluffing up.

Predicting Full Bloom Dates for the Daffodils

The full blooming date for daffodils with a dry March is the third week of April. When there is a wet March, the bloom date is "between" the first and second week of April. With average rainfall for January, February, and March, the date should be the second week of April. Even if the moisture has been normal, dry or wet, when the three-month average temperature is in the 40s, the full bloom date for the yellow daffodils is the fourth week of March, which is four weeks early.

If the average temperature is 48F or higher with normal rainfall

Daffodils bloom two weeks earlier. Lawns green up three weeks earlier. Yellow crocuses are blooming three weeks earlier; silver maple buds a week earlier.

If the average temperature is 48F or more and it is above normal rain

The lawn greens up later because there has not been enough sun. Chives come up eight days earlier.

If the average temperature is cold, like 30 to 39F

Windflowers come out the first week of April. Lawn greens up the last week of March. Yellow daffodils bloom the second week of April.

The second-tier blooms one to two weeks earlier than the blooming date of the full bloom of all the daffodils. That area does get more sun. The dwarf daffodils in the front garden bloom from the last week of March to the first week of April.

Blooming dates for daffodils and crocus February, March and April *(Plain numbers are for April)*															
Occurrence	05	06	07	08	09	10	11	12	13	14	15	16	17	18	19
Mixed daffodil	6	Mr24	Mr31	11	Mr28	Mr20	5	Mr17	9	7-16	9	1	6		
White-yellow						4			5	12	11	Mr12	12		
Yellow daff full	10	16	6-16	11	18	Mr22	16	Mr17	9	14	18	1	12		Mr31
done	24	22	28		26	19	21	Mr23		My2	My9	29	21		
full-wane		6			8		5	6					9		
full-done	14		22			27				18	21	28			
Little yellow daff pond	Mr7	Mr4	Mr2	Mr19	Mr16	Mr9	Mr18	Feb13	-	Feb29	Mr1	Mr13	Mr21		
Little yellow frt	Mr12	-	Mr30	Mr20	Mr24		Mr20	Mr30	-	4	6	Mr17	Mr21		Mr26
Yellow-org cent									6	14	18	17	Mr30		
White crocus	M6		Mr13	Mr19	Mr17					Mr9					
Purple crocus	Mr6	Mr20			Mr17	Mr9	Mr18	Feb17	Mr5	Mr15	Mr10	Mr1	Fe8	M19	Mr15
Yellow crocus first	Mr6	Mr4	Mr2	Mr19	Mr17	Mr9	Mr4	Feb15	Mr5						
second patch	Mr18	Mr18	Mr13			Mr19	Mr18								
Snowdrop						Feb22	Mr4	Jan28	Mr5	Mr18	Mr14	F28	2		

<u>Daffodil Wane</u>: From full bloom to wane is six to 16 days depending on the amount of rain that falls during full bloom.

<u>Pond</u>: Iris in the pond is greening up. Frogs are jumping around the pond. I look closely to see tadpoles paddling around.

Shrubs: The lilac and azalea are forming buds. Now is the time to put fertilizer on acid-loving shrubs. Bushes on the hill are all showing new growth. Lilac is getting buds.

Trees: The silver maple is in full blossom which is resulting in a heavy layer of greenish-yellow pollen. If the weather has been warm, the apple trees may be in blossom. The serviceberry tree's white blossoms are ready to pop.

Vegetables: Chives are six inches tall and the rhubarb is uncurling.

Review of Data

The review of the observed data of every month is looking for consistencies and trends. For birds it is number of overall species, number of residential species that I call the gang, the number of species in the gang, the population of the gang. When and which birds migrate, when they change color, fledge. When and which animals are active. When the insects emerge from their winter habitat. Pond activity becomes noticeable. When do the perennials, shrubs and trees bloom and wane. Once all this life is started and how it progresses depends on the weather. Weather is the monthly amount of rain, year to date amount of rain and snow, temperature, and its occurrence.

Animals: When a warm week occurs chipmunks will be running around, and a lumbering groundhog will cross the hill. One or two rabbits can be seen at dusk and twilight. I sometimes get a whiff of a skunk.

Birds: The amount of rain during the first three months or the average temperatures does not significantly result in a change of the total number of species or the population of resident species. The one exception was in 2009, the rain for the first three months was half the normal amount. Then the total number of species observed went from 30 to 20. Snow does change the populous during a storm by double or triple. It goes back to normal two to three days after a storm.

The number of bird species observed includes the local plus migratory. The local are the residents that come to feeders and those that don't. The migratory are those that come through and stop for a day or so and those that stay for the summer. Depending on the weather locally and afar and unknown factors, the total number of species varies from year to year. Overall, the number has increased from 15 to 30 over 36 years with ups and downs in between. There has been a 50 percent increase in the number of bird species for March from the 1980's to 2024. The increase may be due to climate change in that the northeast temperatures are rising, enticing more species to migrate further north.

MARCH

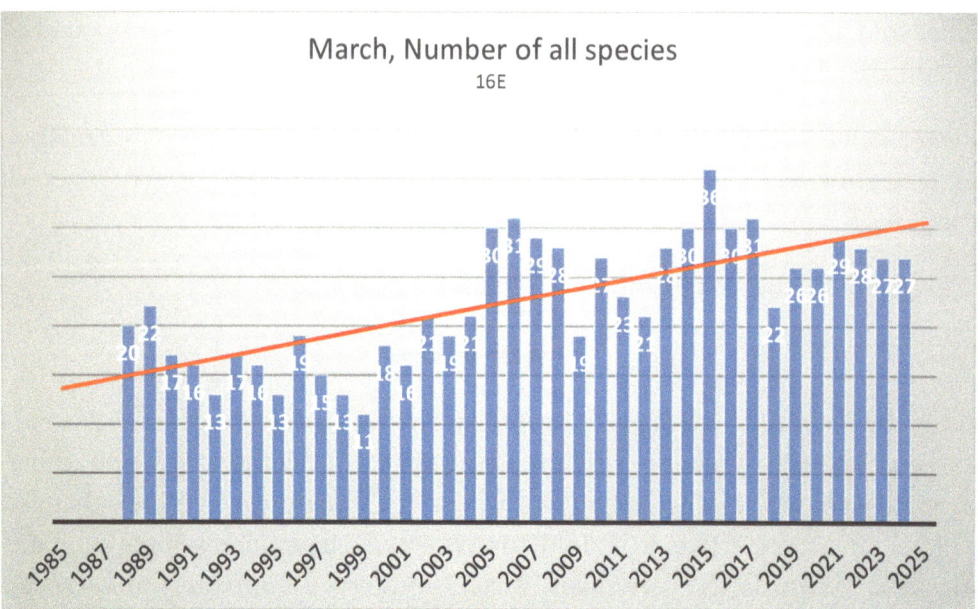

March, number of bird species

The species are divided into resident and migratory. The residents are divided to those that visit the feeders on a regular basis, designated as the "gang" and those observed but do not visit the feeders such hawks or robins.

The number of species in the gang changes from month to month. Over the past 18 years on average there are 13 to 15 species.

For March, going back to 1989 to 2004 the average number of gang species was nine to 10. By 2015 the number increased to 15. Then it decreased to 13 in 2018, but 2024 it was back to 15.

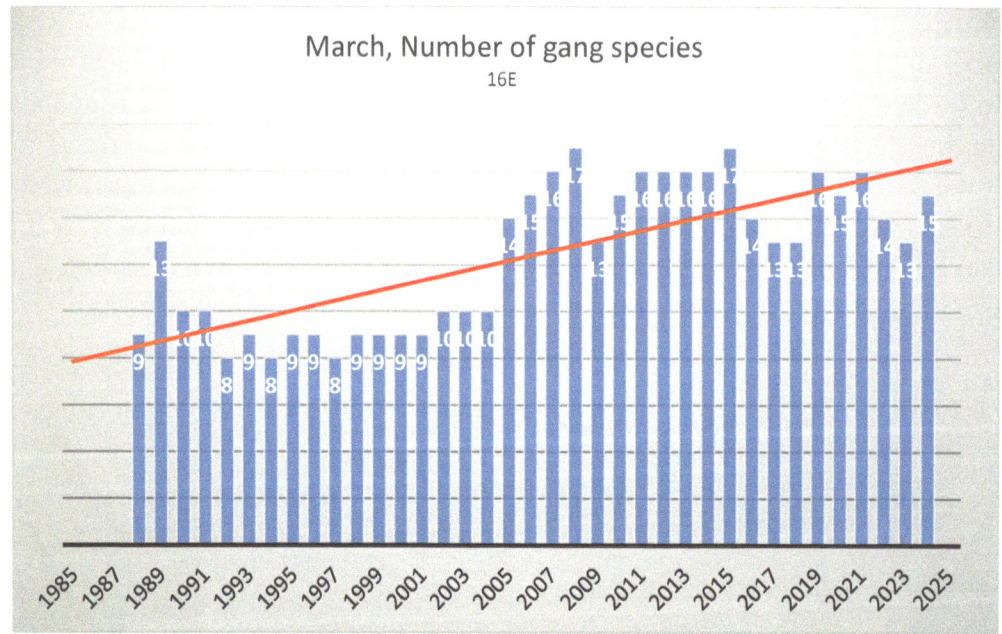

March, number of species in the gang

Over the span of 35 years the gang population in March has gone to a low of 24 to a high of 64. Quite a bit of birds for the yard to sustain.

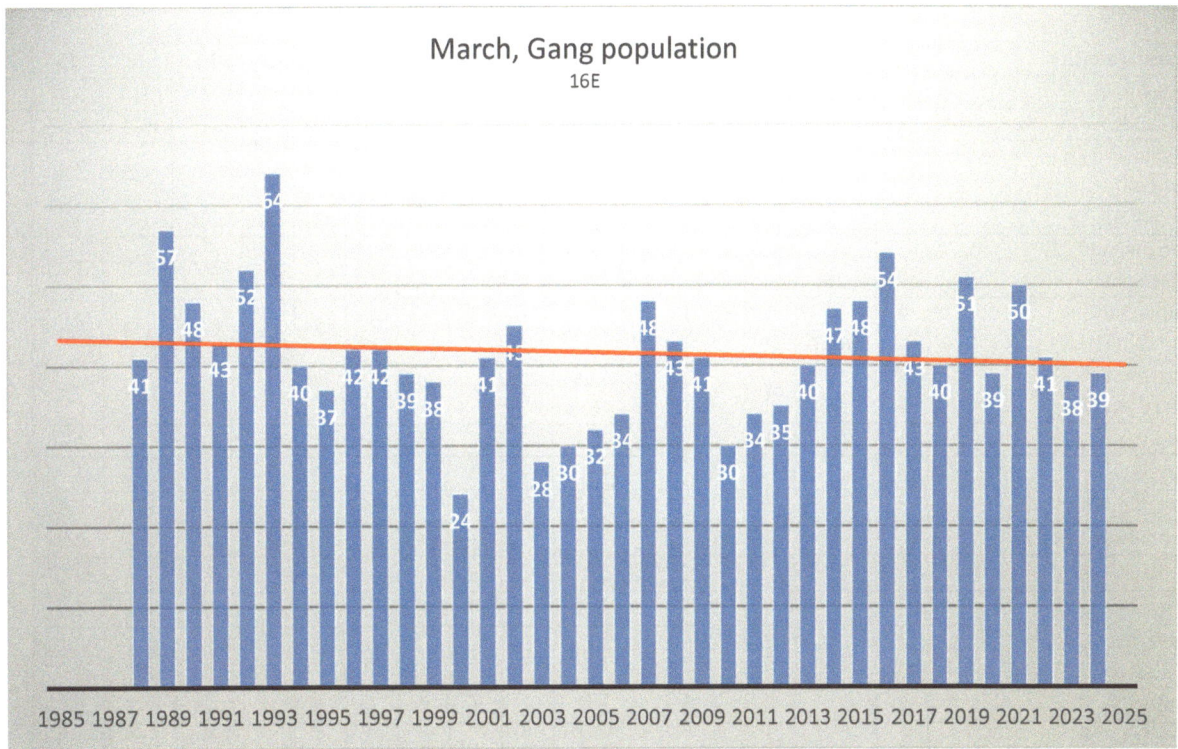
March, gang population, 1988-2019

The weather conditions of the previous months or years such as a drought or excess water affects the population. This 2011 graph shows the monthly number of species and the population.

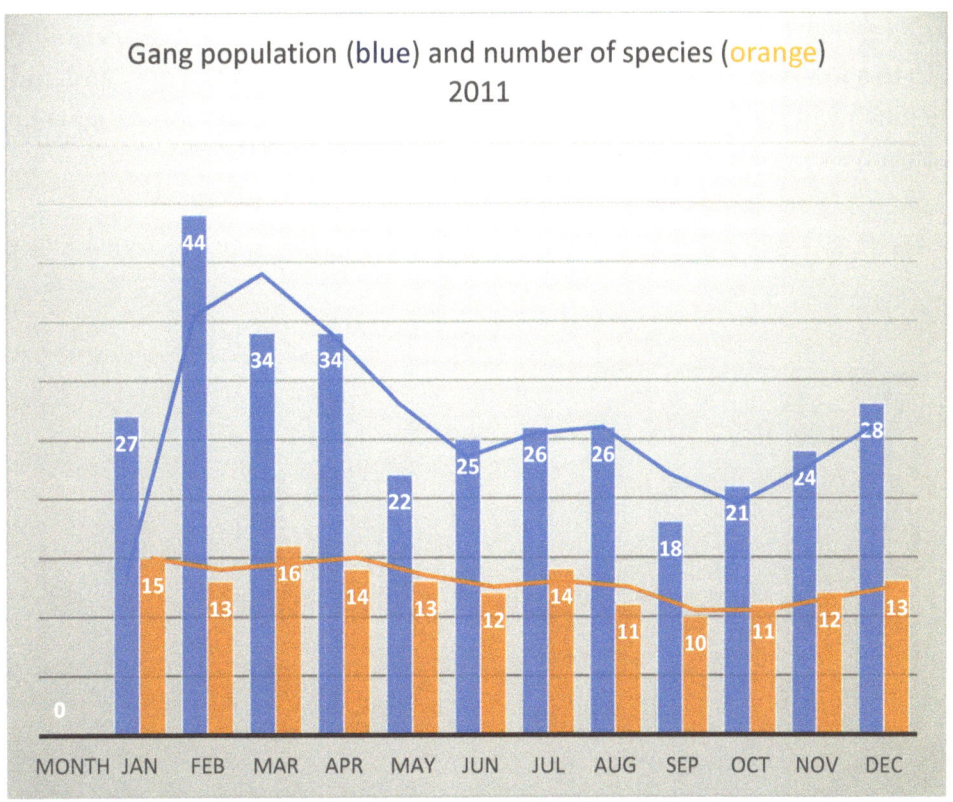
Number of gang species and population for 2011

MARCH

Species/Quantity	Gang for March R=resident, G=gang quantity/date															
	05	06	07	08	09	10	11	12	13	14	15	16	17	18	19	20
Blue jay R, G	3	3	1	8	0	2	2	2	2	1	2	3	1/3	1/3	5/17	3/6
Cardinal R, G	1	2	3	2	6	3	5	5	8	4	5	7	2+1	5/4	4/13	2/6
Carolina wren R Arrival date		1 30	3	2	0	1	2	2	2 20	2 25	2			1/5	1/2	2/6
Chickadee R, G	4	2	3	4	2	2	2	2	2	3	3	1	2/19	1/5	3/8	2/6
Downy wood R, G change color	2	1 26	1	1	1	1	1	2	2 30	2	2	2	2/2	1/3	2/23	1//1
Goldfinch R change color date	1 14	4 16-27	1 30	2 29	2 18-25	1	2 30	2 27	1	1 25	2	2	2/31	1/3	2/1	2/18 30
House finch R, G turns red date	4 29	3 26	4	2	4	4 31	6	6	6	6	4	6	4/19	6/4	7/2	4/6
House spa R, G	1	1	4	6	6	4	2	2	2	4	2	1	4/3	2/4	2/16	3/6
Junco R, G	8	6	12	4	8	3	3	2	3	2	5	5	7/4	3/5	5/4	5/4
Mourning dove R, G	3	2	4	3	4	3	3	2	1	9	11	17	5/7	12/4	12/1	7/1
Red-breasted nut R, G		2	2	1		1		1	1			1	2/7	1/1	1/1	
Red-bellied wood															1/3	2/5
Savannah sparrow R, G			1	1	1	1		1	2	1	1		2/4		1/8	1/24
Song sparrow R, G strips get darker	1 10	1 10	2	2/5	1	13	1	2	16	27	2		1/31	1/1		1/21
Titmouse R, G	2	2	2	1	1	2	2	1	2	2	1	3	2/2	1/1	2/17	2/4
White-breasted nut R, G	1	2	1	1	1		1		3		2	1	2/19		1/1	
White-throated sparrow R, G male change color	3 27	2 26	4	3	4 27	1 30	3 11	4	3 30	8 17	4	4	4/19	3/3	3/4	2/6
Population	33	32	43	43	41	30	34	35	38	45	46	54	43	40	52	39
Total gang species	14	15	16	17	13	15	16	16	16	17	14	13	13	13	15	

The population of juncos for March has steadily decreased over the 30-year period from 10 to five with a high of 15 to a low of two. When upper feathers change to a charcoal gray and under feathers to white, the male junco will start his chasing for a mate doing acrobatically sonic flights up-over-down-around the shrubs.

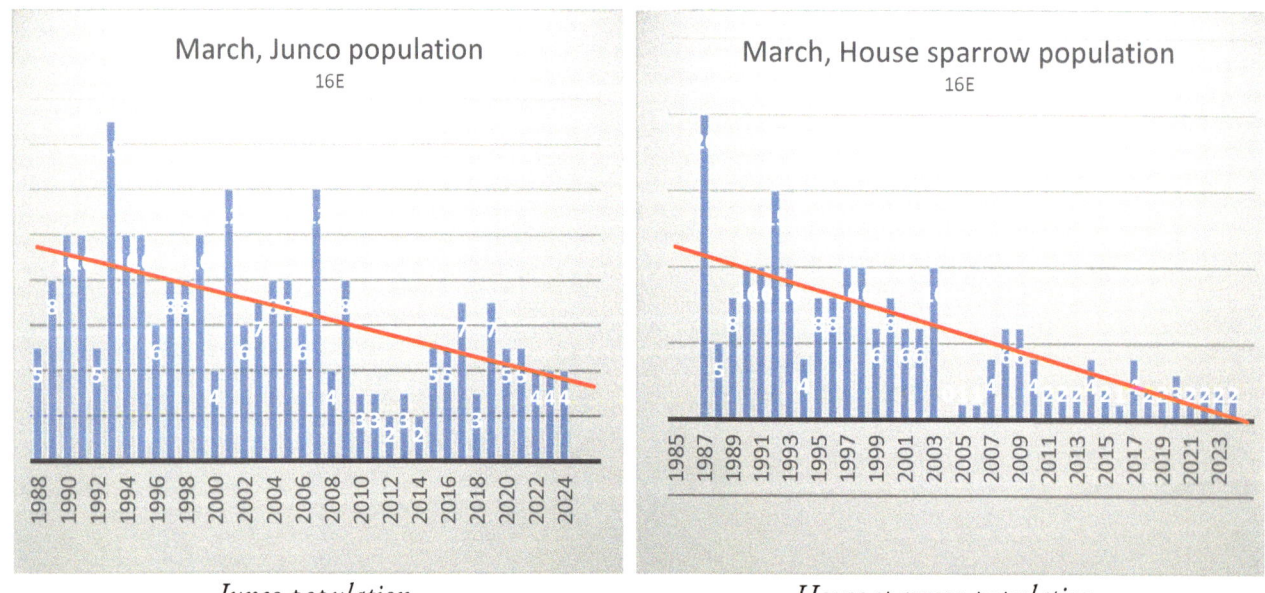

Junco population   *House sparrow population*

The population trend of the white-throated sparrow, house finch, house sparrow, and mourning dove have all shown a decrease over the past 30 plus years.

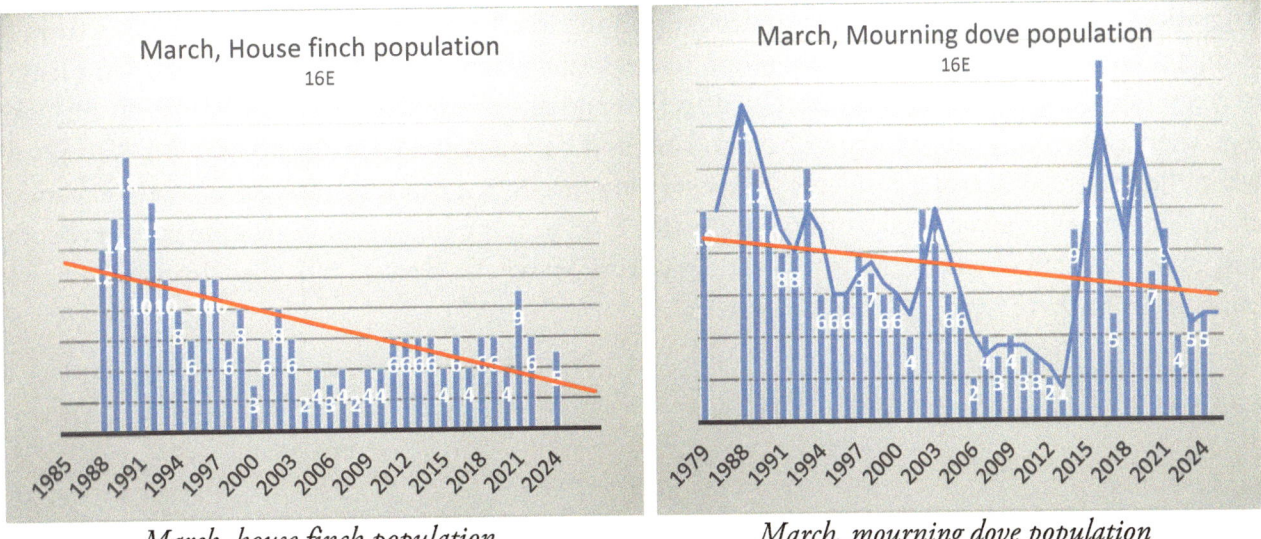

March, house finch population   *March, mourning dove population*

The dove population is affected by how many sharp-shinned hawks are in the area. Cardinals have had a steady increase from two to seven and eight but dropped in 2020.

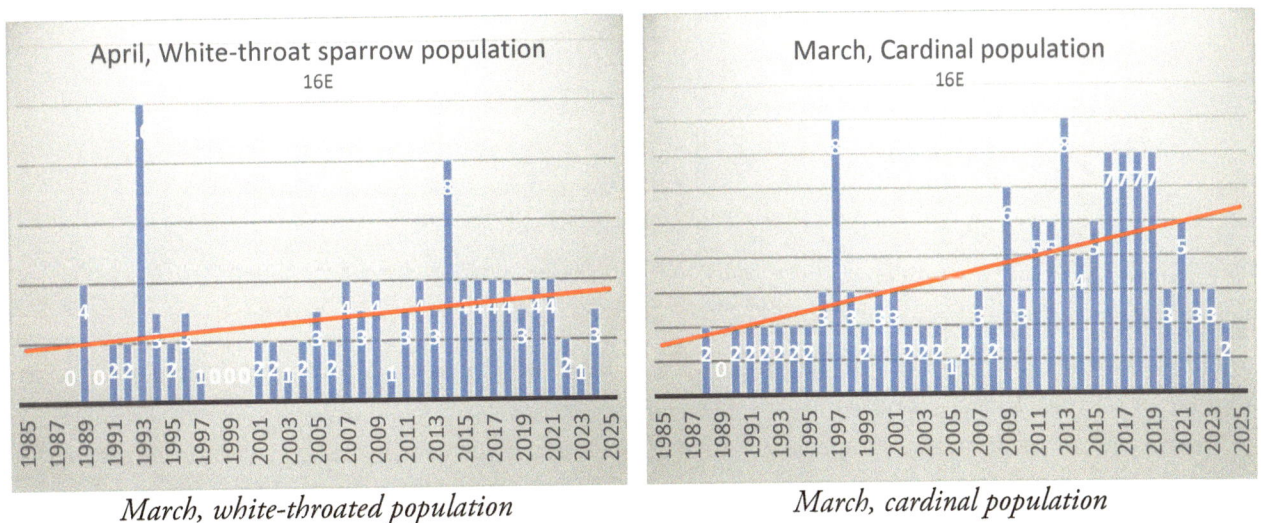

March, white-throated population   *March, cardinal population*

Blue jay population from year to year goes from one to three, averages two. Once in a while, a flock of seven or eight will pass through.

The graphs of a number of species shows an overall view of what is happening to the populations and maybe some anticipation. Over the 30 years, of the seven species, five show a strong decline. Three are permanent residents, the other two are winter residents. Peaks and valleys of the various species populations almost occur during the same year period. Crow population was down to five to eight because of the West Nile virus until 2015, then rose to 20 by 2017. In 2020 I see eight to 10 now and then. Ten miles from here is a large rookery of hundreds.

Fledglings: Blue jay, cardinal, house finch usually brings fledglings to the feeders every year in March. Chickadee and titmouse may bring one fledgling, but not every year.

Migration: The migration of the species depends on the weather from where they are coming from or going to. The built-in signal to start migration relates to the amount of daylight. Once started, the length of migration depends on the weather, the direction and altitude of the prevailing winds, storms, and fog. The weather may cause them to have to land and have a layover, just like planes do. As a result, when they arrive as summer residents or just passing through, it can vary by weeks. Weeks expanded change in arrival time can be disastrous because the food supply of insects or vegetation may have come and gone. I used to observe eight to 10 grackles during March, then after 2007, they dropped to one or none. In 2020, there were 12 flying through.

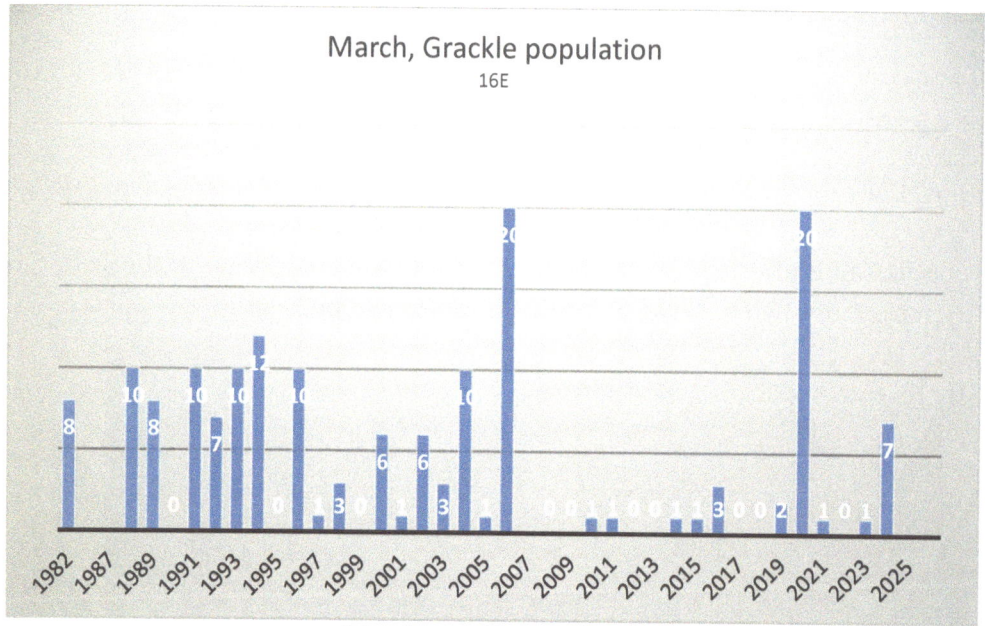

March, grackle population

The migratory are divided into large flocks that fly over of geese, robins or cedar waxwings, and singular or couples like cross beaks, warblers or fox sparrows.

Over a period of 38 years, 31 species of birds have been observed migrating across the area in the month of March. Of those, 10 species have only been seen one to four times. Many birds migrate at night, while I am sleeping. The remaining 21 pass through often, but when in March? Where are they are originating from? Some arrival weeks have moved forward two to three weeks, others moved later by two weeks. Three species arrival weeks have moved from the fourth week of March to the third week of February. Except for saying that a species will be observed in March, anticipation for a particular week is almost out of the question. In 2013, it was reported that different species are migrating further north. Since then, I have been observing new species, such as woodcock and scarlet tanager.

Bed and breakfast birds (birds that only stay a day or a week before moving on) at the end of the month were: 2007, towhee with red eyes arrived a week early and stayed for a week; and 2011, a Connecticut warbler was here for a day. The broad-winged and red-tailed hawks used to arrive for the early spring; in 2020 they were here in February.

Male towhee

Female towhee

Some species were only observed one to four times over the period. The catbird arrived during the first week. The cedar waxwing flock was only observed five times over a 40-year period, each time for only a few minutes during the last week. Fox sparrow arrived from the first week of February to the last week of March, they stayed two weeks before moving on. None have arrived since 2016.

When the flocks of geese decide to go north is a mystery. Coming from the north in the fall migration depends on when the lakes freeze over and availability of food but going north is another situation. How do they know when the ice has broken-up? Maybe they send scouts north. Over the period, flocks would go over between the second and third week. In 2017 there was a very warm November and December, consequently the snow geese did not arrive until January. In 2018 we had a very cold January. Unfortunately, I did not observe or learn when they went north. With construction of the many nearby large warehouse, I no longer observe flocks of geese in my area. I have to go ten miles west to see the over wintering flocks.

Bird arrival or migrating for March (1996 to 2020)							
Species	Quan	Years observed in March without weeks	February	Week 1	Week 2	Week 3	Week 4
Blackbirds	Flock	2002,18		2002/03/18	2013	2004/15	
Broad-wing hawk	2	1994, 2018,19		2005/18/19		2013	2014
Catbird	1	2006				2006	
Cedar waxwing	Flock	1979,82,83,85,90					2017
Carolina wren	1-2	1979,89,90,91 93,96,2001		2019-2020 overwintered			2015
Chipper sparrow							2020
Conn. warbler		2011					2011
Cooper's hawk	1	1994,95,96		2008/10	2013	2005/07	2008
Cowbird	1-4	1988,90,91		2005/06	2015	2015	2015
Duck, mallard	2 to 5	1995,96, 2018	2018 2nd week	2015	2006/12	2016	2008/09/11
Eastern phoebe	1	2006			2006	2015	

MARCH

Flicker	1	1983,89	2018 3rd week	2002/05/14	2006/15	-	2009/15/20
Fox sparrow	1 to 2	2002,03,05		2002/13/16	2005/14/15	2003/07/17	2006/13/14
Goose, Canada	Flock 100's	1986		2015/19/20	2013	2013	2015
Goose, Canada	Flock 1000's	2001		2001/02	-	-	-
Goose, snow	Flock 100's	2006		-	2012	2007	2014/15/17
Goose, Snow	Flock 1000's	2002		2002/07	2015	-	X +120 smaller
Grackle	Flock	1992		1982/83/2007	2015	2011-2017	2016/29
Great blue heron	1 to 2	2005			2015	2013/20	2011
Great horned owl		2006			2011		2006
Hairy woodpecker	1	1989,91		2008 2019-20 overwintered	2013	2017/19	
Junco	5-15	1992		2018/19			
Kinglet	3	1988,90					
Mockingbird	1	1988,89,90		2005/13/14/15/20	2006/12	2006	2017
Pine siskin	Flock	2009		2013		2009	2015
Purple finch					2012		
Redpoll	1	1989					2011
Red-headed wood	1	2001					2006
Red-tailed hawk	1-2	1984,88,89		2005	2006		
Red-winged blackbird	2	2002,03,04,18,19		2002/03/05 2014/15/18/19	1984/2015	2004/17	2010/13/14/20
Robin	1 to 4	1982,83, 2019	2018 3rd week 1985 3rd week	1982/83/2005/10/13/19	1984	2008/17	2007/09/15
Robin	Flock 20 to 30	1979		2012/13/14 2015	-		
Savannah spa							2020
Seagull	1-3	1979,82,83		2013/19	2011		
Sharp-shinned				1984/2018			2019
Song sparrow						2020	
Starling	Flock	1988,89,90	2018 3rd week	2008	2015		2009
Towhee							2020
Tree sparrow	1	2008,10,15,16					2006
White-crown spa	1	1988,90			2005	2014	2005/12
Woodcock	1	2014			2014		
Woodpecker knocking	1	2002			2002	2003	
Total species		**37**	**6**	**24**	**25**	**19**	**28**

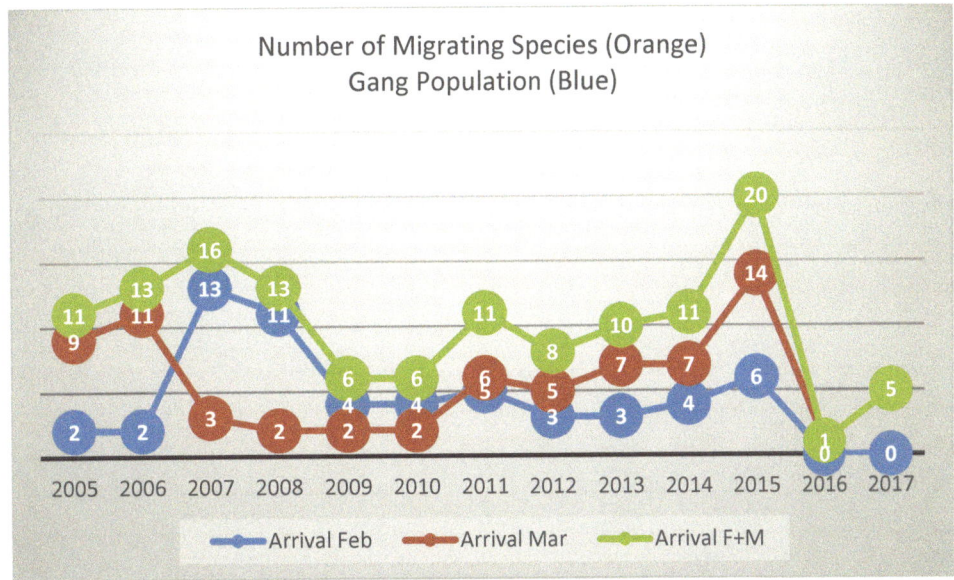

The decline in the number of migrating species in 2009 may be caused from the very dry 2008. We had a very wet year in 2015. The dip in 2016 may be from a warm-dry 2015 November and December, 2016 January blizzard, and warm-dry February. Just looking at the monthly mean temperature of March, when the mean temperature is between 33 to 38F; there is a greater number of migrating species. This showed up 2005, 2006, 2013, 2014 and 2015.

Insects: When the air temperature is 60F the bumblebees will be out of hibernation.

Flora

Lawn: When the average air temperature reaches 40F and the year-to-date rainfall is 10 inches, the lawn will green up at the end of the third week. When the temperature stays in the lower 30's, it will take up to the second week of April for the lawn to green up even with 10 inches of rain.

Perennials: Daffodils are very weather sensitive, especially temperature. They bloom within the period from the fourth week of March to third week of April. Crocus are temperature sensitive. The yellow crocus may bloom on the fourth week of February and as late as the fourth week in March. Sometimes the purple crocus bloomed two weeks later than the yellow, other years it was in reverse. The silver maple blossoms and skunk cabbage breaks through the third week. Windflowers bloom in the lawn and garden the last week of March into the first week of April. The fourth week ferns are up 12 inches, azalea buds are swelling, miniature iris are out. Dogwood buds are opening. The white oak is getting leaves.

Vegetables: Chives are green and rhubarb crowns break through the middle of the fourth week.

So, ends turbulent March.

Index of April

April Overall	61
Weather	61
First Week	63
Second Week	69
Third Week	70
Fourth Week	75
Review of Data	78
Weather	78
Fauna	79
Animals	79
Birds	80
Population of Species	80
Table of the Gang	85
Migration	85
Migrating Table	93
Insect	94
Flora	94
Perennials	94
Shrubs	96
Soil Temperature	96
Trees	97
So, ends April	98

April

Week number for the year of 48 weeks					
Relative Month	Month/Week	First	Second	Third	Fourth
Previous	March	1	2	3	4
Present	**April**	5	6	7	8
Following	May	9	10	11	12

Overall Summary of April

Where March is the awakening month, April is the beginning of the adjustment month. Windflowers are out in the beginning of the month, then the daffodils open, followed by the heavy infestation of pollen from the maple trees, causing allergy suffering. The fiddleheads of the ferns are unraveling further from the captivating earth. The end of the month is the beginning of the azalea show. I might catch a glance of a brown thrasher or a warbler passing through. A pair of downy woodpeckers start making a cavity in a dead centennial maple. The male goldfinch, white-throated sparrow, and house finch are changing to their bright come-see-me finery. The Carolina wren has migrated north just as the male house wren arrives from the south. The hawks are also migrating north and looking for a quick energy replenishing meal. On a warm afternoon a black or yellow swallowtail or a copper hairstreak butterfly may flit from one flower to the next.

Weather

The average high temperature on the 1st is 52F and the low is 34F.
The average high temperature on the 30th is 65F and the low is 42F.
Extreme daytime high for the month: 95F, 1926, followed by 92F, 2010.
Extreme nighttime low for the month: 12F, 2nd of 1923, followed by 13F, 3rd of 1923 and 2013.
The full moon of April is the Pink Moon for the wild phlox, one of the earliest wild flowers to bloom, or the Fish Moon, because that is when the shad swim upstream.

APRIL

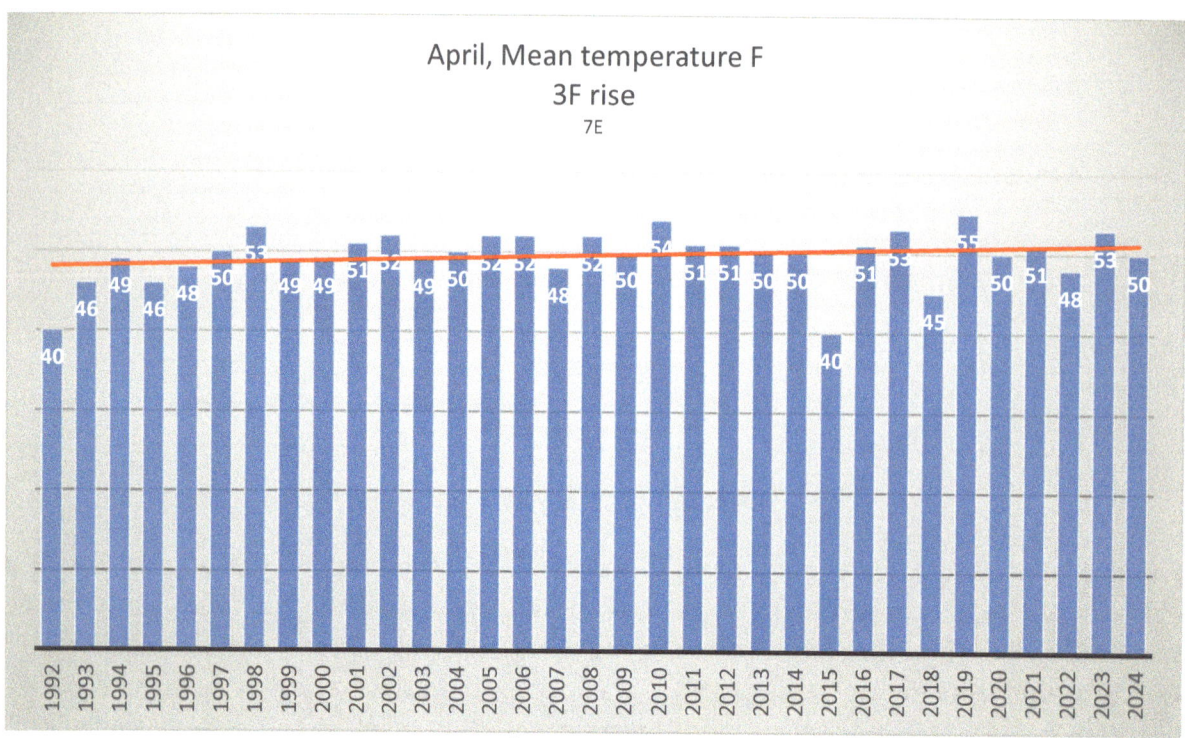

April mean temperature

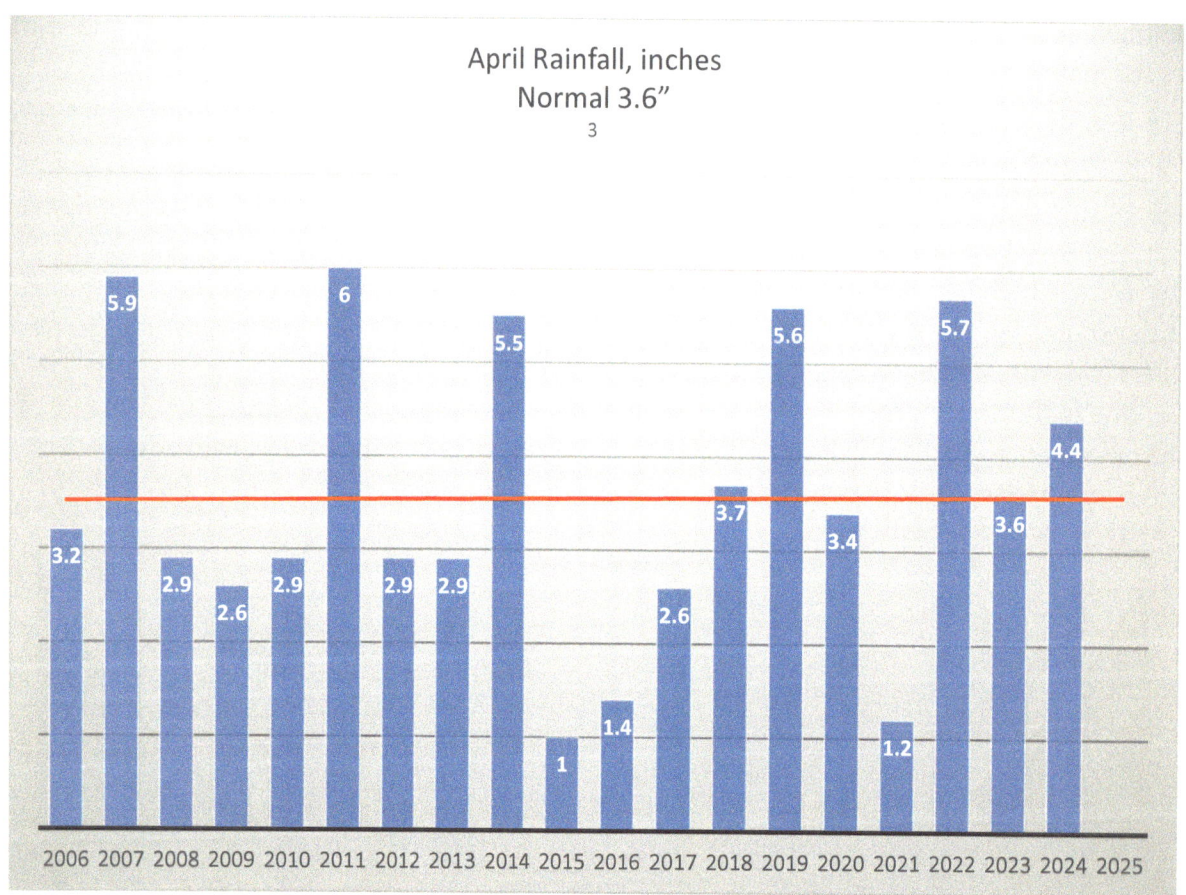

Rainfall for April

Year of 2000's	06	07	08	09	10	11	12	13	14*	15	16	17	18	19	20
Actual month, normal 3.5"	3.18 Drought	5.9	2.9	2.6	2.9	6.0	2.9	2.9	5.5	1.8	1.4 Dry	2.6	3.7	5.6	3.4
Actual year, normal 12.6"	11.7	14.0	16.9	7.1	15.3	18.6	8.0	11.3	16.7	10.8	11.4	13.0	15.6	17.1	12.0

*coldest April Easter in decades

The weather in April dictates what can happen for the entire growing season, including the colors of the leaves in the fall. This is another yo-yo weather month. We can have lots of rain for the first week and then have no rain for the rest of the month. It can be so dry that the ground cracks.

The temperatures can be way above normal the first week then drop to normal, then rise again and then go back to normal. Or the temperature can be below normal, then a nor'easter will bring in lots of rain. Or we can have a red flag (dry weather) warning. Some years (2018 for one) we had a heat wave (five days of hot weather) at the end of the month, that fouls up the growing cycle.

Try this for a month: go from cool to hot (92-95F for a heat wave) then cool to a nice 77F. Thunderstorms and heavy rains come with the high temperature, but changes to snow. Snow can be squalls of three to 11 inches. We have even had one inch of snow on the 29th! Algae forms in the pond, and then we get a frost or heavy fog. When there is a morning dew and not too much of a chill, it is grand to step outside and breathe in the sweet spring fragrance of Mother Earth.

From Lehigh County Extension, information for Bethlehem, PA, March, 2008

Probability of 32F or lower in the spring:

90% March 31st
75% April 7th
50% April 14th
25% April 22nd
10% April 28th

First Week

Spring blooming

Weather: In 2016, 62F with 50 to 60 mph winds, the temperature drops to freezing. Awoken by thunder at 4:30 a.m., I look out to see snow flurries rapidly coming down. The next morning the yellow daffodils that were blooming are frozen; the blossoms have gone slimy limp at the end of the stalks. Their color is now a transparent creamy white.

Animals: Squirrels are eating the buds of the dogwood tree and candy bar. A rabbit is eating the cattail sprouts and the new hosta shoots, plus the lawn.

Cottontail rabbit

Squirrel enjoying a Reese's peanut butter cup

Looking out my front window, I see a squirrel on the other side of the 50 foot asphalt circle with a big walnut in its mouth. I never took noticed how a squirrel hops like a rabbit. I have seen them run and take short hops, but not a continues hop. Well this one hops across the circle, then up my 40 foot driveway, makes a left across my 50 foot lawn, makes a left, goes another 50 feet back into the circle, crosses the 50 foot circle again to my driveway, up the drive again then jumps on to a picket fence and looks around. All this time hopping with the walnut. It paused but did not stop. That is almost the length of a football field. I lost track after that.

Birds: At 7:30 a.m., cloudly, 40F, I first hear, then look up searching, and spot a flock of 100 geese going north.

Starting the first week and throughout the month, various birds arrive, some leave and others for the summer. Brown thrasher and eastern phoebe pass through. Sometimes the catbird and mockingbird arrive (2020, mockingbird arrived the first week of March, 2019-2020 had a non-existing winter, were back 2021-2024) for the season this week, but mostly the fourth week. They are hungry, especially for peanut butter.

Brown thrasher

Catbird

Mockingbird approaching the peanut log

Downy woodpeckers and flickers are knocking out their invitations for a mate on the trees and telephone poles the first and second week. They also use the sides of houses. The male downy is identified by the little red spot on the back of his head. Throughout the month various members of the blackbird family are passing through as individuals and flocks; there are starlings, grackles, cowbirds, crows, and red-winged blackbirds. Male robins are fighting for their territorial rights. Shortly after the fights are settled, I see a couple gathering nesting material.

Birds need calcium to strengthen their eggshells. If they do not eat enough calcium, the weight of the mother bird can crush the eggs. It is suggested by Cornell University to augment the bird feed with shells from grocery store eggs. After washing the shells, put them in the oven at 240F for 20 minutes. Do not put them in a microwave, if you do, your house will smell like rotten eggs. Let them cool and crush them to the size of a penny to a nickel. Blue jays, titmice, and red-breasted nuthatches will devour them.

Crushed egg shells

Crow with pine needles

I spot a female crow with a twig in its beak; the myth says she is looking for a mate. Starlings try to take over house sparrow houses, but lose. Starlings are making a glug-glug sound!

A pair of mallard ducks used to arrive the first to second week. Temperature or rainfall does not seem to affect their arrival time. By the end of the month there may be four to six. Once here, they usually make a circular pattern fly over the circle just before dusk all month long.

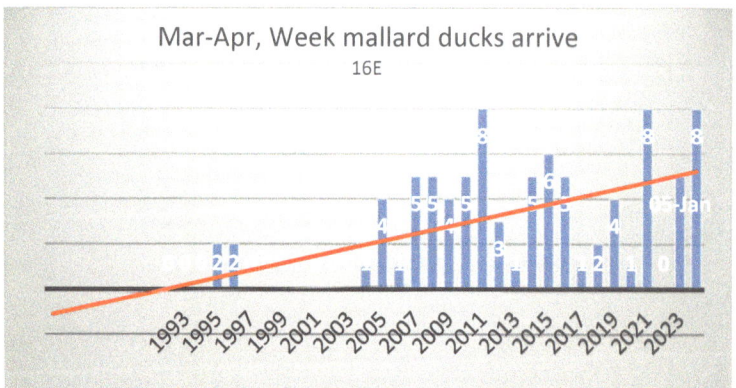
Mallard ducks

<u>Insects</u>: When it is 50F the bumblebees are flying around.

<u>Lawn</u>: This usually is the first week to cut the lawn. Blooming of the yellow bell is significant because between blooming-wane is when turf fertilizer and crabgrass preventor is applied. Then four to six weeks later it is time to apply broadleaf weed preventer.

<u>Perennials</u>: Depending on the temperatures of February-March the windflowers come out the last week of March, but most likely the first week of April.

 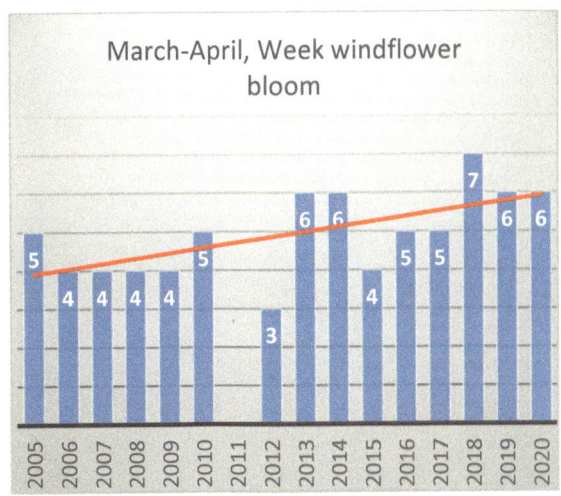

White and blue windflowers bloom the last week of March into first week of April

What I think of as a normal April can go like: hellebore is out, the apples of the Mayapple are forming under their umbrella. The ferns behind the pond are up, a different species arises the third week. Mayapple is poking through, but with below normal temperatures the emerging may be delayed two weeks. A mixture of colors of daffodils is out on the third tier. The little yellow daffodils usually come out the end of this week and bloom for seven to 10 days. One year, this week was so dry that the daffodil stalks were drooping. Even though it was dry, earthworms came out onto the circle.

Grape hyacinth and daffodil leaves in the first week

APRIL

Purple hellebore (Easter Rose)

Cinnamon fern behind pond

Ferns by the pond are up, then depending on the weather, they will unwrap to six to eight inches in one day! In 10 days it may be 12 inches and in 18 days it may be to 18 inches; an inch per day!

Pond: The temperature goes above 50F, frogs are sunning at the edge of the pond.

Shrubs: Yellow bells seem to be very temperature sensitive. When March is warm it blooms the first week, with normal temperatures it blooms the second week and when unseasonably cold, the fourth week. Blooms last 10 days.

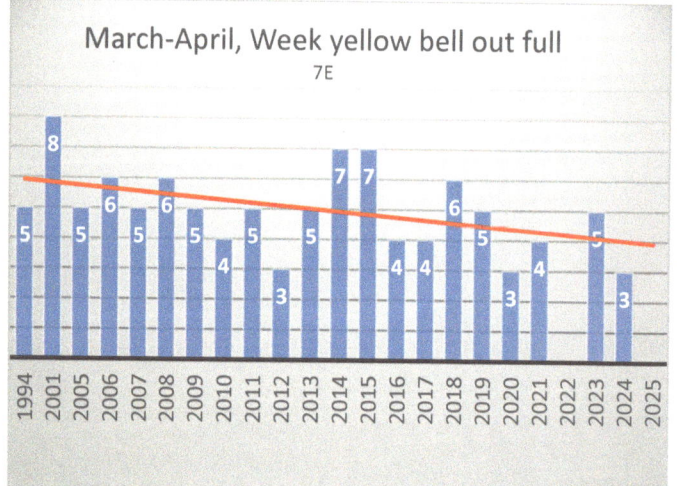

Yellow bell

White or pink clusters form the first week in the third tier japonicas, when the temperatures are normal, warm or cold.

Trees: Spray the apples for scab.

Vegetables: If the thyme plant has survived the winter, green growth now appears at its base. Put 10-10-10 on the vegetable garden and first tier flower garden. Apply 5-10-10 on rhubarb and asparagus.

Second Week

<u>Birds</u>: A cardinal's mating ritual that occurs this week is when the male feeds the female. A redstart stays for a couple of days. Two male house sparrows were having an intense battle over territory or a female, when a sharpy swooped down and took one, ending the battle. Bed and breakfast birds are the brown thrasher and magnolia warbler.

Birds and animals need water, be it a hot summer or a frigid winter. Its beak size and shape will dictate how a bird partakes its drink. The following table lists how fauna drink.

	How birds and animals drink
Cardinal	Dips down, raises head horizontally, then partially up to look around; another raised head high.
Cardinal	After a rain, Card reaches up from a branch and takes a droplet off the end of a rhododendron leaf.
Chickadee	Dips head down, raises it horizontally.
Chipper sparrow	Dips head down, drinks. Brings head up horizontally to look around, dips head down again, brings it up horizontally, then flies away.
Dove	Dips down, keeps head down to drink.
Grackle	Dips head down, lifts up 45 degrees, then swallows.
House sparrow	Dips down drinks, flies away.
Jay	Dips down, slowly lifts head above horizontally, repeats.
Junco	Dips down, immediately raises above horizontally, dips again.
Chipmunk	Puts head down, keeps head down until done (20 seconds).
Squirrel	Puts head down, keeps head down until done (20 seconds).

<u>Insects</u>: A black swallowtail butterfly flits through.

Black swallowtail

<u>Perennials</u>: Yellow iris behind the pond is in bud. This iris also re-blooms in late fall. Solomon's seal is up.

Solomon's seal

Yellow daffodils are in full bloom and wane during the fourth week. Variegated hosta are up the second week.

There are 13 types of daffodil

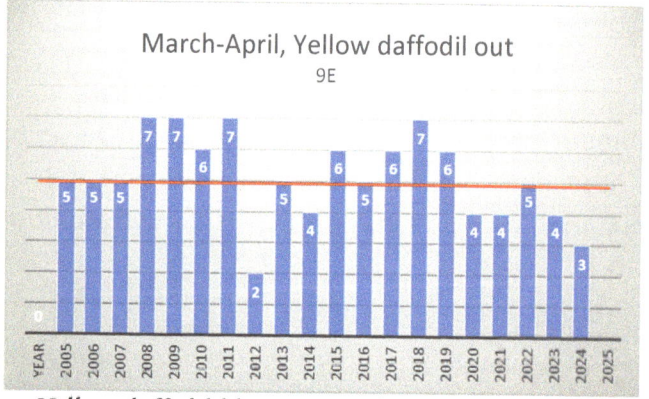
Yellow daffodil blooming week – 2012, 2020 and 2024 were warm winters

The ferns behind the house are in full shade and come up a week later than the ones in the sun. Clematis on the silver maple will be in full leaf. The Mayapple is in full flower.

Pond: The 60-year-old skunk cabbage behind the pond may up 10 inches.

Shrubs: The white lilac is getting flowers. The japonica by the upper gate is in full white cluster.

Trees: The Empire apple gets leaves the second or third week of April, but will not blossom until the fourth week. The Winesap gets leaves a week before the Empire.

Third Week

One evening in 2016, 8:00 to 8:30 p.m., it was clear, calm, 66F, I just sat on my maple tree bench and observed. The only sound was that of the traffic on a distant highway. Then silently the house finch, cardinals, doves, and jays came to the feeders as couples. They fed and moved on, no calling or song. After eating, the doves sat in the branches and softly coo-coo-cooed. A white sulphur butterfly came flitting to a PJM blossom. Silently a pair of robins hopped along the grass to find worms. Soon two male cardinals arrived chasing each other, and then another male and two females came. They all silently fed and flew on. As the sun was setting I spotted the silhouettes of three squirrels high in the tall maple, they were far out on the branches eating the blossoms. They seemed to fly from branch to branch. Another squirrel was traveling along the power line highway with a big nut tightly grasped in its jaw. As dusk came, I made out a pair of downy woodpeckers with their talons gripping the bark of a heavy branch, in a chase circling around and around. They met up and flew on, all in silence. Now it was almost dark, the silence was broken by the songs of a cardinal and robin. The show is over, time to go in.

Weather: This week we can be getting a heavy rain one day and a frost the next. The effect of the cold winter in 2015 was still showing up the third week when suddenly we got two hot days. The second day it got to 84F; delayed perennials came out in hours. The ferns sprang up 12 inches overnight; the ground was very dry. The windflowers opened on the 17th; normally they open in the first week.

Animals: The lawn in the evening looks nice and green and flat; in the morning there is a large lump of grass in the middle of the lawn. During the night a female rabbit has dug a hole, lined it with her belly fur and dropped her kits. Sometimes she may choose the vegetable garden; she can jump over a two foot high chicken wire fence, dig a hole next to the lettuce and drop her kits. What an insult!

Birds: A pair of downy are still drilling out a cavity in the silver maple stump; it takes two weeks. There are lots of chainsaw size chips at the base of the stump.

Downy cavity seven inches in diameter by 16 inches deep

Bright yellow goldfinches arrive from the south. The head strips on the white-throated sparrow have fully changed to bright yellow. Head of the male house finch turns bright red. A pair of red eyed towhee visits for a couple of days.

The anticipation of the arrival of migrating species is interesting. A specie will arrive within the same week year after year. Many years it is a single bird, other years it may be a pair. The average life of a wild bird is three to five years. So how is the knowledge for a towhee to drop by my house passed on to a new relative? Not only that but at the same location in my yard and almost at the same time of the day!

The great blue heron flies northwest early in the morning and returns at dusk. A grackle hits a window as it was being chased by a sharpy. This time the sharpy missed, but another time the sharpy got a grackle in mid-flight.

Great blue taking off from the pond

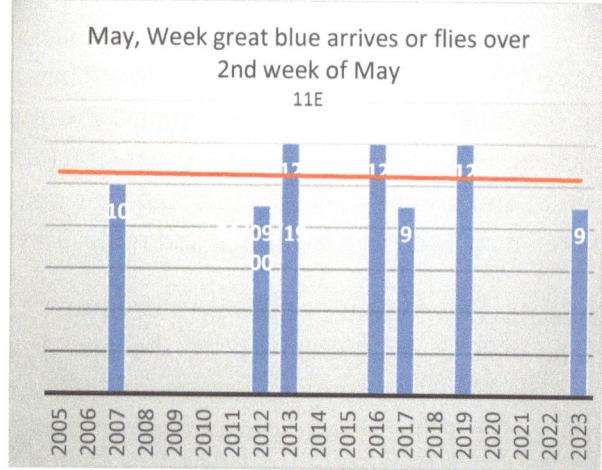

Now that the crows have built their nest, they are occupied by chasing a broad-winged hawk from attacking it.

Insects: There are 15 bumblebees chasing each other. A small copper hair-streak butterfly goes through.

Perennials: Hosta should be in leaf. A Jack-in-the-pulpit is sticking through the ground. Dogtooth violets are up; they take 12 to 20 days to bloom. Royal hosta are up, ajuga, mountain pink, Virginia bells, tulips are in full flower. Fall clematis has new leaves; with a hot March, the leaves arrive the first week of April.

Jack-in-the-pulpit

Dogtooth violet (left) and ajuga (right)

The grape hyacinths along the border of the first tier will be blooming for the next two weeks. The double blooming iris (now and August) are out full. The dark blue ajuga is out by the steps.

TRENDS DUE TO CLIMATE CHANGE

Double blooming yellow iris

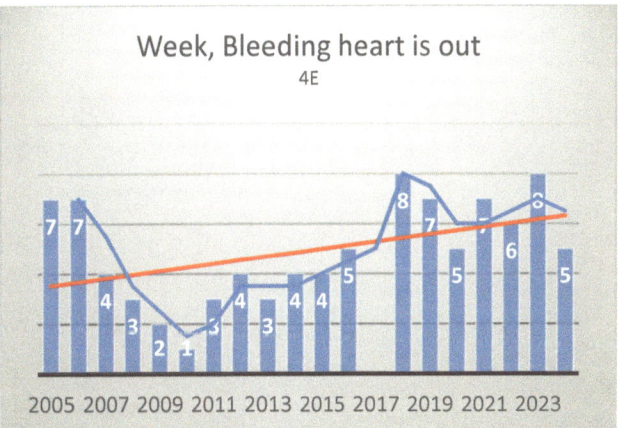
Bleeding heart

Virginia bells

Ajuga

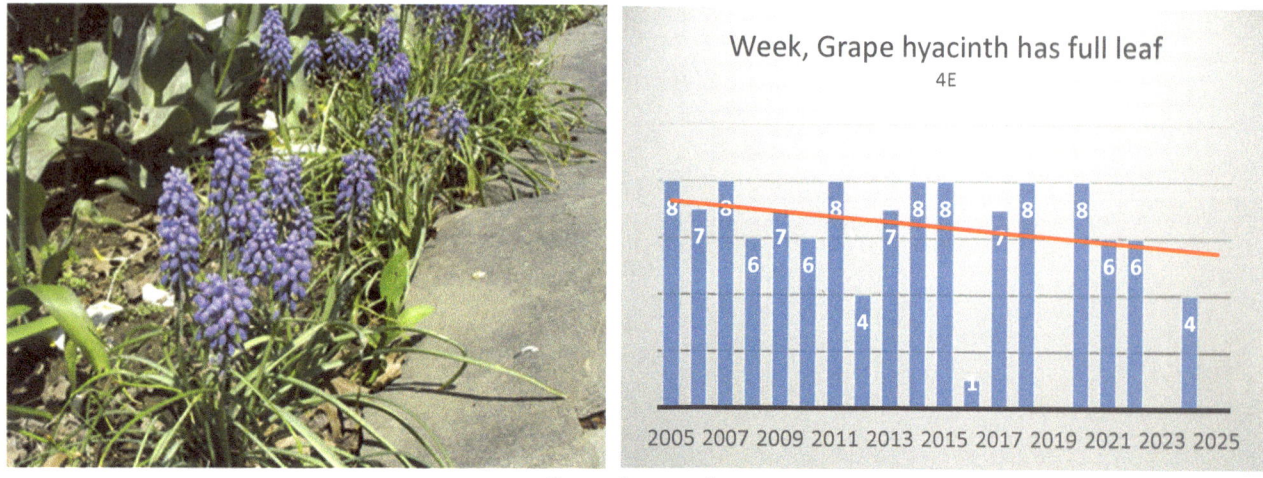

Grape hyacinth

Pond: Ferns by the pond will be fully unwrapped to their full 18 inch height.

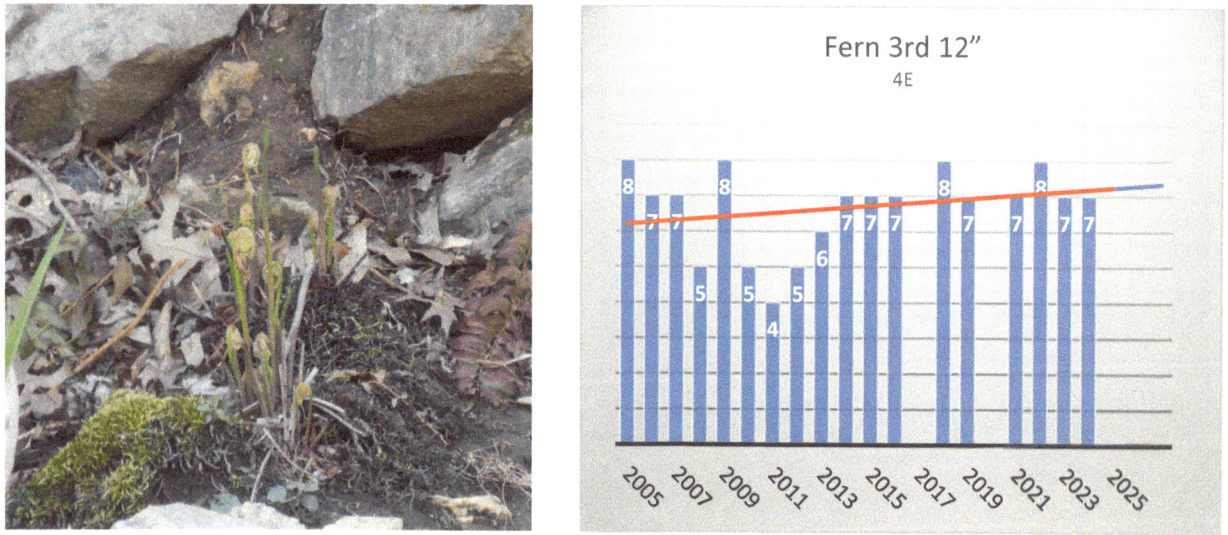

Fern unwrapping by the pond

Shrubs: PJM bloom the third week. When March and April are cold the PJM bloom in the fourth week. Flowers last seven to 10 days before waning, the trendline indicates that they are flowering longer as the years go by. The trend also indicates that they are blooming a week earlier in the past 25 years. The PJM behind the hemlock usually blossoms two weeks after the PJM on the second tier. I do not know if it is due to location or species. From full bloom to new growth is from one to four weeks depending on the weather.

PJM on second tier

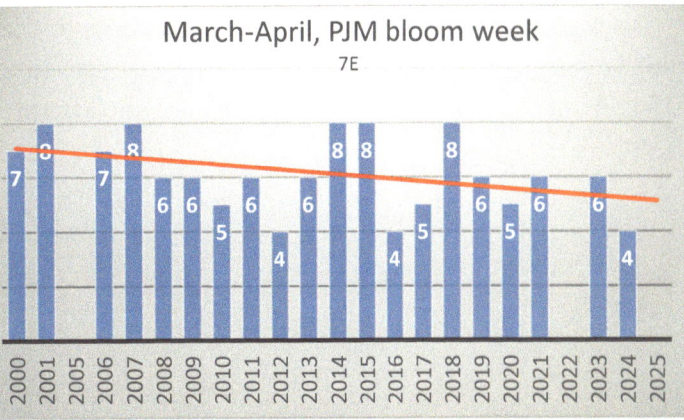
PJM second tier bloom week

Trees: The serviceberry tree is out full. The silver maple is in full blossom. The pollen forms the second through third week, it produces a carpet on the patio. I cannot help walking across that carpet and then drag the pollen into the house. However, it will form the fourth week of March when the temperature is high and the rain is low. The hawthorn tree usually gets leaves the third week. Evergreen growth is usually the third to fourth week in April. In 2018 growth occurred the last week of May!

Fourth Week

The fourth week ferns are up 12 inches, azalea buds are swelling, miniature iris are out. Dogwood buds are opening. The Mayflower is in full umbrella. The poplar and birch are getting leaves. The dandelions are in full bloom. The grape hycinth will bloom for 10 to 12 days.

Birds: The male house wren arrives the last week. It investigates three or four potential houses for his mate. The chipping sparrow arrives throughout April, but most of the time it is the last week. For many years this was the week that the Carolina wren left, but I am not sure what it is doing anymore. In 2019-2020 a pair overwintered. This is the week that the juncos go north.

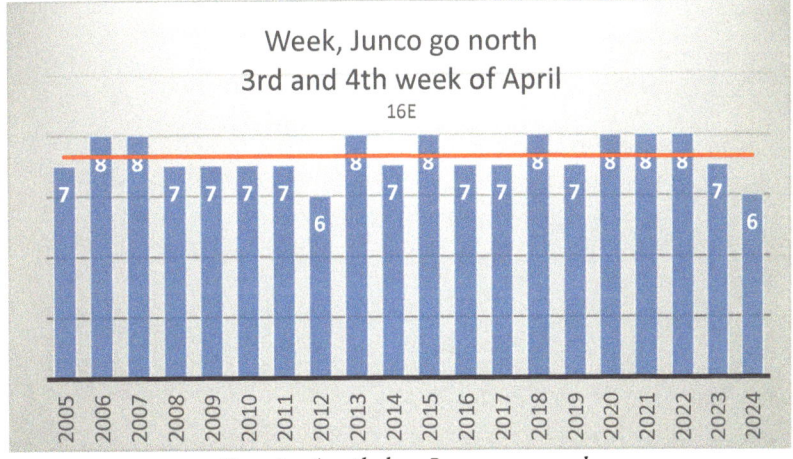
Day in April that Junco go north

To my surprise I had a red-tailed hawk come to the feeder to take eggshells! Another surprise, two blue jays are going after earthworms! They are also singing a different type of song, maybe it is an after-effect from the worms! This is the week for the arrival of the tree and barn swallows.

Barn swallow

I feel fortunate to see barn swallows fly over with their swept-back wings and orange under-breast. I have no environment to attract them. They prefer a bridge, barn or an eve under a large building. Sometimes the tree swallows will take residence in my front birdhouse. They may try to evict the house sparrows who were there first.

Perennials: The ferns on the third tier are a different species and do not break ground until the fourth week; the rate of growth is an inch to an inch-and-a-half per day. They will be 24 inches tall in 24 days!

Mayapple umbrellas

Miniature purple iris

Ferns on the third tier

Pond: Water iris leaves are 10 to 15 inches tall. Below the surface of the pond, the water lily leaves are heading upward.

Shrubs: This azalea is one of two over 75 years old. It comes out in the fourth week of April to the first week of May. It blooms best every other year.

75-year-old azalea

Trees: Korean dogwood forms leaves. Pine tree candles will be lit with pollen. Spruce tree growth will develop no matter if the temperature is normal, high or cold. The birch tree forms leaves and pods unless it is warm, then they form the second week. If the temperature is cool, development will be delayed until the first week of May. Dogwood trees will be in bud, which the squirrels will eat. The poplar will be leafing. Leaves form on the pin oak from the last week of March to first week of May, a six week spread! Pollen from the maple tree buds will be very high during the last week. April 2018 was cold, so the pollen was not released until the last week as an explosion, causing many allergy sufferers a lot of misery.

Vegetables: If the soil is not too wet, it is now time to turn the vegetable garden and plant lettuce and spinach seeds. I take a handful of soil and squeeze. Water should not drip out. If is does it will compact. It should crumble in my hand. Then it is okay turn it over. Plant more lettuce seeds. Pick asparagus four days after they emerge.

Review of Data

Weather: The climatic conditions that affect flora growth in a month depends on the amount of rain from the beginning of the year to date, rain during a month, humidity, the average monthly temperature, the high and low temperature of a month, the cold or hot extreme temperature waves, the position of the jet stream, the amount of sunlight, sun spot storms, and temperature of the soil. The jet stream and solar storm affect weather over a period of two to 12 months. You may wonder why they bloom at all!

If the conditions are right during the first two weeks, when you go out in the early morning and inhale deeply, the air will have a sweet earthly fragrance. Ah, it is very refreshing!

As the ground warms up, fog may form late at night. This can go on into the third week.

Frost is not out of the question on the last week of the month and into May.

In a housing development, the flora is put into a spot dictated by man (location, location, location). Flora consists of the perennials, trees, and shrubs. When there is a warm-wet January-February, week one of blooming can occur the first week of February or a cold-dry March it will push week one to the first week in April. In all cases the chronological order of blooming will basically stay the same.

The two prominate parameters that effect the growth is amount of rain to date for the year and temperature. The amount of rain during the month, did not show as much effect as the progressive amount of rain to date for the year. The mean temperature of April has risen three degrees since 1992. The amount of rain from January through April has been increasing resulting in wetter springs.

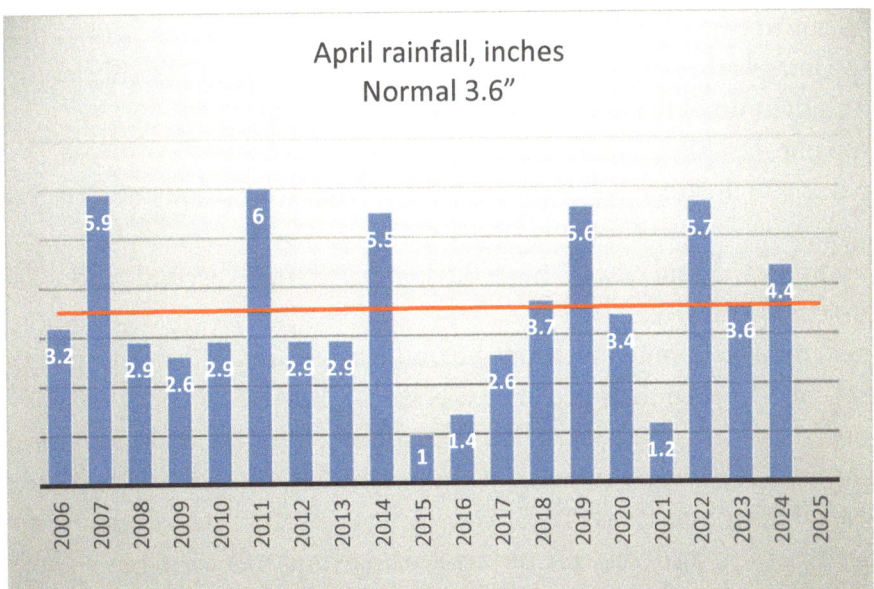

April rainfall 2006 through 2024. The average temperature in April was 50F for the period.

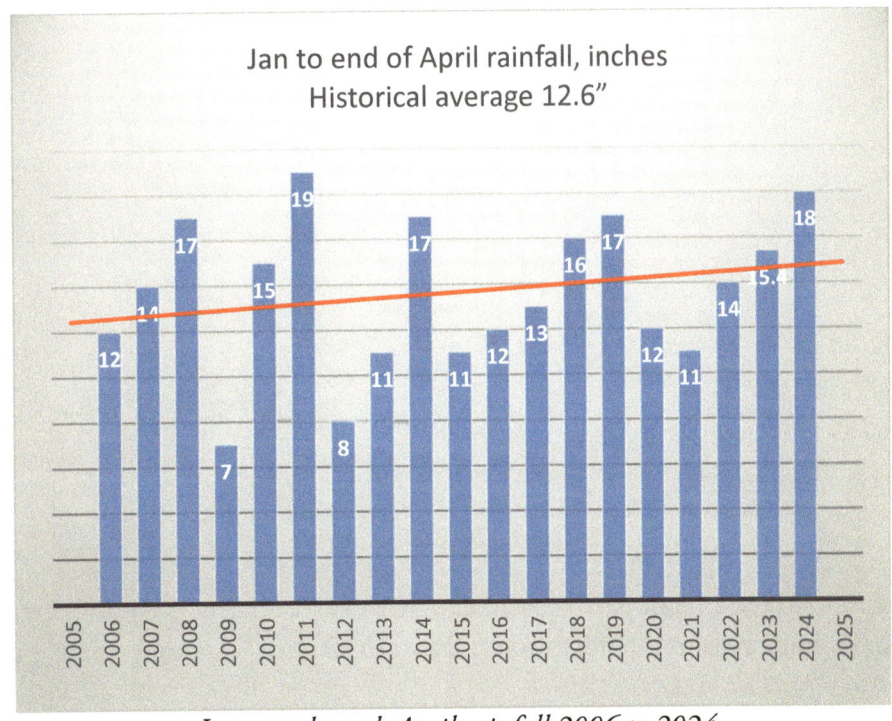

January through April rainfall 2006 to 2024

Fauna

pop = population

<u>Animals</u>: Bat kill, a report in the newspaper says white-nose syndrome fungus is killing big brown bats along the east coast caves. It was discovered in Pennsylvania in December 2008. It first surfaced in New York state in 2006. There could be a rapid extinction in seven to 30 years. Bats in the wild live 30 years, they have one pup a year. They come out of hibernation in April. In 2016 and 2017 I observed only one bat during the summer and none in 2018. In 2019 to 2023 there were two.

Chipmunk, pop one to two, come out of semi-hibernation in March. They run around eating spilled bird seed. Like the squirrels they have chases. One male is trying to gain territorial rights. The winning male, with his tail straight up in the air, then chases after a female.

Groundhog, pop one, last observed March 26, 2024.

Opossum, pop one, are nocturnal and not seen very often. The last one observed was the first week of March 2024.

Rabbit, pop two to four. Their kits are born during the first and second week. Usually the kits are running around on the fourth week.

Skunk, pop one. You usually smell their left-behind fragrance in the morning in the last two weeks of the month. Last time I smelled their presence was spring 2024.

Squirrel, pop one to five, are doing three things all month. One, the male is chasing the female, they stop, do their thing, rest for five minutes, then they continue the chase. When they have successful interaction, the second thing you see is they are taking leaves, cattail stalks, and evergreen branches up the tall trees to make a nest. Third, they are out on the dogwood tree branches eating the young buds.

Squirrel with cattail stalks

Vole, pop one to two. I see them dart around the rocks surrounding the pond. Strange, soon I see many voles, then for some reason the next month they are all gone! Are they victims of hawks, fox or starvation?

Birds: This is a list of resident species; some are permanent, others are winter ones and others are summer. The list is for the April's population and activity. Where I found a correlation with a species relative to flora it is noted by *italic print*. The *italic* is also shown in the list of migrating species.

Blue jay, pop one to four, starts to sing in the first two weeks.

Canada goose, pop is 20 the first week, then drops down to one or two.

Cardinal, sings loudly the first week, continues to sing throughout the month but not as loudly. Pop two to seven for the first two weeks, then down to two to three. They may bring one to three fledglings the first two weeks. Young males fight for territorial rights during the second to fourth week. They may have mange throughout the month.

Carolina wren, pop is one to two throughout the month, they sing loudly the second week. Some years they seemed to leave the third week and then others, or the same wren, arrived the fourth week.

Catbird, pop one to two, arrives between the 27th-30th. They then become members of the gang. They especially like smooth peanut butter.

Chickadee, pop two to four, is stable throughout the year.

Crow, normal pop of two to four, it increases to 10 to 12 when they flock together to chase broad-winged or red-tailed hawks from the area. I have seen a female with a twig in her beak on the third week. This could be for nesting or I heard a tale that it is a signal that she is ready to mate.

Downy woodpecker, pop one or two, mostly one. The male starts tapping on a telephone pole the first week and taps throughout the month. From 2016 they have resided all year as a pair

Mom downy feeding her son

Goldfinch have been overwintering since 2012; now and then they come to my feeder in one or two pairs. My neighbor has a thistle tube separated 30 feet from other feeders outside of a second story window and has five to six finches come for breakfast and dinner regularly. With a normal average temperature in April, the overwintering male goldfinches start to change color from drab green to gold during the third week. When there is a very warm March, the color change starts the first week of April. Likewise, when there is a cold first week in April the change takes place during the fourth week. The males start to sing for a mate the third week. The fully color-changed finches returning from the south arrive the last week of the month. *The change of color occurs a week after the Empire apple gets leaves.*

Goldfinch

House finch, pop fluctuates from year to year from two to four to six for the month. The species stays as couples throughout the year. Unfortunately, this species can get conjunctivitis throughout the year. To attract a mate, the head of the males turns to a light red the third week of the month.

Head of the house finch turns red

Male house sparrow

House sparrow, pop can be two to six. It is very aggressive in getting and keeping a house.

Junco, pop varies from one to six over the years. The first two weeks they chase each other in fantastic acrobatics across, up and over and around the shrubbery. Then on the third week they go north, which is the *week that the Virginia bells are blooming, the same time that the house wren is arriving.*

Mourning dove, pop one to six, about the third week the unabashed bulls chase the hens.

Coffee klatch of 12 mourning doves

Red-bellied woodpecker, pop one, first arrived as a winter resident in 2013, then became a winter resident. In 2020, a pair overwintered.

Red-bellied woodpecker with a bad hair day

Red-breasted nuthatch

Red-breasted nuthatch, pop two, was a winter resident for many years. Starting in 2012, *it or they arrived a week after the Mayflowers were up.* Last sighting was the last week of 2014, then they came back as a resident pair in 2016.

Sharp-shinned hawk, pop one to two, is here all month. Out of nowhere it usually dives in for breakfast an hour after the gang first arrives. It misses many times while going for a dove. Sometimes it hits the back window or the house itself. It dusts itself off and shakes its head and goes for another try.

Song sparrow, pop one or two, is very elusive, but its song is always refreshing. I observed it gathering nesting material the fourth week. *It usually does not start to sing its wonderful song until the first day of spring.*

Song sparrow

Titmouse, pop one to two, likes to sing during the first two months of the year.

Titmouse

White-breasted nuthatch

White-breasted nuthatch, pop two, started to overwinter since 2018.

White-throated sparrow, pop one to four for the first week, then the pop decreases during the month. The whitish-gray throat and dull yellow head stripe change to pure white and bright yellow any week during the month.

White-throated sparrow is gray breasted in the winter

Table of the Gang

April Species/Quan.	Gang members and quantity for April quan/date R=resident, G=gang													
	05	06	07	08	09	10	11	12	13	14	15	16	17	18	19
Blue jay R,G	1	4	2	2	1	1	2		1	2		4	2	2/7	7/24
Cardinal R,G 9/20mange	2	4	3	1	6	4	2	2	6	2	2	4	3	2	4/5
Carolina wren R, G			2	2	1	2	1	2		1	1		1	0	
Catbird G arrive date			1	1 27	1 26		1 28	1 23		1 8			May	1/28	1/24
Chickadee R,G			3	3	4	2	2		2	2	2	2	2	1/2	
Chipper spa G	2		3	3	4	2	2		2	2	1	5	1	1/2	1/2
Downy wood R,G	1	2	2		2	2	2		1		1	2	2	0	2/2
Goldfinch R,G change color April date March date	2 25	5 21	4 24 30	5 16 29	5 16 29	2 12	1 30	1 5 28	2	3 15 25	1 9	1 3 28	5 11	1/10 18	2/105
House finch R,G	2	6	4	6	4	2	4	2	4	3	2	4	5	6/4	2/4
House sparrow R,G	3	3	4	4	4	4			2	5	4	2	2	3	2/7
House wren		1/24													2/24
Junco chase date	4	3	6	3	4	1	1	1	2 29	1	1	2 3,7	3 2	7/10 Mr19	2/4
Mockingbird													1	1/2	
Mourning dove R,G	3	4	3	2	4	2	2	3	2	6	7	6	4	2/4	2/2
Red-bellied wood													1	1/22	2/9
Red-breasted nut													1	0	1/2
Robin															2/4
Savannah sparrow				1	2							1	1	1/1	
Song sparrow	1	3	2	1		1	1	1	1	1	1	1	2	1/1	1/2
Titmouse R,G	2	2	2	2	1	2	1	1	2	2	1	1	1	2/2	1/2
White-breast nut Arrival date	1		2	2			1			1 24	1	1	1	0	1/4
White-throated sparrow change date													4	4/2	4/18 5
Total gang	11	10	13	13	13	13	14	9	12	14	13	14	19	16	18
Population	23	37	43	38	43	26	23	14	27	34	23	36	42	38	46

Migratory Bird Species During April

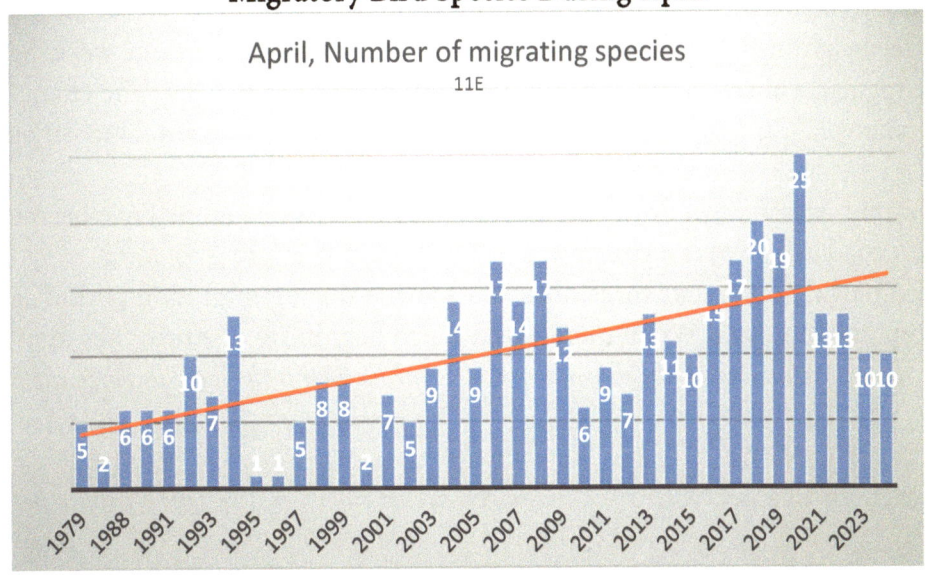

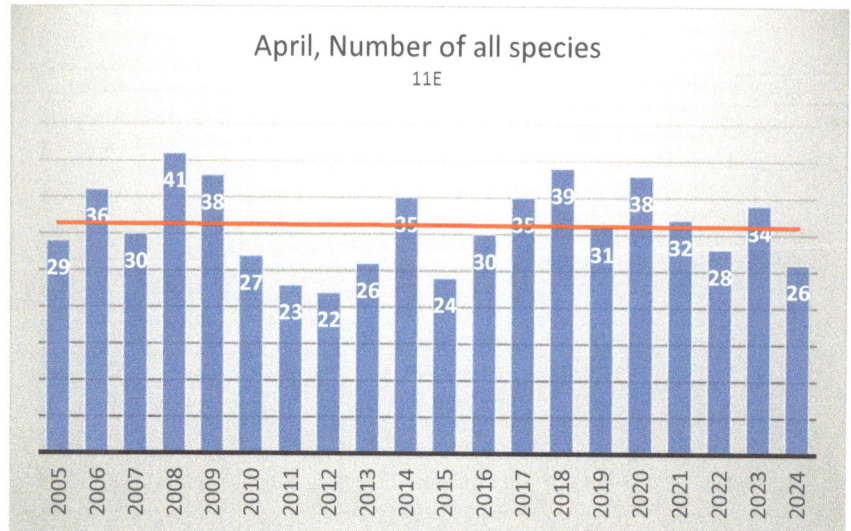

Over 19 years, the quantity of total species has changed from 22 to 41. Twelve were only observed once or twice over the period. Some migrating species have bed and breakfasted for two to three days. Others I just happened to look out and see them get a drink and some seed before moving on.

Weather affects the quantity of migrating species. Not knowing the weather condition of where the birds are migrating from, but reviewing the data, the higher the average temperature for March and April the greater the number of species. Likewise, the greater the amount of rain the higher the number of species. Temperature seems to have a greater affect than moisture. The more moisture and temperature, the more insects there will be available for the migrating birds. Over the past 25 years the average temperature of March has increased by five degrees and that of April by three, resulting in buds and insects developing sooner. This can be a positive in that food is available when birds arrive early or bad if the food has already waned upon arrival.

There does not seem to be a firm correlation of the count when the migrating birds arrive or when males change from dull to bright colors with April's temperature (average-high-low).

The data did show in some species when there was a warm or cold March, the migrating count changed accordingly in April. The same thing with the amount of rain. There are 23 migratory species that have been observed over a 10 year span during April. Other species may have an absence of five to eight years between sightings. For most, the sighting usually is within the same week. Some migrating species that are regularly observed in April may not be observed until May. The changes are possibly due to global change, the environment changes surrounding my yard, and my own garden's environment.

Barn swallow, pop was a small flock that arrived the fourth week. None have been observed since 2007 when a new road bridge over a stream was built a half mile away.

Broad-winged hawk, pop one or three observed soaring during migration periodically during the fourth week from 2006 to 2022. When there was a warm beginning of April, they arrived the second week. *The migration correlates when the mountain pink is full, which is the same as when the big white azalea are in full bloom.*

Brown thrasher, pop of one.

Canada geese go north a *week after the windflowers are in bloom or when the Virginia bells are up.*

Canada geese

Chipper sparrow, pop one to four. Like clockwork they usually arrive between the first and second week of April, which is the time that *the white hosta are coming up.* It has a very high pitched trill.

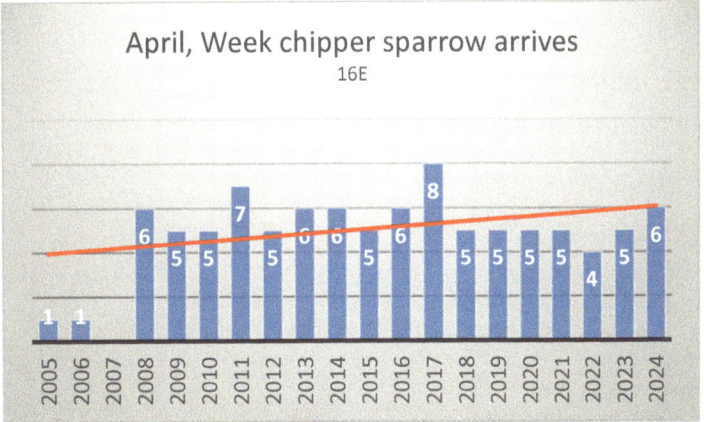
Weeks chipper sparrow arrives

Cowbird, pop one to six arrive first week, and will be here for the month. The female lays her eggs in a song bird's nest. They arrive when *the mountain pink is full and the Mayapple is full leaf.*

Eastern phoebe, pop one, was last observed the first week of April 2024.

Flicker, pop one, arrives the first week giving its very distinctive call.

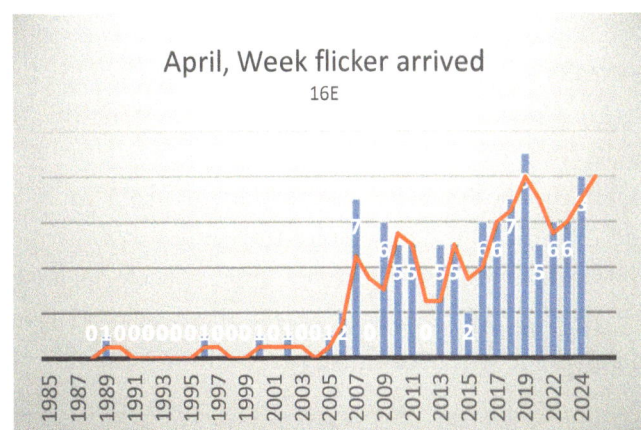
Weeks flicker arrives

Fox sparrow, pop one, arrives in March. Pop may increase to two in the third week. They may continue north on the fourth week, but this is not a firm observation.

Fox sparrow

Grackle pop was large migrating flocks before 2004, since then the first week pop is one to 10! *Migration occurs when windflowers are out and when the poplar tree gets leaves.*

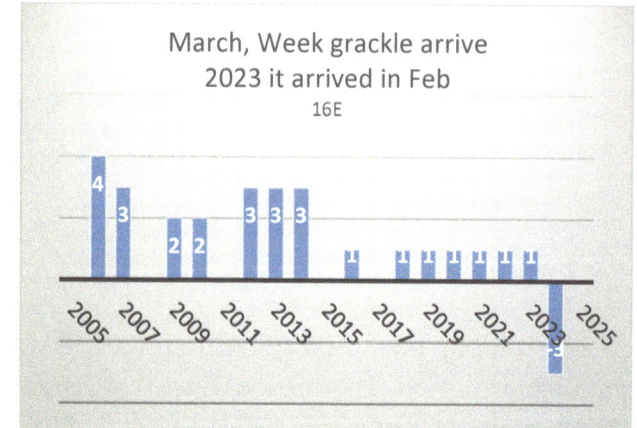

Grackle

Great blue heron, pop one. This graceful six-foot wingspan bird flies over regularly. Now and then it sneaks a landing by the pond for brunch. This has been going on for over 25 years. It reacts to the slightest movement in the kitchen and is gone in a flash. Arrival has gone from the third week of April to the first two weeks of March!

Hairy woodpecker, pop one, arrived in March, it will move on later but the winters of 2018-2019 and 2019-2020 it overwintered.

House wren, pop one male, until two weeks later when the female arrives. He usually arrives when the *Virginia bells are blooming* (the first week of May). The trend shows he is now arriving the last week of April. He will investigate two to three houses for her choosing.

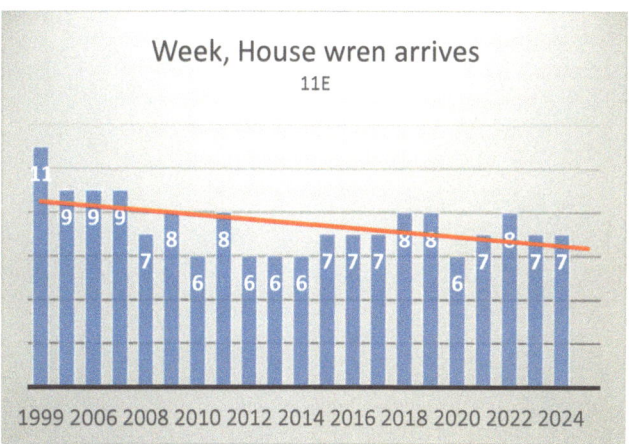

House wren sings its heart out

Mallard duck, pop two to six, arrive the second week.

Mockingbird, pop one. Many years this bird was absent. In 2016, for the first time in many years, there are two mockingbirds at the suet and the peanut butter log. On March 7, 2020, they arrived after one of the warmest Februarys on record.

Mockingbird   *Female purple finch*

Pine siskin, pop small flock of four to six, was last seen in April 2015.

Purple finch, pop is flock of three to five.

Red-breasted nuthatch, pop one to two. Until 2010 the red-breasted nuthatch has been arriving two weeks after the Mayflowers were up (the last week of April). Starting in 2012, *it arrived a week after the Mayflowers were up.*

Redpoll, pop one, the last time a flock passed through was the third week of 2011.

Red-winged blackbird, pop of one to three. Like other blackbirds, there used to be big flocks that came through in March, now we are down to a pop of one or three, with the exception of March 2016 when a flock of 10 stopped at the feeder for five minutes. To me, from my youth to now, the song of this bird truly represents spring. The striking flash of its red epaulets is also deeply engrained as spring. In 2018 they arrived during the first week of April. Since 2016 one or two are overwintering. When we had the January 20, 2016, blizzard two were among the grackles that arrived at the feeder in the middle of the storm.

Red-winged blackbird

Robin taking a bath

Robins, after their initial arrival as a flock of 10 to 15 in March, the pop goes to two. The males will fight for territorial rights. After the battles, the winner will sing loudly. The winner and his mate will then start to build their nest. This usually occurs during the first week of April. The others will disperse to different areas so there is only one couple in the yard. They start to *gather nesting material a week after the windflowers are in bloom.* With the unusual warm winter especially during February 2020, there was a flock of eight in the yard on the 8th.

Rose-breasted grosbeak, pop usually is one male and female the first week of May. In 2005 a couple arrived on the 28th of April. In 2006 and 2008 two females came through in the last week of April. In 2017 four females arrived April 12th, the male arrived May 2nd.

Female red-breasted grosbeaks

Sharp-shinned hawk, pop one to two. The sharpy is active all month trying to get meals to feed its young; most times it is going for a dove. In 2020 it was active in March.

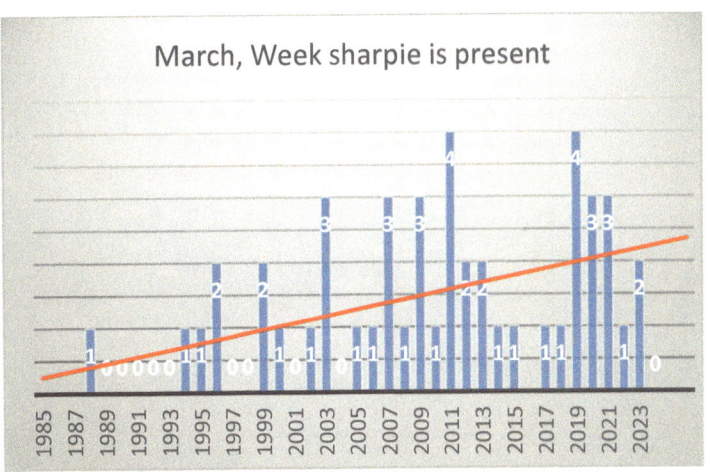

Sharp-shinned hawk

Starling, pop one to two, arrive in March in flocks of 10 to 20. They then raid all the feeders before flying on. In April, there may still be one to two left, but they too will move on.

Starling

Towhee, pop one to two, usually arrive the last week of April. They like to scratch around the leaves. The male arrives one to two weeks before the female.

Tree swallow, pop two to six, like a standing birdhouse on an eight foot pole. They used to arrive during the fourth week, until the house sparrows took over the house during the winter. The house sparrows refused to move out when the tree swallows arrived.

Turkey vulture, pop one to four. In 2010 five landed on the roof of the house, spooky. They return the first to second week when there is a normal temperature, but delay the return to the fourth week when there is a cold first week. With the mild winters; they have been overwintering since 2016. It is nice to see them soaring on a clear, crisp, winter's day. Maybe this is due to climate change. 2023, they were landing on the lawns!

Turkey vulture

    White-crowned sparrow, pop one, passes through going north the third or fourth week of the month. Usually it is only here for one or two days. It looks very much like the white-throated sparrow; the two white stripes on the top of its head are very pronounced. It was last observed in 2024.

    White-throated sparrow, pop three to six. During the winter their throat feathers are a dull gray and the head stripes are a light yellow. Come April the throat feathers turn to a clean white and the head strips become bright yellow. When there is a cold March and beginning of April the colors change during the second week. When there is a warm beginning of April the change takes place in the fourth week. *The change in color correlates when the grape hyacinth are in full bloom.* They migrate north during the fourth week.

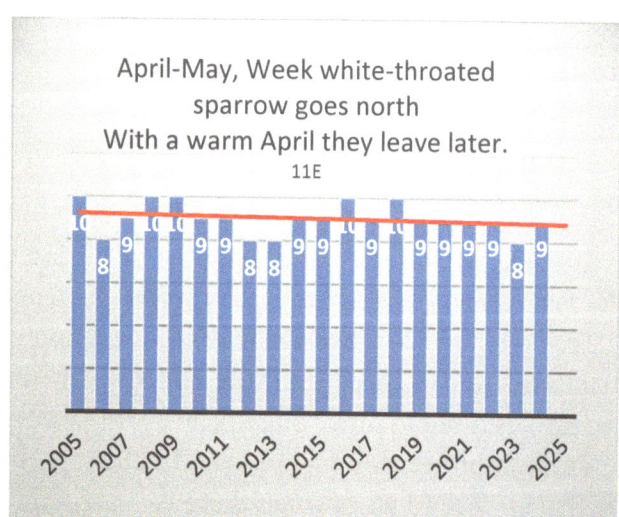

White-throated sparrow

Yellow warbler, pop one, was observed April 24, 2007. Then one passed through May 3, 2017. This is the same time as *when the bleeding hearts and PJM are in full bloom.*

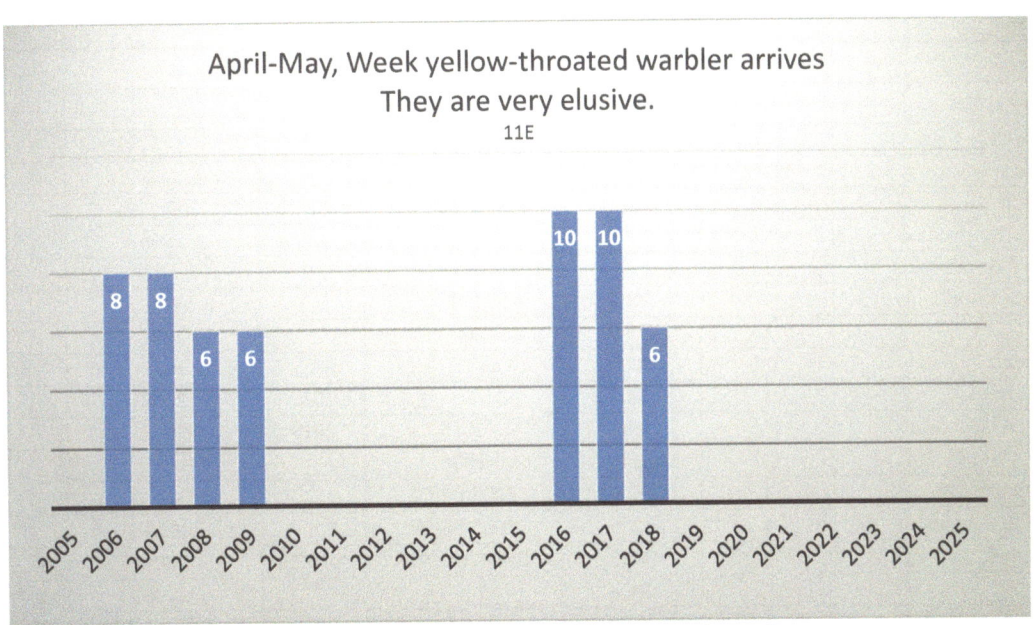

Weeks number yellow warbler passes through

Week migrating species were observed in April						
Species	Quan	Year first observed in April	Week 1	Week 2	Week 3	Week 4
Baltimore oriole	1-2	1994				2002/09
Barn swallow	2-4	2005			1996	2005/06
Black-white warbler	1	1996				
Broad-winged hawk	2	2004				2004
Brown thrasher	1	2006		2016	2019	
Catbird	1	2005	2005/06			2008/18
Cedar waxwing	Flock	1979,94,97				2001/17
Carolina wren	1-2	2005	2005/06			
Chipper sparrow	1-2	1992	2005/06/09/18/19			2017
Cooper's hawk	1	1998	2013/16	2017		2010/14
Cowbird	1	1992	2018			
Duck, mallard	2-5	1996	1996/2018			
Eastern phoebe	1	2005	2005/07		2008	
Flicker	1	2002			2007	2002/19
Fox sparrow	1-2	2005	2005			
Goldfinch changes color		1979	2006/12/13/16	2010/17	2008/09/13/14	2005/07
Goose, Canada	Flock 100's	2009	2009			
Grackle	1-4	1993	1993/2005/06/07 09/10/18			2017
Great blue heron	1	1984	2010/12/13/16	2017	2001/08/11/14	
Great horned owl	1	1995				
Hairy woodpecker	1	2012	2012			
House finch head changes red			2006/13/18		2015	2008/11
House wren arrives	1	1979	2012	2010	2008/15 2017	2009/11/18/19

Indigo bunting	1	2003			2003	
Junco	1-4		2005/06			
on the chase						
leaves			2005/06			
Kinglet	1	1992	1994			1995/2017
Magnolia warbler	1	1979	1994, 2008			
Mockingbird	1-2	1992	2008/16/18			
Pine siskin	Flock	2009	2009			2015
Purple finch	1-6	2006	2009/14/18	2017	2008/16	2006/13
Rose-breasted grosbeak	1-2	2005			2008	2005/18
Red-bellied woodpecker	1	1993				1993/95
Red-headed wood	1	2008	2008			
Redpoll	1	2009	2009		2010	
Red nuthatch	1-2	2005	2006/07/13/17			2005/09/10
Red-tailed hawk	1	2008		2017	2008	
Red-winged blackbird	1-2	1992	2018	2019	2009	2001
Robin	1-2	2006	2006			
Savanah sparrow	1	2005	2005/06/07		2008/09	
Sharp-shinned hawk	1	2005	2005/06/07		2003/09	2011
Song sparrow	1	2005	2006/16/17/18	2013/14	2005	2009
Starling	1-2	2005	2006		2005	
Towhee	1	1992		2008	2005/07/13/17/19	1992/2009/14
Tree swallow	4	2007			2016	2007
Turkey vulture	1-2	2007	2008/09	2013/15	2007	
White-crown sparrow	1	2006	2007			2006
Wood thrush	1	1989				2017
Woodpecker knock			2006			
Yellow throated warbler	1	1995				1995
Yellow warbler	1	1995			2008/09	1995/2006/07

Some migrating species arrived two to three weeks later in 2010 (chipper sparrow, fox sparrow, house wren). The chipper sparrow left two weeks early!

Insects: The appearance of the insects depends on the weather. When the afternoon temperature rises to 55F the bumblebees emerge from their below-ground hive and become active. This is when they appear to be mating. Then when the temperature rises higher, to 60 to 70F, is when white sulphur, blue hairstreak, orange copper, black swallowtail and maybe a yellow swallowtail butterfly will flit through. A hundred worms facing south, may appear in the circle in the morning after a heavy rain the night before.

Flora

Perennials: There was a three week spread of when growing activity started in perennials from 2004 to 2019. With trees and shrubs the spread was from two to eight weeks. April 2018 was much cooler than normal; many shrubs predictions were a week or two late.

TRENDS DUE TO CLIMATE CHANGE

Perennials April Week numbers March 1-4, April 5-8, May 9. Moisture is year to date														
Occurrence	05	06	07	08	09	10	11	12	13	14	15	16	17	18
Temp April	norm	norm	norm	norm	norm	warm	norm	norm	norm	norm	cold	norm	norm	
March	cool	norm	norm	norm	norm	high	norm	hot	norm	cool	hot	hot	hot	
February	norm	warm	cool	norm	norm	cool	norm	norm	norm	cold	cold	hot	hot	hot
Moisture, April	dry	norm	wet	soak	V dry	high	soak	crack	dry	soak	dry	warm	norm	soak
March	norm	dry	crack	soak	crack	crack	crack	crack	crack	soak	norm	norm	norm	soak
February	soak	norm	crack	soak	crack	wet	norm	crack	norm	soak	soak	soak	crack	
Temp+rain		norm	norm	norm	nor	hot	norm	hot	norm	cold	cold	hot	norm	cold
Little yell daf, bck	1	1	1	3	3	2	3	Feb13	-	1	1	2	3	
Little yell frt	2	-	4	3	4		3	4	-	5	5	5	4	5
Grass greening	4	3	4	4	3	4	2	2	6	6	5	3	4	6
Windflower	5	4	4	4	4	5		3		2	2	4	6	6
Mixed daff	5	4	4	4	4	3	5	3	6	5	6	5	5	8
Cut lawn		5	5	6	5	4	7	4	4	7	7	7	7	7
Yell daff full	6	6	5	5	5	5	5		5	7	7	5	6	8
Myrtle out	6		5	6		5	?	3	6	6		3		8
Virg bells up	5	6	6	5	5	5		-		6	6		6	6
Fern pond up	5		7	5	5	5	5	3		6		6		7
Fall cle 3rd leave		5			7	5	8	4	5	5	5	3	5	8
Whit Frt Host up	8		7		5	6	6		5	5E	6	4	5	7
Fern 3rd up	8	7	7	5	5E	5	8	5	5E	6	7	7		7
Dog tooth full	7	7	8	5	8	6		1	6	6			5	8
Bleed heart full	7	6	8	6	6	5	7	8	7	8	8	5	8	8
Min purple iris	7	7		8	8	7	8		8	8		8	8	8
Jack-in-pulpit up		6		6	8	8		5		8		5	-	None
Grape hyat full	8	6	8	6	7	6	8	4	6	8	8	1	6	8
Virg bell full	8	7	8	7	7	7E	7		7	8	7		5	8
Dandelion full	8	7	8	5	-	8	8		7	8	8	3		7
Mt. pink full	8	7	8	6		-	8	3	7		8	5	6	
Mayapple full	8	8	8	8	8	6	8	4	6	8	7	8	8	8
Iris yell 2nd out	8	8E	9E		8	7	8	2	8			8	8	8 bud

The following table shows a listing of blooming of perennials, trees, and shrubs in chronological order.

Perennial, shrub, tree initial bloom growth with normal rain and temperature Number 1 is the first week of March, number 8 is the last week of April.					
Perennial	Week	Shrub	Week	Tree	Week
Little daffodil back	1				
Little daffodil front	2				
Grass greening	4			Silver maple bud	4
Fern up	5				
Windflower	5	Yellow bell full	5		
Mixed daffodils	5				
Virginia bell	5				
Dogtooth violet	6			Empire apple leaf	6
Yellow daffodil full	6				
Bleeding heart full	7			Serviceberry full	7
Dandelion full	8	Japonica full	8	Oak leaf	8
Hosta up	8	PJM 2nd full	8	Poplar leaf	8
Yellow iris out	8	Mt. laurel growth	8	Empire apple blossom	8

Shrubs: Japonica by the back gate gets white clusters in the second week no matter the temperature or moisture. The three in the front get cream clusters in dry, wet, normal or cold tempratures, however, if March is very warm they will bloom in the first week.

Japonica clusters

Soil Temperature: The normal conditions do not occur very often. Either it is too hot or cold or wet or dry. As an attempt to determine what causes the plants to emerge or bloom at different dates I had taken the soil temperatures at a four-inch and 12-inch depth. I felt this may be more important than the air temperature or precipitation. Unfortunately I did not take the data at the same time nor at the location of the plants, so I had to estimate for time and location. As a start I took the data of 10 plants of when the new growth started, then went to the soil data to determine what temperature growth had started. The results showed that many of the new growths for different years occurred close to the same soil temperature. The most notable finding was that the start date from one year to another was as much as eight weeks.

Across the span of eight years some soil temperatures are eight to 12 degrees higher than what looked to be the norm. Looking closer at the data showed that most times this occurred when the air temperature was at an extreme of 20 or more degrees above norm.

There was no correlation in days from initial growth to full bloom. That is because after the initial growth all kinds of weather occurred from hot, cold, snow, dry or a deluge. I don't how to use this information except that I now know that initial growth happens when the soil reaches a certain temperature.

TRENDS DUE TO CLIMATE CHANGE

| Perennial activity from soil temperature at 4" depth corrected for location of plant est= estimate | | | | | | | | |
| Date/soil temperature F | | | | Normal air temp for April = 52F | | | | |
Plant Mean March F	2013 37	2014 32	2015 33	2016 40	2017 38	2018 37	2019 37	Change in weeks
Dogtooth violet up	Apr 10 62 (air 81)	Apr 13 54	-	Mar 25 56	Mar 7 54	Apr 13 56 est (air 77)		5
Fern 3rd tier up	-	Apr 14 54	Apr 19 70 (air 85)	Apr 17 51	-	-		0
Mayapple up	Apr 9 62 (air 82)	Apr 13 54	-	Mar 30 50	Apr 11 54	Apr 13 56 est		2
Grape hyacinth up	Apr 16 61 (air 69)	Apr 25 60	Apr 21 66	Mar 4 50	Apr 13 50	-		8
Daylily up	Apr 9 62	Apr 8 54	Apr 16 59	Apr 22 57	Mar 2 54	Apr 9 62		8
Bleeding heart up	Apr 7 62 (air 64)	Apr 1 50	Apr 4 52	Mar 17 52	Apr 10 52	Mar 21 60		4
Hosta up	-	Apr 2 53	Apr 11 59	Mar 3 53	Mar 3 53	Apr 16 57 est		7
Fall clematis leafing	Apr 19 64 (air 77)	Apr 18 55 est	Apr 20 66 (air 84)	Mar 18 53	Apr 7 55	Apr 25		6
Windflower out	Apr 1 58	Apr 8 54	Apr 11 59	Mar 25 56	Apr 2 55		none	3
Rhubarb up	Apr 11 54	Apr 3 52	Apr 1 52	-	Apr 1 53		none	2

A four foot difference in elevation can have a difference in soil temperature The temperature on the third tier is three to four degrees warmer than the first tier.

The soil temperature at a four-inch depth remains the same from 7:00 to 10:00 a.m., then increases four to five degrees from 10:00 a.m. to 2:30 p.m. At 12 inches it only changes one degree. The temperature in the raised garden at four inches is about 5F different than in the ground. At a 12-inch depth, there is only zero to one degree difference. There is no general correlation between the air temperature at the four-inch depth and air temperature. Air temperature fluctuates too much for the soil temperature to track the change.

Trees: Dogwood blossoming is weather sensitive. The front and back dogwoods bloom the same week. The third week, the dogwoods in the front and back form buds with normal or hot March and April temperatures; with a cold March or April buds form the fourth week. It takes eight to 10 days to go from buds to full flower, only five days when the temperature is very warm. They bloom the fourth week, except 2010 was a warm March-April and they bloomed the first week of April. Both 2014 and 2018 had a cool March, blooming occurred the first week of May. The birch trees get leaves the third to fourth week.

Silver maple budding usually occurs the first week of April. With a warm March budding may be the first week. Budding and blossoming usually occur a week apart. When there is a cold season there can be a two to three week separation between bud and bloom. There is a one to two week difference from full blossom to flower drop. The second week, the silver maple forms buds. It will be the third week with a cold February. It takes another one to two week for leaves to form. The sugar maple blossoms one to two weeks after the silver maple.

The serviceberry tree gets its leaves one week before flowering. Leaves come out the second week. The trend is toward the first week. It was warm in 2012 and the leaves came out the last week of March. When April is cold, flowering will be delayed until the fourth week in April or the first week of May.

The poplar leafing is strongly dependent on the weather. In 2006, 2007, 2010 and 2012 it leafed the first week of April. Other years it was mostly the last week of April.

So, ends April.

Index of May

May Overall	100
Weather	100
First Week	102
Second Week	106
Third Week	112
Fourth Week	116
Review of Data	123
Fauna	123
Animals	123
Birds	124
Gang	124
Table of the Gang	124
Migration	126
Insects	128
Flora	128
Canopy	129
So, ends May	130

May

Week number for the year of 48 weeks					
Relative Month	Month/Week	First	Second	Third	Fourth
Previous	April	5	6	7	8
Present	**May**	9	10	11	12
Following	June	13	14	15	16

Overall Summary of May

May is when parent birds bring fledglings to the feeders, the white-throated sparrows leave to go north, and the female house wren arrives. There may be as many as 29 to 30 species of birds migrating for summer residence or just passing through. Hawks soar in the warm air of thermals. Pollen is getting higher from the maple trees, the highest in the second week; the Mayflower are in bloom. Mold is high the third week. Bumblebees come out of the ground when the temperature reaches 50F. The bunnies are eating tender, lush grasses. The ferns still unravel through the captivating earth. Azalea will be blooming throughout the month. Dogwood and apples are in blossom, the candles have formed on the pine boughs. The tall deciduous tree canopy is completed. The skunk cabbage and cattail sprouts are up. There is a multitude of insects all over. As the earth warms up, the air will smell fresh in the morning. During the last week, the bearded iris open. Purple iris, roses, and brilliant yellow evening primrose are out. Rhododendron and mountain laurel are in full flower. Conditions may seem right to plant annuals. Beware do not plant too early, as a frost may still occur. Hummingbirds return during the last week and it may be warm enough for the lightning bugs to rise around 8:30 p.m. At the end of the month some cool weather crops like lettuce and broccoli are ready to pick.

Weather

The amount of daylight on the 1st is 13 hours, 57 minutes, on the 31st it is 14 hours, 48 minutes.
The average high temperature on the 1st is 67F and the low is 43F.
The average high temperature on the 31st is 76F and the low is 52F.
The extreme daytime high for the month is 96F in 1946 and 2002.
The extreme nighttime low for the month is 28F in 1947.
The full moon of May is called the Flower Moon because of the many flowers that are in bloom.
At the beginning of the month the sun rises at 6:00 a.m. and sets at 8:00 p.m., at the end of the month the sun rises at 5:35 a.m. and sets at 8:25 p.m. April has yo-yo weather, but May is normally nicer. The morning air can smell sweet; there may be a dew on the lawn.
The temperatures can go over 90F during the last week. Except for a few years, every May has had a 90F or plus reading since 1925. More than once we have had a heat wave in the latter part of the month. At the other end of the weather spectrum, we can still get a freeze during the second and third week. We can get downpours that cause major flooding or it can be bone dry. Now and then we get a tornado

warning. In 1985 a tornado stuck 20 miles west of Allentown. Since the 2000's thunderstorms have been developing the third and fourth weeks. Wind gusts of 40 to 70 mph can happen. Many days there is a wind of 25 mph.

October 2015 was cold, then November and December were hot resulting in rhododendrons blooming! The final result, we had very few flowers in May of 2016 and the canopy completion was very late, the 31st of May. To continue with the crazies, 2017 had one of the warmest daytime temperatures on record, 95F (not the warmest mean temperature).

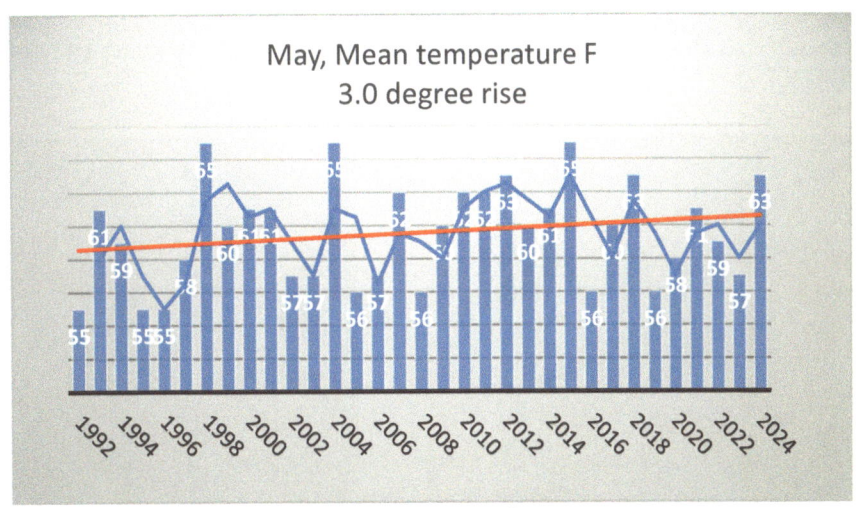

Mean temperature for May

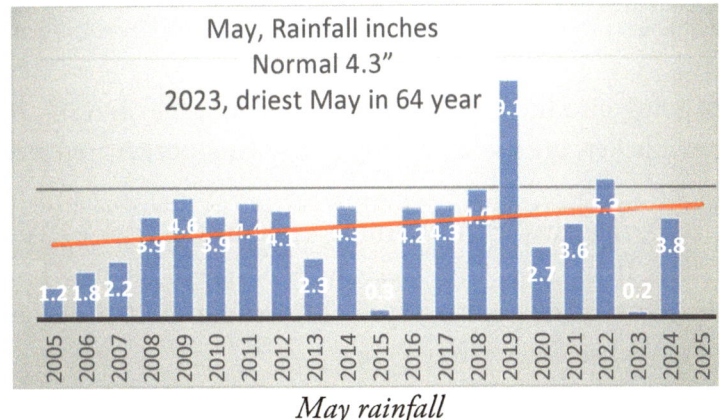

May rainfall

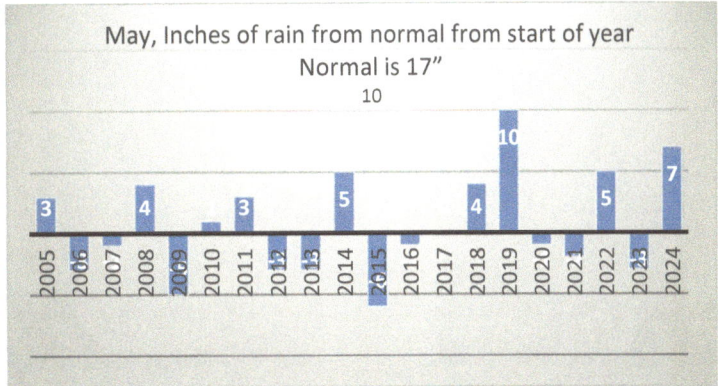

May, inches of rainfall from normal from start of year

In 2009 and 2015 monthly rain, dry springs, was minus five and six inches. In 2014 we had a wet spring and a drenching in 2019.

First Week

<u>Weather</u>: The sun is out, when suddenly it gets overcast and there is a downpour for 15 minutes with thunder and lightning and then a brilliant sun. At 6:00 p.m. a storm releases marble size hail for 20 minutes including thunder and a tornado warning. Typical for May. Pollen is heavy in the air

<u>Birds</u>: The white-throated and white-crowned sparrows go north; many years they leave the last week of April.

House wren

White-crowned sparrow

During the first week going into the beginning of the second, the arrivals are the Baltimore oriole, catbird, chipping sparrow, flicker, great blue heron, hairy woodpecker, red-bellied woodpecker, and tree swallow.

Rose-breasted grosbeaks arrive early in the morning from the last week of April to first week of May. Since 2016 the females arrive first. In 2018 no males were observed; in 2019 both were observed early May. 2021 and 2023, they arrived last week of April. 2024, none spotted; When you spot a grosbeak, look around, the blue indigo is usually nearby.

Male grosbeak

Female grosbeak

Red-breasted grosbeak arrival and departure in May														
Year	2005	06	07	08	09	10	11	12	13	14	15	16	17	18	19
Arrival		6	4	2	7	-	vac	-	6	2	-	-	1	Apr 28	1
Departure		20	9												4

May 12th – indigo bunting

I hear a Baltimore oriole before seeing one, usually it is perched at the top of the poplar tree. The male continuously calls for a mate. When paired, I will only have the pleasure of its song for a couple of days before they move on. Except in 2020, for the first time they nested here.

Male Baltimore oriole

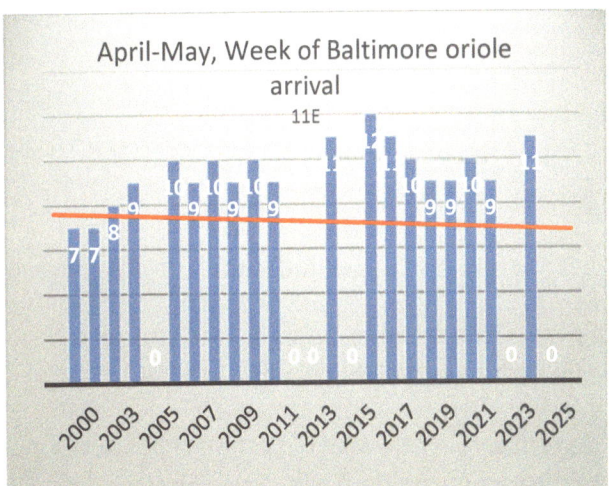

Oriole arrival week

Baltimore oriole arrival and departure day in May, quantity of one																	
Year	2005	06	07	08	09	10	11	12	13	14	15	16	17	18	19	20	24
Arrival	10	4	9	6	10	5	vac	-	2	7	18	11	9	3	7	5	none
Departure	20	19	-	20	19			30	-	14	26	-			9	stayed	

I usually hear the catbird before seeing it the first week of May. It has arrived for its summer residency. Many years the male arrives first, from 2015 they arrive as a pair. In 2017, I have a pair in the front and another in the backyard, the twain does not seem to meet. The backyard catbirds will be taking nesting material into the shrubs on the hill. Two catbirds take a bath under the lawn sprinkler. I do not have bluebirds because they want an open space like a meadow. When an open area is available, the bluebirds arrive as early as February, mostly late March. Some local areas have them all year.

The male house wren usually arrived as a solo from April into May; in 2019 they arrived as a pair. White-crowned sparrows are going through from the south from late April to the first week of May. The flicker arrives the first week.

Grackles will take whole peanuts; they also eat dry dog food. A pair of cardinal land on a dogwood branch, the male bends over and kisses the female.

Father and son cardinals are at the feeder, the son follows the father's every move. Robins chase jays and cackle at anything that comes anywhere near their nesting area. A flock of 10 jays goes east. Three jays chase a crow that has something in its beak. A young robin with a speckled breast and a short tail is doing short flights around the yard.

Young robin

A sharpie misjudges and slams into the back window. It shakes itself off and flies on.

Insects: On a warm evening as the sun sets, a swarm of little bugs can be seen dancing in the golden beam as it passes through a graceful bough of a pine tree.

Perennials: There is a lot of carryovers of developing flora from the last week of April. The perennials are rapidly coming to life. Yellow iris, mini-iris, and bleeding hearts are in bloom this week into the beginning of the second week. When the average temperature of April is in the mid-50's the Mayapple is in full leaf in May. Hosta are up six to 10 inches but growth will be delayed 12 days when the average temperature is only 48F. Grape hyacinth and tulips are done, yellow bell foliage is growing fast, miniature iris is still blooming. Some of the azaleas are in full blossom. The clusters on the japonica have turned from pure white to semi-white. Cattail sprouts are almost three feet high. The holly bushes are in white flower. The pink lilac blooms the first week, its blossoms last four to 13 days. Bleeding heart are full. Poplar tree flower blossoms fill the driveway. This is a dirty tree; it blossoms in the spring then drops thousands of tiny samaras in November and December. Cinnamon fern is in flower, so is some coral bell.

Pond: Most pond plants do not start sprouting until the second week of June and then they take off. The arrowheads are four inches by the 21st and go to 12 inches by the 30th. Pickerel weed has blue flowers by the 31st.

Cattails have a rapid growth rate, about an inch per day or a foot in 12 days. They sprout at the beginning of April; they are seven feet by the middle of June. They form an eight-inch male spike at the top and a 10- to 12-inch female punk below the spike.

May 31st – cattails and evening primrose

On the 7th there were a few frogs croaking, by the 21st six were croaking from 4:30 to 5:00 a.m. and soon there were 14 small ones. Tadpoles are swimming and surfacing for a gulp of air. Frogs are doing their mating ballet; I'm surprised that the females do not drown.

Doing their thing

Trees: Holly and tall birch are leafing. The small birch is developing seed clusters, dogwood and serviceberry are in blossom.

It is snowing dogwood pedals this week. The pedals turn brown and slippery and I track them into the house.

White dogwood

Japanese maple

The pink and white dogwoods are in full blossom. The clematis on the maple tree is out full. The holly bushes have heavy growth.

Pink dogwood

<u>Vegetables</u>: Plant lettuce the first week. When it comes up, it may have to be protected from frost. It is ready to pick the end of the fourth week.

Second Week

Weather: Daytime temperatures are in the 70's F.

Animals: Squirrels are taking branches to the top of the spruce to build a drey.

Birds: Grackles are at the feeder; another one goes under a bush with a whole peanut and cracks it open. They also eat waffles. Looking out the window watching a female grackle by the watering pan I noticed she had something white in her beak; she dunked her bill in the water and flew off. Five minutes later she was back, this time swinging on the perch by the globe filled with split peanuts. She kept pecking at the nuts; the broken pieces fell to the grass. After a few minutes she dropped to the grass to collect the broken pieces. Then with a beak full of broken nuts she flew back to the watering pan, dunked the nuts and flew off to her nest. This was repeated most of the day. Sometimes she would go directly to the watering pan, take a drink and fly to the nest.

Grackle chipping at the split peanuts

Gathering the pieces

Dunking the split peanuts

Getting a drink

Swallowing

Ready to go back to the nest

Two male cardinals are hissing at each other. The hissing sounds just like two cats. White-throated sparrows are still around, the junco has gone north.

House wren emptying sparrow's house

A house wren is emptying a sparrow's house. The house sparrow is ready to attack. This battle will go on for a couple of days; the sparrows will keep their house.

On the 9th of 2017 I hear a different song calling from the poplar tree. Then the familiar song of the oriole. Much to my delight, next to the oriole was a scarlet tanager, the first one I have observed since living in the circle for 48 years.

Scarlet tanager

Catbird eating an orange

I glanced up from the laptop just in time to see a yellow warbler leave the feeder, it was right on schedule. Mid-morning a pair of Canada geese honk as they fly over.

Tree swallows are in the house on the pole in front of the house. However, house sparrows vacate the swallows. I put rubber plugs in the entrance holes to keep out the sparrows. The sparrows proceeded to make an entrance hole into the particle board siding. So be it, they win.

Yellow warblers may pass through and stay two days. Even though the house sparrows are aggressive and not the prettiest, an import from England and multiply more than rabbits, they do a fairly good job in getting white moths in mid-flight.

Two male goldfinches look at each other, then both fly up like helicopters six inches apart going at each other for territorial rights. Twilight, 5:20 a.m., a male cardinal sings its heart out for 20 minutes.

At 7:30 to 8:30 a.m., even though it is raining hard with 50F, the full gang comes for breakfast. The Carolina and house wrens continuously sing, trying to outdo each other.

A red-tailed hawk comes in from nowhere and snatches a grackle in mid-air, showing-off to a watching sparrow. At 6:00 p.m., 77F, I look up toward the north. At approximately 2,000 feet a kettle of eight raptors, eagles or broad-wings are flying wingtip to wingtip in a spiral. This continued for 10 minutes until the kettle floats silently out of sight.

Tree swallows are circling in the evening. On a warm evening I may see a bat the second or third week. Starlings, robins, and catbirds helicopter up to the suet feeder.

Catbird helicopters up to suet

Fruits: Blueberries are in white blossom.

Perennials: Jack-in-the-pulpit is in flower. The double blooming yellow irises by the step iris is still blooming. They will wane in seven to 10 days. Double blooming because they bloom in the spring and then again in late fall. Cinnamon fern is in flower. At the end of second week put down 10-10-10 on flowers, aluminum sulfate on hollies. Mayapples develop flowers. Woodruff is out; blooming is delayed to the third week when there is a cool April with an average temperature of 48F.

May 13th Hosta up, columbine out, ferns up

Cinnamon fern

May 9th – Jack-in-the-pulpit

May 13th – Mayapple

Astilbe are fully up. Bearded iris has tall buds. The big Hosta is almost in full leaf, daylilies are two inches high. The honeysuckle is in full flower.

Shrubs: The rhodie on the hill blooms for six days. The honeysuckle and sand cherry shrub are in blossom. More azaleas are out. They do not all bloom at the same time. It takes a month from the first species to the last to come out full. This week the deep pink by the upper gate, red on the third tier east, light pink in the front yard, and light red by the driveway are out. Spray hemlock hedge with horticultural oil to prevent the woolly adelgid insect. Spray with Isotox for spider mites at 10-day intervals on evergreens and azalea, also time to fertilize the lawn. The white lilac blooms the second to third weeks. At end of the week, the orange Mollis azalea is full. The 70-year-old big white biannual on the second tier is out full.

May 11, 2007 – 70-year-old azalea bloom week
(22 days)

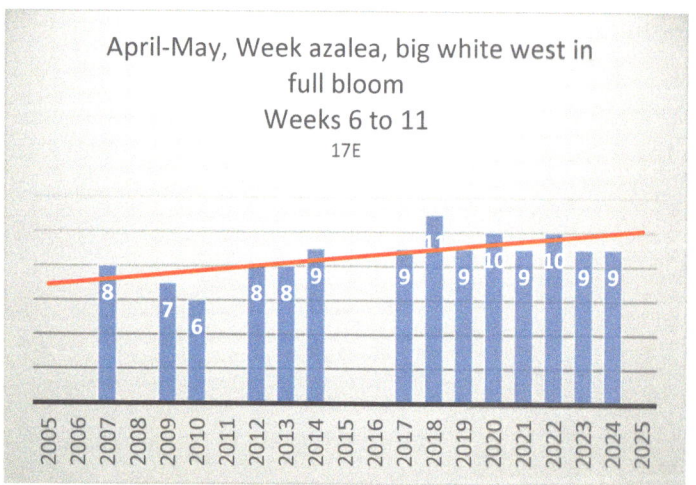

Trees: Maple samara or key (as a boy on Long Island we called them pollywogs) are twirling like a flock of little helicopters. With a little breeze they go up and across the sky before descending for their landing. Hemlock gets new growth.

Maple samara, nice to track into the house

Full tulip poplar flower

The pine trees have new candles during the second to third week. The hawthorn tree forms flowers. The hemlock hedge and large spruce trees develop new growth. Small evergreens are showing growth.

Late afternoon of May 16, 2018, we get high winds, heavy rain, and thunder. The canopy had just been completed two days before. The next morning, I find poplar tree buds spread all over. Normally it does not bloom until the end of the month.

<u>Vegetables</u>: Put 17-16-28 on the asparagus. Plant dill and parsley.

<u>Fruits</u>: On May 13th the Empire apple is in full bloom. A rope from the chimney is to support the branches. I use rope rather than wooden props to the ground to prevent squirrels from climbing into the tree. The brown roof flashing is to prevent squirrels from going up to the apples.

Rhubarb is ready for picking the end of the second week. I may be lucky enough to pick four to five cups of rhubarb. It gives you a good spring tonic.

Empire apple supported by rope around the chimney

Rhubarb

Third Week

<u>Animals</u>: A pair of squirrels is up in the serviceberry tree eating away the berries; too bad I don't get any because serviceberry make a good pie. Young oak branches that are snipped off by squirrels for a drey are falling to the ground. A chipmunk watches two squirrels chase each other up and down and around the trees and bushes.

A squirrel in the maple yells and chatters as an old black cat slinks by.

<u>Birds</u>: At 7:00 a.m. a pair of majestic, graceful, great blue heron go north on their daily flight from their nesting area. On the trip back, they were being chased by a crow! A grackle cracks open a whole peanut. A pair of robins on the lawn cock their heads looking for worms. A blue jay dropped down from the maple toward the main feeder but landed on the ground because a grackle was on the feeder. The grackle chases a dove from the feeder. I look closely under the bushes to find a fox sparrow which was passing through. After many years absent, a red-headed woodpecker comes for a bed and breakfast on May 16, 2017.

Red-headed woodpecker

Fox sparrow

Robins and cardinals sing 10 minutes before sunrise. I open the door to the sweet fragrance of morning air. Sunrise, crows in an unorganized flock fly west. Early morning catbird, robin, song sparrow, and house wren are singing; mid-morning the Baltimore oriole is singing.

A woodpecker is making a cavity in a broken branch on the poplar tree. Song sparrows are nesting in the tall arborvitae. Two male robins are in a territorial-mate battle. After the battle, the winning male preens itself on a dogwood branch.

Song sparrow runs along the rocks by the lawn, its nest is in the arborvitae. It chirps as it goes, then chirps louder as I get closer. I spot one blue jay feeding on a fledgling sparrow. Catbirds, house sparrows, and robins like to take baths in the birdbaths, especially while the sun shines.

<u>Fruits</u>: It is time to spray the apple and pear trees.

<u>Insects</u>: During the third week I should see black swallowtail and greater fritillary butterflies.

Depending on the temperature, I may see the first Japanese beetle on the roses or apples. Gnats are all over my face; time to go in. I have been told to wear a tall hat; they will go there rather than your face! Skin-So-Soft helps. Ticks are also prevalent. Tent caterpillar webs form in the apple trees.

<u>Lawn</u>: This is the last week to put down lawn seed.

<u>Perennials</u>: The ostrich and Boston ferns are full; they are temperature sensitive. When there is a cool April, they are not in full fronds until the fourth week. The deep purple and white columbine are in bloom. The trumpet vine gets new leaves. The Royal Hosta on the third tier are beginning to leaf. Solomon's seal is in flower. The Korean dogwood blooms. Lavender clematis is out on the maple. Mountain laurel by kitchen has red buds. The blooming period of grape hyacinth, 15 to 24 days, starts the third or fourth week of April. The cooler it is while blooming, the longer the period of bloom.

Bearded iris is out the third and fourth week (when the monthly temperature is 66F). Little yellow flower, double purple-white columbine, and pencil iris are out behind the pond.

May 24th – bearded iris out

Columbine among the fern

May 27th – clematis

<u>Pond</u>: When needed, muck out the pond the third week.

Water lily leaves come to the surface

Moss by a rock in the pond

Pencil iris

Shrubs: Trim the arborvitae, yellow bell, and burning bush with the electric shears. Cut off winterkill branches from the shrubs. After cutting a 50-foot extension cord a couple of times, I learned to add a six-foot extension cord between the cutter's plug and the main extension. It is easier to replace the six-foot cord than splicing the main cord.

Mountain laurel cluster

Trees: Pollen dust from the silver maple is all over, three-inch-long tassels drop from pin and white oak. After three weeks, dogwood blooms are done.

May 9th – white oak tassels

Vegetables: Again, with a false warming, I planted some of the vegetable seeds too early. I replanted because they all rotted. Need to wait until the soil temperature is a steady 50F at four inches. It is hard to resist; wait, wait, wait! Now is the best time of the year to plant fruit trees. Chives are ready to add to an omelet.

Fourth Week

May 31st – yard

Weather: The weather in April controls the growth in May. One year I was weeding in the first week, three years later started on the fourth week. The same goes when putting in annuals. Usually buy 24 wax begonias and 24 marigolds. Must be careful with the fertilizer. If you put down too much 10-10-10 you will have good looking plants, but few flowers. Superphosphate would be the better thing.

Animals: A kit squirrel is eating silver maple samaras. A rabbit eats the pansies and zinnia. A chipmunk and a rabbit are nose-to-nose while eating safflower under the feeder. A munk finds a bare area of soil, it makes an indentation. It gets in and rolls over in it for five minutes, then gets up and scurries off! The chipmunk is up a dogwood eating the flowers.

Birds: At 4:55 a.m. the loud repertoire of a robin wakes me up, it stops, might as well get up. Open the front door to get the paper to the fragrance of a skunk's overnight visit. Other than that, the air has a fresh, sweet smell. Spot four starlings peeking out of the owl house. A grackle tries to chase a bull dove from the main feeder; the dove raises its wing, the grackle leaves. A jay eating split peanuts is chased off by a grackle, a few minutes later four jays return, the grackle leaves. The jays also did this to a dove when the dove chased a jay from the feeder. A jay chases a male cardinal around the yard! Robins are chasing other birds away from their nest area. The poor dove is chased by a sharpie, it bangs into the window, and there are lots of flying feathers. At the end, the sharpie had missed.

    A catbird takes peanut butter, it then rubs its beak on a steel ring to get the butter off the side of its beak. Grackles also like peanut butter. A house finch takes a safflower seed, chews in its bill in a rolling motion for two seconds, and then swallows. A chipper and song sparrow pick up a seed, swallow, and repeat again. Goldfinch takes two or three thistle seeds, chew, swallow, rolls out the husk from its beak while upside down on the thistle tube perch and then repeats. Catbirds eat the safflower one at a time. Mourning doves just vacuums the seeds like there is no tomorrow. The male cardinal feeds his fledgling at the feeder, so does the male house finch. A young grackle is flying around being a real pest to its parents. Surprise, a house sparrow takes a crumbled eggshell; usually it is just for blue jays and titmice. A catbird likes peanut butter and grape jelly. White with brown streaks breasted young robins are chirping, "We want worms."

TRENDS DUE TO CLIMATE CHANGE

We want worms

Unusual, a robin eats safflower. A pair of bright red cardinals chase each other for territorial rights. Chipper sparrow, song sparrow, and house wren are all getting insects for their young. The chipping sparrows have their nest in the yellow bell bush.

Pair of house finch bring their fledglings to the feeder; the fledglings have their little feathered horns. I can hear a young sharpie yelling from the nest high in the spruce tree, "Where is the food?" I am on the lookout for the great blue heron to land in the pond for a meal. It detects the slightest movement in the kitchen and takes off immediately. House sparrows are copulating on a dogwood branch. When finished they fly their separate ways. Teenage cardinals are changing to adult colors.

In the 1980's I had six to eight hummingbirds active at a large trumpet vine; in 2018 I feel lucky to have one. Still only have one in 2024. They arrive the last week May to the third week of June.

Hummingbird

<u>Insects</u>: While cultivating, spot a daring jumping spider patiently waiting for a victim to be snared in its web. Bumblebees are in the lupine and rhodie flowers. Interesting to watch a bee land on a lupine; the weight of the bee on the pedals open to reveal the stamen with the nectar. The bee goes to the nectar, and when it leaves the pedal closes. A sitting fly is rubbing its front legs together in a sun beam. Thousands of small ants and flying ants swarming on the sidewalks are gathering outside the

underground nest getting ready to migrate. In the evening observe swarms of little bugs in the sun beams under the boughs of the evergreens. A yellow swallowtail erratically sails through the garden in the warmth of the afternoon. As it warms up in the afternoons white and yellow-sulphur butterflies' flit erratically over the flowers, never stopping for more than a moment on anything. They will dance in front of each other for an instant before continuing their journey to nowhere, but to themselves. At 9:00 p.m. the lightning bugs rise. A nice way to end the month.

White sulphur, alias cabbage moth

Lawn: First cutting of the lawn is during the first week; spread herbicide for the crabgrass the end of the second week. In 2020, with the warm winter, the first cutting was April 1st, April Fool's Day!

Creeping Charlie, wild strawberries, and purple violets are doing well on the hill. Purslane and red, yellow, and white clover are all spreading in the lawn. Red thread disease is appearing on the grass. Mushrooms are poking up in the lawns.

Mushrooms from the lawn

Perennials: Blue bachelor buttons, coral bell, bleeding heart, Stella de Oro, rose campion, and red-hot poker come out. The pink clematis and Klondike are in full bloom.

Stella de Oro

The evening primrose reaches peak bright yellow; it blossoms for 18 days. The plant is a little over a foot high with 10 or more half dollar size buttercups on each stem. They close down in the evening and open up the next morning. After they finish blooming by the third week, the plant starts to expand its surface roots. This is a nice plant, but it is prolific.

Evening primrose, bees love yellow

Bachelor buttons

Shrubs: Azaleas are out during the last three weeks of the month. Prune flowering shrubs (forsythia, azalea, hollies, PJM, and rhodies) as soon as possible after the flowers fade. This is to prevent disturbing next year's forming flowers. It also gives the plant better direction for growth. By the fourth week the yellow bell and front shrubs need trimming. Holly bushes are in blossom and have a sweet fragrance. The roses bloom about a week after the bud forms.

Blooming of the Azaleas: The azaleas bloom in the same sequence year after year. The blooming week basically stays the same. They do appear to be temperature sensitive. They will bloom a week later when the average temperature in April drops from the mid-50's to the high-40's. The length of blooming is different from bush to bush, again they may be temperature sensitive.

I attempted to make blooming patterns from all the azalea data to determine: What date did the plants come out? Was that date consistent year to year? Did the blooming date depend on the color of the bloom? Did the color of the winter leaf indicate what color the flower would be? What was the length of time that the plants stayed in bloom? In the 1980's and 1997 the azaleas opened nine to 18 days later. Is this change due to global change or some other weather condition?

The length that a plant stays in full color is constant but is not the same for all the species. Those in the front of the house, it seems that no matter when the plants opened in May, they all wane by the last week of May or first week of June. The length of full bloom goes from seven to 31 days.

Azaleas		
Winter color of leaves	Length of bloom in days	Color of flower
No leaves	14 to 16	Cardinal red, Mollis orange and pink, light pink
Light green	15	White
Green	20	White
Dark green	10, 21, 31	White, pink, salmon, red, light lavender
Green-red	25	Front of house next to holly
Maroon	16, 10, 31	Red, brilliant pink, red-orange, salmon
Dark maroon	10, 12, 17	Pink-orange, red, salmon, lavender, white
Orange-red	10	Large pink
Orange	22	Bright red

The location of the plant in the yard does not seem to make a substantial change when the plant will bloom. From the color of the winter leaves, which go from light green to dark maroon, is not an indicator of what color flower the plant will produce.

I did not find a blooming pattern by flower or leaf color. I developed an anticipation table of when they may bloom and length of bloom. The table shows by date, color, and location when 19 azaleas come into full bloom and days in bloom. The dates change with the weather conditions, but the progression stays the same.

Azaleas			
Day in May when in full bloom	Color	Location	Days in bloom
2	Light lavender	Front window (3)	21
4	Orange	Vegetable garden	15
5	White	By tall Chinese lantern	20
7	Red	By tall Chinese lantern	15
7	White	By tall Chinese lantern	15
7	Red	East of PJM	25
7	Brilliant pink	Split rail fence, third tier	31
8	Bright red	Third tier, east, small Chinese	22
8	Lavender	Front, by cut off holly	17
11	Red	Behind hemlock	10
12	Orange	Third tier, west by path	15
14	White, old	Third tier, west, bi-annual	14
15	Pink-orange	Under dogwood by pond	10
15	Salmon	Third tier, east, small Chinese	10
16	Cardinal red	Front, east	14
17	Red-orange	Hill by bench	10
20	Salmon	First tier, west, by bench	10
23	White	Front, east	12
27	Lavender-white	Third tier, east and front	6

TRENDS DUE TO CLIMATE CHANGE

Following are pictures of the opening on a particular day in a given year, (then the average day the bush blooms and length of blooming).

May 9, 2007, April 29, 2017
(May 7th, 15 days)

May 10, 2007
(May 15th, 10 days)

May 12, 2007 – Mollis

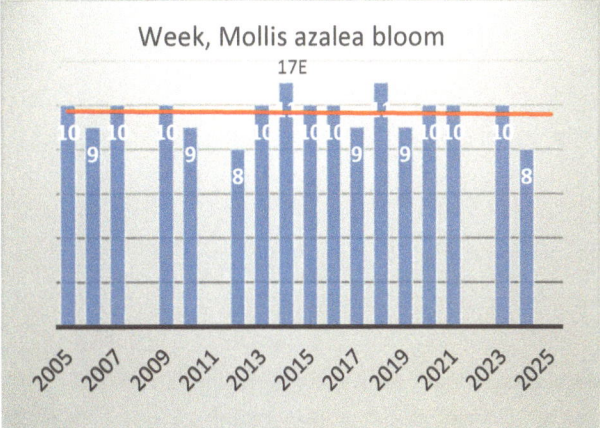

May 21, 2007
(May 15th, 10 days)

May 21, 2007
(May 17th, 10 days)

MAY

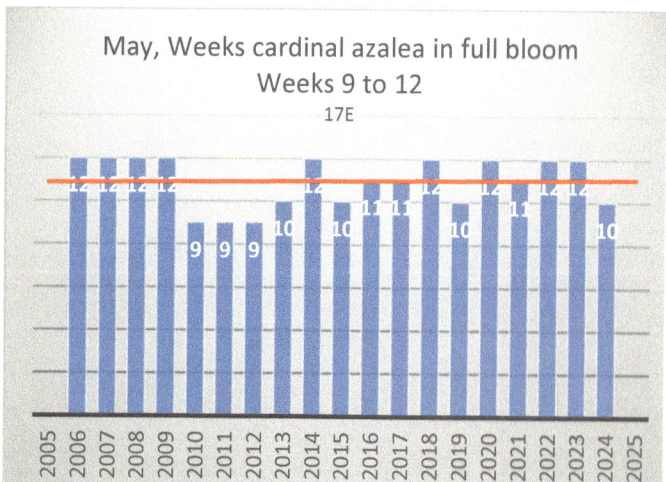

May 24, 2007 – Cardinal azalea

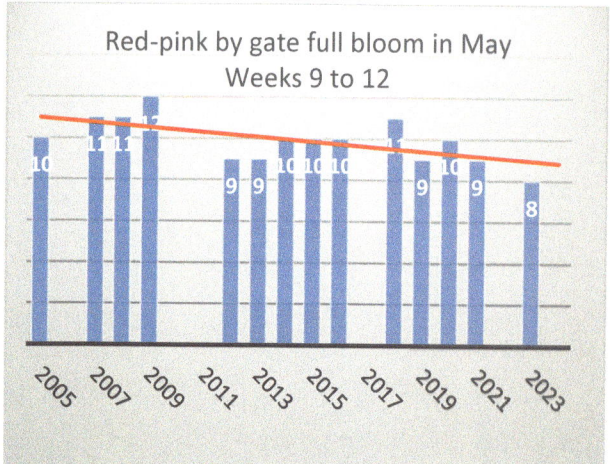

Azalea by gate

Rhododendron: The large pink rhododendron seems very temperature sensitive; it bloomed from the first to last week of the month. Usually trim the rhododendron, japonica, and holly at the end of the fourth week.

May 27, 2007

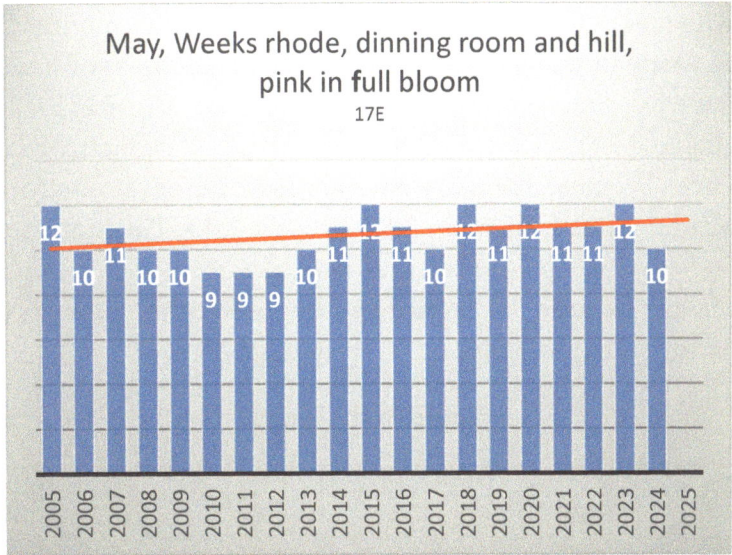

Rhodie bloom week

<u>Trees</u>: Pedals from the dogwood are all over. The spruce trees drop brown cones that formed last fall.

<u>Vegetables</u>: Some years I have cultivated the soil the beginning of the first week; it is best to wait until the end of the fourth week. When a ball of soil crumbles and falls apart it is okay to plant. No, it is not okay if the ground is not warm enough. Buy plants (tomato and herb) early to harden them, set them out in the sun during the day and then put them in the garage at night. Plant pole beans and Swiss chard only after the ground is above 50F at a depth of four inches. Wait for it to happen! Strawberries are in flower. Plant burpless cucumber, squash, and melon seeds. Plant the seeds on the north side by the trellis so their tentacles will seek the arms of the trellis. The fruit will then hang long and straight, also, the fruit is cleaner, and the critters may not eat them. Don't let the hanging fruit get too big, it puts stress on the vine. Work rotted manure or compost into the soil before laying the seeds. Sprinkle or spray the ground with Sevin before putting in the seeds and then sprinkle the covering layer of ground with Sevin. If the squash and melon do not get at least six hours of sun, they will not do well.

Picking green and red leaf lettuce that was started from plants. Broccoli plants are a foot high; the chipmunk is keeping them short. They eat the strawberries; neither netting nor mothballs help. I also watched one climb 20 inches vertically to get at a feeder. Chives go into flower the end of this week.

Review of Data

Fauna

<u>Animals</u>: Buck rabbits are chasing the hares in the beginning of the month. The hares drop the young in a hole in the middle of the lawn or jump over a two-foot chicken wire fence to deposit them in the middle of the vegetable garden! The young bunnies can squeeze through the tiniest gap to munch on lush lettuce and perennials.

Pairs of chipmunks can appear throughout the month. I smell or see scat of skunks in the third or last week. Raccoons raid the bird feed the third and fourth week.

Birds: It is enlightening to find out how many species are observed in a month. On an average there are 32 species during April. Many times, during the day, unidentified flocks of small birds skim across the top of the canopy, no less what is missed while I slumber. Between 2010 and 2012 there was a substantial "dip" of quantities to the low 20's. From 2005 to 2019 the number fluctuated between 22 to 39. The reasons for the changes can be, most likely, my environment and/or the environment they came from. Looking at May, the trend is going from 36 in 2005 to 30 in 2019. The dip to low 20's of 2010-2012 is still apparent.

The gang species changes because the catbird and house wren are now residents for the summer. The white-throated sparrow and junco have or will be going north.

Table of the Gang

Species/Quan.	05	06	07	08	09	10	11	12	13	14	15	16	17	18	19
Blue jay R, G	3	3	3	1	3	4	2	2	4	2	2	3	4	2	4
Cardinal R, G	2	3	3	3	3	4	2	2	4	2	2	2	4	2	4
Carolina wren R, G	1		1	2	2	1	1	1	2 +3	1	2	0	0	0	
Catbird G arrive	2/2	2/3	2/4	0	2/6	3/3	4/1	2/2	1/1	2/2	3/2	2/2	2/1	2/1	4
Chickadee R, G	2	2	2	2+2	2	2	1	2	1	2	2	2	1	2	2
Chipper sparrow G	2	0	2+2	0	2	2	0	2	2	2	1	2	1	1	2
Downy wood R, G	0	0	0	1	2	1	1	1	1	1	1	2	2	1	2
Goldfinch R, G change color date	3 6	3	3 1	3 4	6	2	1	0	3	2	2	2 9	4	3	
House finch R, G	4	2	4	4	0	6	4	4	2	4	4	4	4	6	4
House sparrow R, G	4	4	8	8	0	4	0	2	4	2	2	4	2	4	3
House wren G	2/1	1	2/7	2	1/5	1/3	1/	2/5	1/14	1/5	2/1	2/7	2/1	1/1	2
Mourning dove R, G	3	1	8	4	4	4	2	3	4	4	3	4	7	4	4
Song sparrow R, G	2	1	2	2	2	1	1	1	0	1	0	1	1	0	1
Titmouse R, G	3	2	2	2	2	0	2	1	2	3	2	1	1	2	1
White-breasted nuthatch													1	1	
White-throated sparrow G											2	1	1	2	4
Total gang species	13	11	13	13	12	13	13	13	13	13	14	14	15	14	16
Population	33	25	44	38	31	35	22	25	34	29	34	32	37	34	40

The total number of gang species for the month is basically 13. Some years it may drop to 11 or go to a high of 16. The type of species is constant. The reason that the number of species goes from 13 to 16 is that two species that normally migrate in April stayed here in May. The basic number of species almost stays the same throughout the year.

Number of birds within the species is also constant for May. The chickadee, sparrows, titmouse or downy stay between one or two. The dominance of population of one species over the others only occurs now and then; the prominent species are the house sparrow, house finch, and mourning dove. Total number of birds that the yard sustains ranges from 22 to 44 with an average of 33 birds. They eat a lot of seeds and peanut butter. The 2010-2012 dip appears in the chart for quantity in the gang. The dip keeps showing up month after month in the different charts.

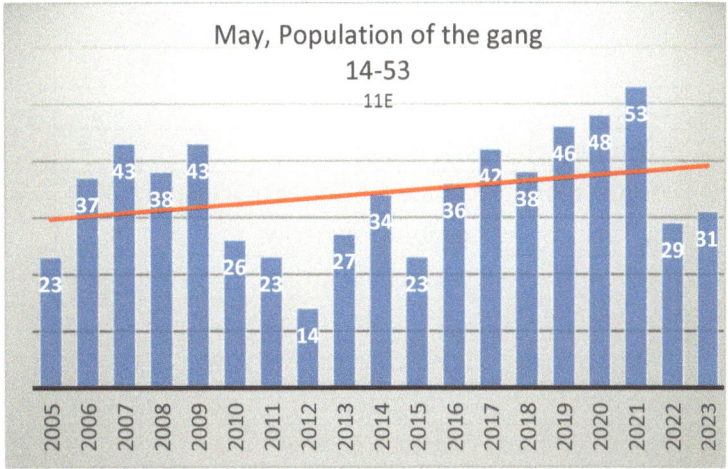

Population of the Gang: May brings the fledgling to the feeders. The number of fledglings within a species basically stays the same. The gender of most fledglings is easy to identify by their feathers. The gender of fledglings of the mourning dove, house finch, and goldfinch are difficult to identify.

Reviewing the dates of the returning summer residents, the catbird is almost consistent during the first week of May. In 2024 it returned the first week of May.

When and how many fledglings are brought to the feeders? For many species, the male feeds the young as well as the female. The fledglings are constantly crying and yelling for food, "Me, me, me." This squawking is not only done by the fledgling of small birds, but also the large raptors. Raptors are much noisier and when there are three fledglings, it will be quite noticeable (loud).

When parents bring young to feeders in May		
Species	How many young	Week
Blue jay	2	3rd
Cardinal	2	3rd to 4th
Carolina wren	3	1st
Catbird	1	1st to 4th
Downy	2	4th
Goldfinch	2	2nd
Grackle	1-2	2nd to 4th
House finch	2	1st to 4th
House wren	1	3rd
Robin	2	2nd to 4th
Sharp-shinned	1-3	4th
Starling	1	4th
Titmouse	1	1st

As the parents need food for the fledglings, the pecking order becomes apparent and not necessarily at the feeders. The birds like to grab, pluck, snatch a plump, fresh protein insect rather than a dry seed, but the seeds sure do disappear. The house wren does a great job in reducing the bugs.

Pecking order in May
Catbird chases hermit thrush
Titmouse chases house finch
House sparrow chases white-throated sparrow
Catbird chases chipping sparrow
Jays chase crow with chick jay
Crows chase broad-winged hawk
Song sparrow chases chipper sparrow
Robins chase jay and grackle from fledgling area
Pair of catbirds chase three jays
Grackle chases dove and a crow
Pair of jays chase pair of catbirds

<u>Migratory Passing Through (Bed and Breakfast)</u>: May is very active for the migration of the 57 species observed from 2005 to 2024. Many are not repeaters, only to be observed once or twice. Some become summer gang members, others just have a bed and breakfast for a couple of days. The number has been on the increase and arriving earlier.

It used to be that the song of a red-winged blackbird would herald spring as the flocks would go north. Flocks of cowbirds, grackles, red-winged blackbirds, and starlings, all in the same basic family, pass through. The flocks used to number in the hundreds, with the environmental changes, a large group, but not flock size, pass through in May. Strange, that one or two red-winged blackbirds were observed during or after a heavy snow in February. Today the numbers in a flock are larger in the urban areas, but still not as significant as 25 years ago. They pass through the first and second week. None were observed in 2024.

Do I observe a change in arrival time caused by climate change? Yes and no, some species arrival week has not changed over the 16 years. Third week arrivals, in 2016 and 2017, I observed a specie that I had not seen in 40 years, a woodcock. Another species I had not seen for 20 years was the red-headed woodpecker and one completely new specie, a scarlet tanager. In 2014 and 2016 a wood thrush stayed for three days.

From 2005 to 2024, 12 species arrive a week earlier. The white-throated sparrow and purple finch are leaving a week later to go north.

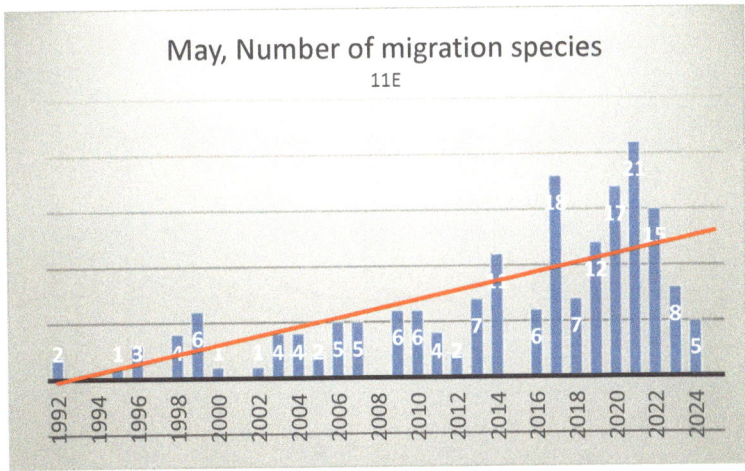

May, number of migrating species

April of 2016 was extremely dry. The flora came up or bloomed on schedule. The birds realized that the food may not be in ample supply when they would arrive because the number of fledglings was low.

In 2006 the red-bellied woodpecker arrived the last week of May; in 2017 it was arriving the first week. Starting in 2019 it was overwintering.

The arrival of the white-breasted nuthatch was the first week of May. In 2009 and 2015 when May was exceptionally warm and dry, they arrived in the third week; in 2019, they started to overwinter.

A real catch-22, the cowbird does not make a nest. She lays her eggs in a songbird's nest, resulting in less songbirds. Now with less songbird nests available there are less cowbirds. In 2006, they arrived the second week of May, in 2016 arrival was the second week of April. None were observed in May of 2018-2019. One was observed in March 2020.

Broad-winged, red-tailed, and Cooper's hawks soar high above around the billowing clouds against a blue-blue sky.

After reviewing the migration in May, how does this compare to the migration of April? From 1992 to 2019, the number of migrating species in May went from one to 12.

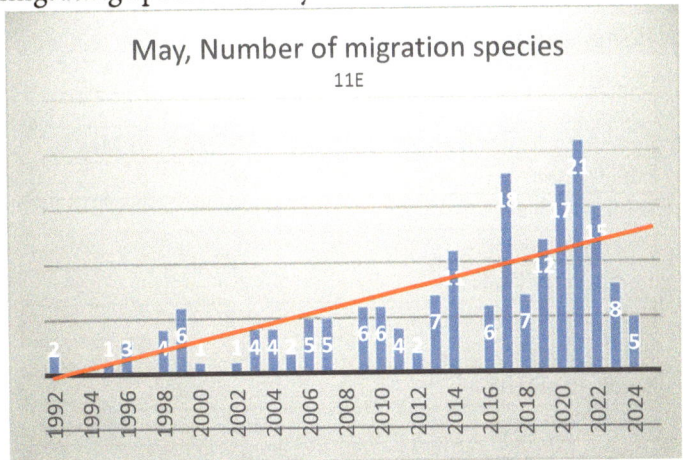

May, number of migrating species

The April chart starts in 1979. Migrating species went from six to 20, almost doubling of what happens in May. The 2010-2012 "dip" is there.

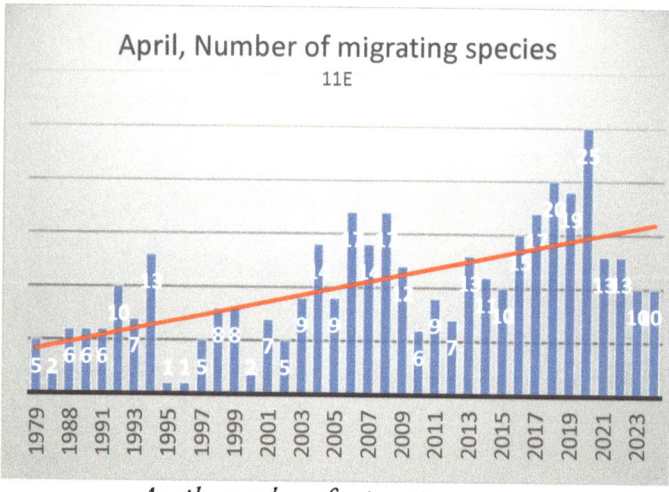
April, number of migrating species

All these species must migrate. It is built into their system-hormones; amount of sun; when insects, seeds or fruit will be available. As the climate changes, the availability of the flora is changing. So how do birds know when to change their migration departure to correspond with the flora?

<u>Insects</u>: Throughout the month when the daytime temperature goes above 55F, the soil just below the ground surface will warm up the hibernating bumblebees. By the afternoon they will be buzzing about.

Flora

<u>Progression of awaking flora</u>: Three characteristics that were reviewed from the data were the progression of awaking between species, the progression changes with the weather, and trends of awaking from year to year.

From 2006 to 2018, the progression occurs the same weeks when the average monthly temperature of April and May is between 53 and 55F and the yearly rainfall is above 13 inches for May. The reason of using April's temperature as well as a portion of May's is that many species awaken during the first two week of May, and their awaking is affected by April's temperature. The weeks of the progression is:

Progression of flora	
Species	Week in May
Poplar leaves	1st
White oak leaves	1st
Dogwood flowers	1st
Serviceberry flowers	1st
Holly, front gets flowers	2nd
Spruce growth	2nd
Hemlock growth	2nd or 3rd
Maple pollywogs drop	2nd
Maple pollen heavy	2nd
Hawthorn flowers	3rd
Canopy complete	2nd
Poplar flowers	3rd

Depending on the air and ground temperatures, and rain (local climate), the leafing of the tulip poplar tree can be as early as the first week of April to the second week of May. When this occurs, the whole progression stays the same, but shifts up or down the calendar. This makes for interesting anticipation.

The poplar tree releases last year's remaining seedpods a week after leafing; very messy. The pods do a nice job of clogging up the rain gutters.

Canopy: The trend of the completion dates of the canopy of the deciduous trees appears to be flat. However, both 2014 and 2016 had a cold-dry April which pushed the completion of the canopy to later dates. Removing 2014 and 2016 gives a different picture. The 2005 canopy completion was the beginning of the third week, by 2017 the completion was the beginning of the first week. Almost a two-and-a-half-week change. A strong indication of climate change.

In 2016 the completion was the 31st of the month, completely out of the trend. The temperatures of the first four months were higher than previous years. The month of May was nine degrees lower than previous years, but that should not have been a major factor since the canopy completion was usually at the beginning of the month. The amount of rain for the year going into May was almost normal, but March and April were very dry preventing the canopy from developing.

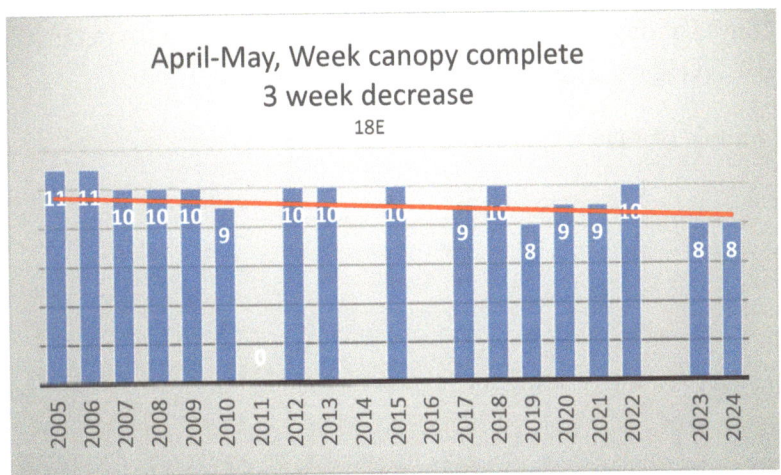

Canopy completion (no data for 2011)
With the removal of 2014 and 2016, shows a strong downward trend

The canopy completion corresponds to other shrub and tree blooming dates.

Many years rhododendrons bloomed on the same week as canopy completion. However, some years the rhodie bloomed one week later, which did not help in anticipation for the canopy. The blooming of the Cardinal azalea occurs the same week as the canopy completion. The dropping of maple samara starts during the third week of May to one week later than the canopy completion. It seems that the majority of samara get tracked into the house.

On average, dogwoods bloom the last week of April or the first week of May. With a warm April, such as 2012, blooming week can be the first week of April. For many years the dogwoods bloomed two weeks earlier than canopy completion. By 2024 there was only a three-quarters of a week difference.

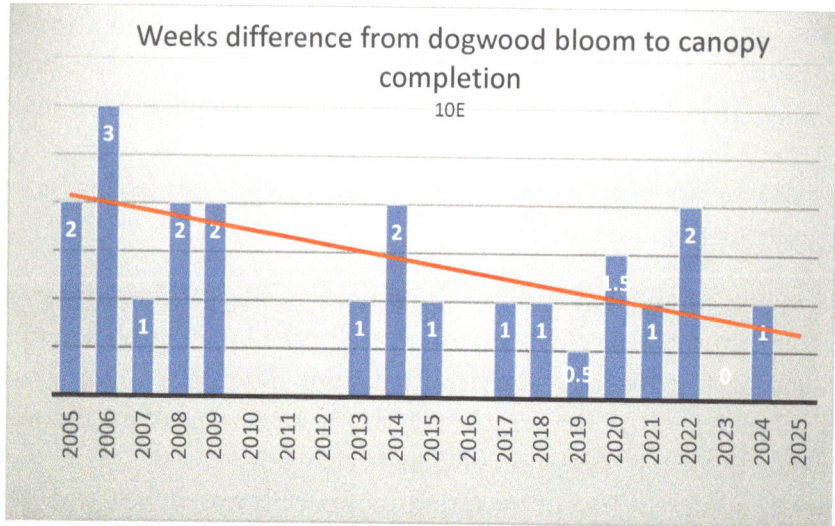

Difference from dogwood bloom to canopy completion

The flowering of the dogwoods has three phases; full blossom, wane when the blossoms turn brown, and petal fall (third week of May). Normal period of blossoming is two weeks. When March and April are warm and dry the blossoms may last three or more weeks. When the year to month is extra wet, the blossoms only last a couple of days. Most seasons, the time from bloom to petal fall is three to four days. With a dry May, it takes five days.

So, ends the busy month of May.

Index of June

June Overall	132
Weather	132
First Week	134
Second Week	140
Third Week	144
Fourth Week	149
Time Out	154
Review of Data	155
Flora	155
Perennials	155
Shrubs	157
Pond	157
Fauna	159
Birds	159
Table of the Gang	160
So, ends June	162

June

Week number for the year of 48 weeks					
Relative Month	Month/Week	First	Second	Third	Fourth
Previous	May	9	10	11	12
Present	**June**	**13**	**14**	**15**	**16**
Following	July	17	18	19	20

Weather

The amount of daylight on the 1st is 14 hours, 54 minutes; on the 30th it is 15 hours.
The average high temperature on the 1st is 78F, the low is 52F.
The average high temperature on the 30th is 81F and the low is 60F.
Extreme daytime high for the month (100F for the following days and years): 6th 1925, 20th 1923, 27th 1966, 29th 1934, and 7th 2011.
Extreme nighttime low for the month: 39F 1st of 1945.
The full moon of June is called the Strawberry or Rose Moon.

Overall Summary of June

The amount of daylight at the beginning and end of the month is about the same and the longest of the year. The fauna loves it because it gives them the most time to collect food. Afternoon temperatures can soar to 100F. The 3H's (hazy-hot-humid) form and so do heavy evening thunderstorms. Now and again we will have a tornado warning. Rhododendrons are out in full glory. With the 3H's comes fungus on the roses, phlox, and lawn, mushrooms spring up on the lawn. Time to spray horticultural oil on the apples. Gnats are out in earnest making it very unpleasant to work outside, however, things need to be done. Plant beans, cucumber, tomatoes.; thin out beets, carrots, and lettuce; harvest the first heads of broccoli. Later in the month trim the arborvitae, holly and rhododendron. All month-long perennials will be out in full glory and water lilies will emerge in full pink and white.

Squirrels bring their pups from the drey to the yard to be taught the tricks of survival and nature of humans (me).

In the evening, I look forward to listening to the croaking of the frogs, and at dusk see the erratic flight of brown bats going for thousands of insects. Then at 8:55 p.m. observe the rising of lightning bugs from the lawn; at 9:00 p.m. I am chased inside by the taunting attack of the female mosquitoes.

Weather

Mold and pollen are not really a weather event, it is the wind that distributes both that can make you feel uncomfortable to miserable. Pollen comes from the growth of the trees and grass. This is the month that the tree pollen is medium until the second week then goes to low the third week; grass pollen was high for the first week then drops to medium until the third week, then drops to low. Mold was medium for the first week then goes to high until the third week then it drops back to medium. It does not have to be hot for the mold to form, just wet. When the 3H's are forecast, it is time to put fungicide on the lawn. The formation of mold depends on the temperature-humidity. It could be high the third week and then moderate during the fourth, or it could be moderate during the first to third week and then be high from the third week to the end of the month. Sometimes the 3H's are so bad that when you walk out the door you immediately start to perspire.

Outstanding weather in June: 1938 the wettest with 10.5 inches of rain; 1989 was the warmest; 1990 had 80 mph winds; 1991 there was no rain during May-June. In 1999 there was no rain in April-May affecting growth of June's flora; 2002 and 2005 received 8.3 inches rain; 2006 it was so hot I could smell the evergreens on the hill (nice, but a bad condition); 2014 very hot and humid. June is the beginning of hurricane season; it goes to the end of November. On the 1st of June 2011, we had a tornado warning between 3:00 and 8:00 p.m. That did not happen, but we did get half-inch hail and very heavy rain. Most of the time the temperature is pleasant, in the 80's. Going back to 1990's, almost every year there was at least one day in June with at least 90F. In 2011 it went to 100F. From 2000 to 2012 these 90-degree days occurred the first two weeks of the month, from 2013 to 2018 they are occurring the last two weeks. Heat waves occurred from 1984 to 2003. After that there have only been one or two days of 90F each month. Climate change? However, the extreme hot and cold records are still from the 1930-1940's.

Percentage of extreme hot and cold days that occurred by decades in June.

Hot		Cold	
Decade	June	Decade	June
1930-1940	50	1925-1940	27
1950-1960	23	1950-1960	33
1970-1980	10	1970-1980	23
1990	7	1990	10
2000-2010	10	2000-2010	7

In 2012 the spring temperature was 5F above normal. In 2016 the temperature was also high and the ground was very dry causing the deciduous canopy to develop late. We can have a thunderstorm at 9:00 a.m.! It only takes 20 to 30 minutes for a small cloud to mature into a thunderstorm. Most thunderstorms seem to occur when the temperature reaches 91F.

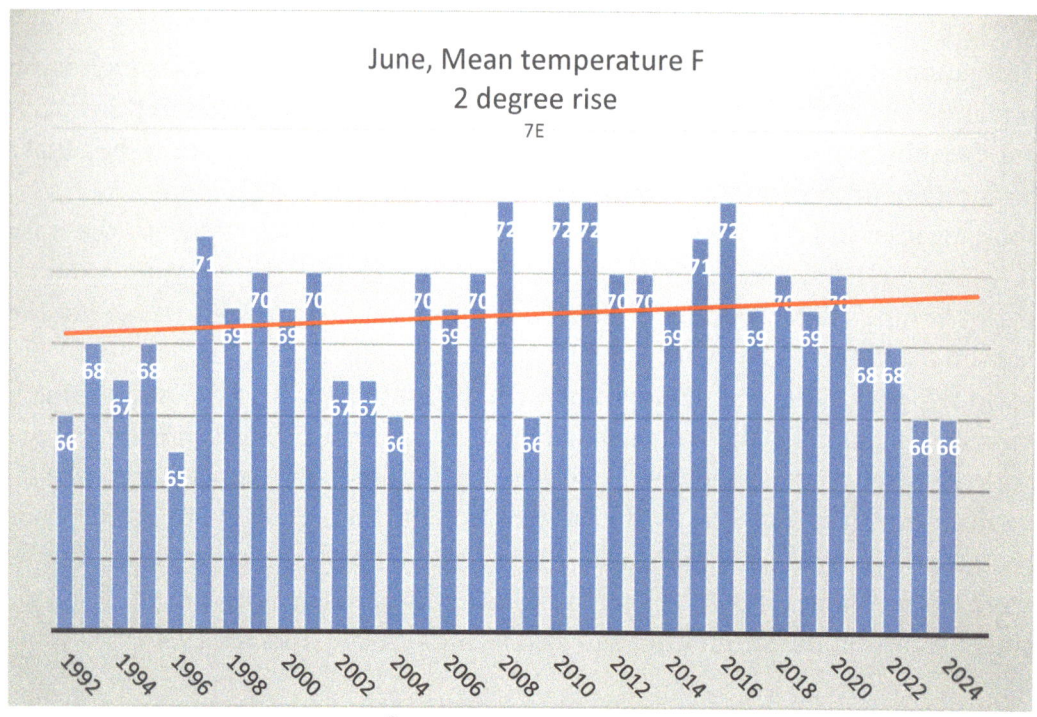

June, mean temperature

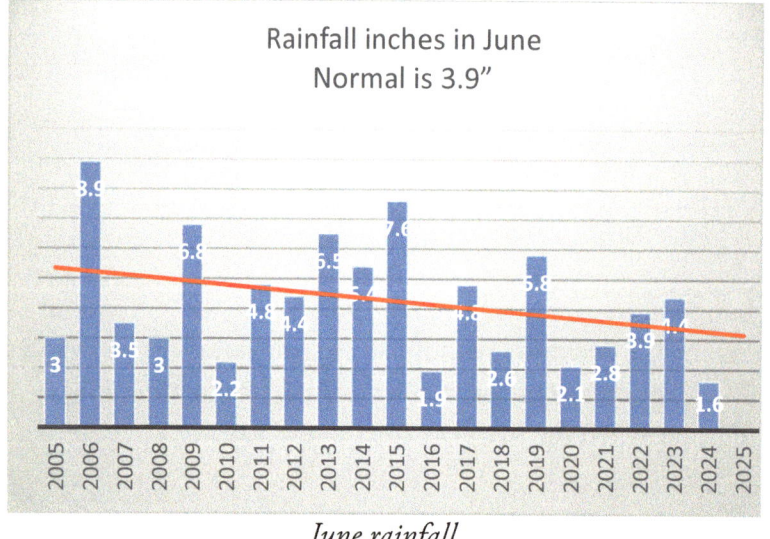

June rainfall

First Week

At 4:45 a.m. I am woken up by robins and cardinals singing and frogs croaking. The male house sparrow feeds his young. At 8:45 a.m. again robins singing loudly, but abruptly stop at 9:00 a.m. With a hot afternoon, a heavy thunderstorm will form and hang around. The air has a great fragrance (ozone) when it is over; it will be cool for a brief time before the humidity builds up again. A downy helicopter up to the suet. I am looking at the perennials when a frog springs up a foot from the ground to get a bug that had just landed on a flower. A female song sparrow is a surrogate mother feeding a young cowbird. A robin is eating suet, which is unusual. A catbird flies onto the feeder and chases the robin. Then a cardinal chases the catbird. It eats some seed and then flies off. A yellow swallowtail flits through at 2:00 p.m. The great blue heron flies over at 3:00 p.m. Two male cardinals go at each other. I have a little clay birdhouse in the lilac bush. I put it there for house wrens, but to my surprise a pair of chickadees have taken up residence.

Evening, a tree swallow goes over for high flying insects. Frogs start croaking at dusk. The bats erratically scour the air for insects. At 8:45 p.m. the lightning bugs rise from the lawn, going higher than the house.

<u>Weather</u>: Almost every year, the first day of this week it is 90 to 100F and humid. At 12:00 noon it is 67F dew point with 86F air temperature, ugh! At 2:00 p.m. it is now 90F, further ugh! The humid days give the right atmosphere to produce mushrooms in the lawn. Tree and grass pollen are high, mold is moderate.

<u>Animals</u>: In the animal world there are young chipmunks, rabbits, and three squirrels all curious and mischievous.

In 2017 there were not many hawks in the area, consequently there are rabbits all over. Three seven-inch-long bunnies squeeze into the vegetable gardens and devour the young plants and proceed to devour the new growth on the perennials. At night, the hare jumps over the chicken wire fence on the raised garden to have a feast. They also like the seeds of the poplar tree. I think I will put the poplar seeds around the raised garden and let the rabbits fill up on them. Late in the night I hear a raccoon trying to get at the bird seed in the metal garbage cans.

<u>Birds</u>: The bird gang is here. Robins and cardinals are eating suet, not together, at separate cake cages. The main feeders are serving five doves, a pair of catbirds, and a young cardinal. Many species of birds bring the fledglings to the feeders. Adult robin, downy, catbirds, cardinals, and crows are feeding their fledglings all day long. A young house sparrow flaps its wings as its father feeds it peanut butter. I am continuously refilling the holes in the peanut butter log. Chickadees catch small moths for their young. Tree swallows leave and house sparrows take over the house. Flocks of red-winged blackbirds and starlings go north. I may see a hen turkey with a clutch of 10 scurry across a road. Sharp-shinned and Cooper's hawks are continuously in the feeder area to snatch a meal. Late afternoon, a hawk or owl has landed in the silver maple; for five minutes crows, jays, and robins are all yelling until it leaves. A cowbird raises its wing to a jay who says, "So what?" A jay dances around a dove while both are on the feeder. The jay leaves and is replaced by a male cardinal, which is then replaced by a female house sparrow. A chickadee chases and gets a moth. A great blue heron is in the pond. It detects me, spreads its five-foot wingspan, and flies off.

A male downy is waiting for the house sparrow to leave the wire cage black ball filled with split peanuts; two house finches beat it to the ball, so the downy flies to a suet house. Then a catbird goes to the ball; then the catbird chases the downy from the suet. Then it chases a female house finch with its beak wide open. The catbird then helicopters up to the suet. Then a blue jay chases the catbird from the suet. On and on it goes. I put up another wire cage, this one is green. I fill it with the same split peanuts. Weeks go by, no songbirds go to the green ball. I'm going to paint it black, maybe red!

Young house sparrows occupy three houses. There are two song sparrow nests, one in the front yard and the other in the back. Crows have one youngster. There are pairs of cowbirds, doves, chickadees, house finches, downy, blue jays, and chipper sparrows. House wren was gone for a week, but then returned.

House wren

A female cardinal has mites, resulting in mange. A bull mourning dove puffs up his breast, and then struts toward a young hen. She turns up her beak and flies off. A jay chases a robin from the main feeder, bully. A grackle brings two fledglings to the feeder.

Put whole peanuts on the main feeder; a blue jay takes the nuts before I get back into the kitchen.

Jay with nuts

Then two grackles come in for the whole peanuts. A male grackle puffs up its chest to show off to the other how big he is. The others shrug, like "show off."

In 2008 and 2018 I found a dead catbird on west side of the house. No signs of the cause.

Insects: A little moth is fluttering around; it is spotted by a chickadee who launches from an above branch and snatches it up. While gardening, I feel an itch and look at my hand to find a tick.

TRENDS DUE TO CLIMATE CHANGE

Ant with moth wing. What strength!

Silver spotted skipper on a zinnia

<u>Perennials</u>: Butterfly weed has green blossoms. Bearded irises are about done. Red hot poker has big blooms. Many perennials are done blooming by the end of June. Zinnia from purchased plants are out.

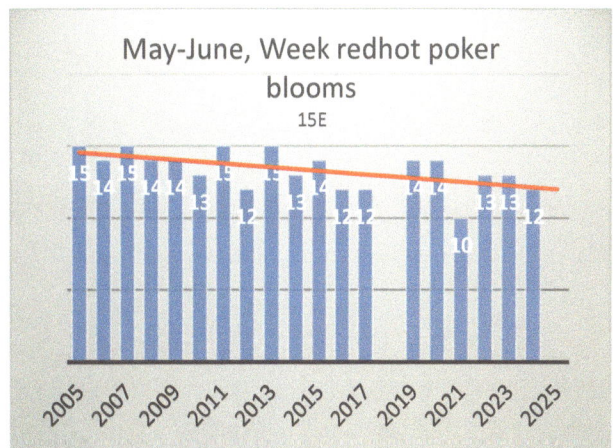

Red hot poker

The astilbe spikes form in early June, rise to over a foot and produce their feathery white, pink, and red florets. Some of the flowers last to the first week of July, but most are gone by the end of the month.

Astilbe

The rose campion by the steps is out full. The blossoms last 16 days. It is self-pollinating and moves further east each year as the wind progressively blows the seeds until there is no soil for the seeds to germinate. It took 20 years for this to occur. Then they were gone. Globe thistle is three feet high.

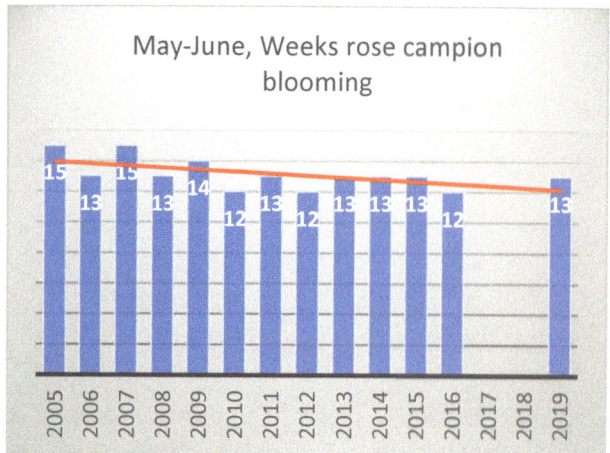

Rose campion

Globe thistle, globe forming

The tough stalks of the cornflower rise to three feet and develop their flower of pink petals with the rust colored button in the middle. Rock sedum behind the pond is draping gracefully over the stones with its dainty yellow flowers. When the flowers wane, the sedum will become tan and bedraggled. The green dragon (part of Jack-in-the-pulpit transformation) is coming up.

<u>Pond</u>: Around the pond there are four frogs, I don't know how to pair them off. Then again maybe they are not pairs. Now how is that for spring romance? The blue pond iris is waning. The pickerel plant in the pond has white flowers. After a heavy rain it is spread out flat like someone stepped in the middle of it. I tie it up and cut off the spent flowers. The water lily bloomed on the 5th. Pencil irises behind the pond are done.

Lilies

Pickerel weed *Frog in the shade*

<u>Shrubs</u>: Mountain laurel blooming June 1st.

Mountain laurel is blooming

<u>Trees</u>: Flower petals are also falling off the poplar trees. Red and white dogwood may still be in blossom.

Fruits: To have larger fruit, I thin out the apples to three to four to a bunch and cut off suckers. The apples have good fruit. Strawberries should be ready to pick. There are green berries on the blueberry bushes.

Vegetables: A morning inspection of the vegetable garden reveals that voles have nipped the snap bean buds. Put Sevin to kill and prevent insects on the roses and vegetables. Broccoli is ready to pick. Cultivating the herb garden, I come across a big bumblebee's nest under the thyme plant; I quickly back away.

Second Week

Animals: A chipmunk is having a breakfast of my cucumbers! Two other munks are on the chase. Chives are around various plants to discourage rabbits from eating them. A squirrel takes nesting branches to the top of the tree. A groundhog is on the hill.

Birds: It is 7:30 a.m., a clear morning of 70F, things are tranquil. A blue jay and grackle sit next to each other on the feeder. Another jay goes for a smorgasbord of sunflower seeds, split peanuts, suet, and a drink of water. I hear other grackles grac-grac-grac at each other. A house wren trills at the top its lungs. A flicker is calling for a mate. A dove has gathered nesting material in its beak. A young grackle lands on a young cattail stalk. The stalk gently bends over causing the grackle to slide off. The stalk re-straightens.

Then the pecking order routine. A cardinal chases a catbird from the feeder, then a dove chases the catbird. Pairs of cardinals and robins go after a sharpie that was trying to get at a young robin that was in the arborvitae. Chickadees have two fledglings sticking their heads out of the front birdhouse.

Pairs of cardinals are hissing like cats at grackles which are getting too close to their nest. A pair of house wrens has settled in the cuckoo-clock birdhouse on the front porch. The true benefit of having wrens is they continuously bring fat, juicy insects to the fledglings in the house.

More mom

Opened the front door to a bright morning. The clear song of a Baltimore oriole fills the air. Tree swallows fly over getting insects.

Young house finches are at the feeders. The young finch has two horns of brown feathers about a half inch tall. They disappear as it matures.

Mom feeding junior house finch

Junior house finch who has horn feathers

Young starlings are very noisy; a rare spotting of a wood thrush passing through. Observed a downy eating the seeds of a red-hot poker. This week the downy brings two fledglings to the feeders.

Downy getting red hot poker seeds

Insects: It is a warm and windy evening. I like to sit and observe the yard but get a severe gnat attack and I quickly retreat to the indoors. These attacks also occur in September and October.

Lawn: A wet and humid condition brings excessive numbers of mushrooms all over the lawn. Some years fairy rings develop that are six feet in diameter.

Fairy ring

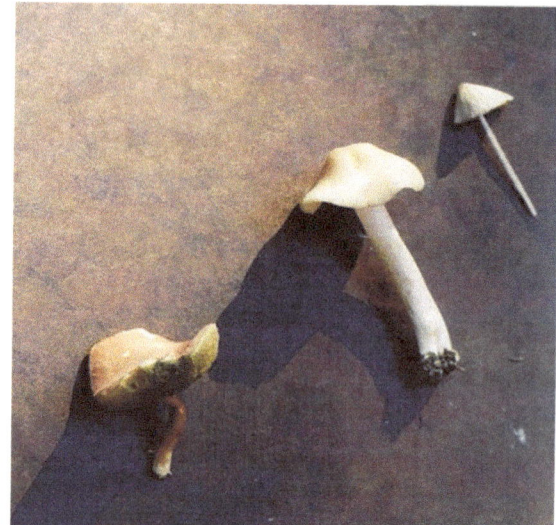
Two types of mushrooms

Now that I told you that we get mushrooms in June, guess what? Because of the same conditions a second war is declared, fungus! The fungus war is worse than the weed war, because this one kills the lawn. At least with the weeds something is still green at the end. The strategy with this war is that you should know when to spray before the other side declares war. Watch the weather conditions for when it is going to be hot and humid. The fungus turns the grass brown. The fungus not only affects the lawn but also flowers like roses and phlox. Put spray or granular fungicide from fourth week of May through third week of June. There is another fungus called red thread. It is easy to detect, just look down at your shoes. If the creases of your shoes are coated with red powder, your lawn has red thread. Its appearance means that the lawn is nitrogen deficient and needs water.

Another occurrence in June is that the grass produces mold that affects our health. The mold starts in the middle of May and peaks at the beginning of June and then declines to none by the first week of July.

If we have a bad dry spell in June, at the end of the month the lawn will go into dormancy and turn brown. When this occurs do not walk on the lawn or cut it. The tall weeds continue to grow. You may relinquish to cut weeds as a cloud of dust follows behind you. Sometimes instead of getting a brown patch you will see brown rings in the lawn. They arise from the beginning of June to the beginning of August. With a hot, dry, and humid summer the lawn will have large brown spots; put down fungicide. When the fungicide is not effective the spots will grow and spread right through the summer to the beginning of September.

You can have a good-looking lawn through the third week of August, then we get three weeks of humid 90F, and all of a sudden, spots. Fungus has arrived. The name of the game is to put down fungicide, even if the 3H's are not present.

Another war is the grub war. First it was crabgrass, then goose grass, then weeds, followed by fungus and now the grub war.

Grubs are the larva of the Japanese beetle. They are white and a half-inch long and a quarter-inch in diameter and are usually curled up like a "U." They eat the roots of the grass. The lawn turns brown and when you pull on it the lawn comes up like a rug. This occurs during the first week of June. The other indicator that grubs are present is you find that your lawn is dug up in the morning from visiting skunks. On the other hand, if the lawn is brown and you rake hard and it does not come up, then the problem is probably lack of water.

There are two ways to get rid of grubs. One is to spread a poison which directly kills the grubs or two is put down a bacterium called milky spore. This is a bacterium that gets into the intestinal system of the grub and kills it. It takes two years for it to get established but it lasts 15 years or more.

<u>Perennials</u>: The yellow flowers of low sedum are showing. Daylilies and astilbe have buds. Feed roses alfalfa meal and compost. Put bone meal around bulbs. The daylilies are ready to open. The evening primrose is still bright yellow, their flowers are full of bumblebees extracting nectar. The umbrella-shaped flower of the yarrow is pure white; lavender loosestrife and yellow potentilla is blooming. Dainty-stemmed pink coral bells are out by this week and finished by the fourth.

Coral bells

<u>Pond</u>: In the morning when I come down the stairs and look out the kitchen window, I see the great blue heron stepping out of the pond as it finishes breakfast. I'm looking at a flower whose blossom is 20 inches above the ground. The blossom is slowly rocking back and forth as a bee is extracting nectar. When wham! From below a frog jumps straight up and swallows the bee. The frog falls back to earth and the stem sways until it comes to rest.

<u>Shrubs</u>: Remove flowerheads of daffodils and tulips. Pinch back the mums, columbine is in seed, clematis blossoms are on the wane. The bleeding hearts are fading. The mountain laurel is in full pink, it will wane in seven to 10 days. Yellow bell (forsythia) has heavy growth.

Rose bushes are out full; peach-colored, lavender, light pink, dark pink, and light yellow.

June 10, 2010 – roses

<u>Trees</u>: Hang 15 CD discs on the Empire apple to discourage birds from attacking the young fruit.

<u>Vegetables</u>: The asparagus fern are three to four feet tall. Broccoli is in flower, cauliflower is forming, patio tomato is forming buds, and the grape tomato has buds. The cucumber and pole beans are up. Sprouts of lettuce are up three inches; beets up half inch, thin them with a scissor. Cut off the lower branches on the tomatoes for air circulation. Plant dill and parsley. What I thought was carrots coming up turns out to be dill!

Third Week

<u>Weather</u>: On the 19th in 2009, 96F, the amount of rain from April 1st to June 18th (39 days) is 12.54 inches. Then we get another 1.4 inches. It bruises apples and peaches, rhubarb leaves are shredded, tomatoes are yellowed. Hay is ruined. Vegetables have fungus. Strawberries rotted from too much rain. Crops were unsuitable to sell. Birds are eating during the heavy rainstorms. Many fledglings die.

<u>Animals</u>: The rabbits on the hill are jumping up and down; I do not know if this is due to the heat or internal stimulation. Two years in a row I find a dead adult skunk in the garden. No physical injury is apparent.

TRENDS DUE TO CLIMATE CHANGE

<u>Birds</u>: In 2006, 10 bird feeders are active. Catbirds have a nest in the arborvitae, they chase grackles from the area. The catbirds meow at me when I walk by. I observe a catbird pulling an earthworm from the soil! A defending house wren chases a squirrel from its house area. Then there was a cardinal and robin working in unison to chase a sharp-shinned hawk from getting a young robin which was in the arborvitae. Unusual, a cowbird raises an aggressive wing toward a blue jay! Another unusual observation is to see a pair of grackles waddle across the lawn. A young, large speckled breasted robin is hopping on the lawn looking for worms. A grackle is going after a young catbird, when suddenly the adult catbirds attack the grackle, who quickly flies off.

In 2016 lots of fledgling activity. Activity started the last week of May and rapidly increased by the middle to the third week of June. I had a hanging platform safflower feeder by the kitchen window, but with the threat of severe weather I kept taking it down for concern the wind would bang the feeder into the window. I replaced the feeder with a plastic tube that holds split peanuts. What a good move, birds that did not eat safflower were now continuously by the window. It got so active that I put up two tubes. I also tried to keep the peanut butter log full. Everything was fine until the chipmunks found how to get on the tubes.

Female downy

Mother catbird

They love the peanut butter log

<u>Insects</u>: A little blue azure butterfly flutters around. Black and yellow swallowtail butterflies flutter through from the first to last week of the month. As I walk across the lawn, again the gnats rise to attack my ears; it takes away any pleasure of being in the yard. Lightning bugs begin to rise in earnest during the third week but are the most abundant during the fourth week.

Swallowtails

Spot an eighth-of-an-inch long green cricket.

Look closely, the hopper is in front of you

Perennials: The two-foot spikes of the veronica are producing their blue fuzzy flower. As time goes by the stem gets taller and flowers continue to emerge at the top of its growth; bumblebees are continuously pollinating the blue spikes. The veronica blooms for seven to 10 days. When the flower fades, it leaves pea-size seedpods on the stem.

Veronica

The male spikes have formed on top of the punks of the cattail. It is time to cut the red hot poker stalks.

Bergamot can be pink, lavender, and red. It has a square stem; the leaves were used as a tea by the colonists after the Boston tea party.

Bergamot

The trumpet vine red flowers are almost trumpeting. The hummers love the nectar. The hummers also do their mating dance in front of the blossoms. The seeds from the blossoms develop all over the lawn. The trumpet vine has a deep taproot and is difficult to eradicate. Squirrels are a little help because they eat the blossoms.

Trumpet vine blossoms forming

Butterfly weed and tiger lilies open this and the fourth week. By the end of the third week the tiger and daylilies have blossoms on the three-foot stems.

JUNE

Orange tiger lilies

White tiger lilies

The butterfly weed, alias Railroad Mary on Long Island, is bright orange. This flower attracts monarch butterflies.

Butterfly weed

Yarrow and red lilies open this week and bloom for a week.

Pond: Dragonflies hover over the pond in the late afternoon. In a blink of an eye they change position over the pond to hover again. A pink water lily is dazzling the pond. With the 3H's, algae starts to form in the pond's water. It is not bad enough with the 3H's, but I'm attacked by gnats again.

Shrubs: Trim the rhodie, Japanese maple, and holly. They should have new growth in two weeks.

Trees: It can get so hot that I can smell the aroma of the evergreens.

Vegetables: Bouquets of green flowers are forming on top of the rhubarb stalks. Strawberries are ripe; who gets them first, me or the chipmunks? Munks are also eating the beet seeds. To deter the munks, I lay a 2 x 4 board over the seeds until the seedlings break.

Fourth Week

<u>Weather</u>: The humidity is causing the split peanuts in the feeders to go moldy.

<u>Animals</u>: A chipmunk makes a dirt patch in the lawn, then for five minutes keeps rolling over, then gets up and runs away! Rabbit eats Solomon's seal leaves. The cute bunnies, young and old, are eating my broccoli, zinnia, portulaca, spinach, and Brussels sprouts! Like the squirrels in March, two chipmunks are on the chase all through the gardens as well as between the furniture on the patio. They race right through my feet! The back one catches up to each other. They proceed to tumble over and over each other, stop, and continue their chase.

<u>Birds</u>: A female surrogate song sparrow is feeding a young loud cowbird.

Cards and robins sing for 15 minutes at 5:30 a.m. and then again at 6:00 a.m. In the evening they start at 8:40 p.m. and stop abruptly at 9:10. Half an hour each time.

Robins, catbirds, and jays like to take baths on hot days.

Blue jay taking a bath

Catbird ready to step into the bath

Robin watching me

A goldfinch is feeding upside down on the thistle feeder. Another battle, I am gardening when suddenly there is a racket of birds yelling all around me. A pair of robins, a pair of jays, a young jay, and a catbird are all in battle. Somehow the young jay got too close to the robin's nest setting off the adult robins. The male robin had jay feathers in its beak. When I moved, it caused the birds to scatter and things calmed down.

Goldfinch eats upside down

A titmouse goes after its reflection in the kitchen window. A pair of grackles chase a jay from the feeder, a teen cardinal is changing from beige to red. A grackle is at the feeder getting a split peanut, when suddenly from behind another grackle grabs the peanut from its beak and flies off as the awestruck one looks on. Brotherly love! Another grackle stops to have a piece of fresh orange. A crow with a baby grackle in its beak has four adult grackles chasing it.

A great blue heron is seen high in the evening sky; its wings appear to be in slow motion as it goes south from its early morning northern trip.

Great blue heron

Robins land on the rooftops and sing their up and down scale evening song, proud and clear. As they finish, their silhouettes can be seen flying into the spruce and maples for the night's roost. But then they change their call to a different chip-chirp, like a final "good night." The night song of the song sparrow can be heard among the spruces.

Fruits: Raspberries are turning pink. The Empire apples look good and have a pink blush, but the leaves have curled; time to spray the fruit and vegetables with Malathion. The blueberries are ready for picking; it is a race between the birds and me.

Empire apple

Insects: A black swallowtail butterfly sails through but stops to sip some nectar from a zinnia. Japanese beetles are out this week, so are more gnats. At 8:50 p.m. I look down the see the lightning bugs rise from the lawn. First, there is one or two and then 30 to 40. They were only a few inches over the grass. As it gets darker, they are a few feet over the grass, they rise 10 or more feet. Males hover in dance two to three feet above the ground, but some go as high as 30 feet. I don't know if they turn off their lights when they have attracted a female. The female mosquito attacks viciously at 9:00 p.m. just as the sun sets, but they usually reduce the attack by 10:00 p.m.

Lawn: The lawn needs nitrogen, fertilize with 20-4-10 or 20-2-0. Lawn weeds of yellow wood sorrel, white clover, purslane, knotweed, and wood sorrel are prevalent.

Purslane, used in a salad in the 19th century

Yellow wood sorrel

Perennials: Depending on the weather, instead of the third week, the end of the month is when the globe thistle, water lilies, trumpet vines, veronica, and bergamot are blooming. Loosestrife, very aggressive, is a great pollinator. It blooms the first through the third weeks.

Purple loosestrife

The amaryllis is not a perennial; however, it has been overwintered for 14 years, sort of a perennial. I put a baby's crib blanket over them with a space heater in between the plants. The temperature is kept around 50 to 60F. I water them sparingly. Come April, I uncover them and give them more water. By the end of May they are ready to be set out. The amaryllis go into bud the third week and blooms to the end of the fourth week.

Amaryllis that I have overwintered for 19 years will bloom into July

Blanket, coneflower, and spider plant flowers open. Daffodil foliage is turning to dull green. Wild thistle parachutes are floating in the gentle afternoon breeze.

Annual note: In 2015 all impatiens got a fungus and died in days. Garden centers did not sell impatiens for two years.

Spray fungicide on roses and phlox. Spray horticultural oil on hemlocks, apple, phlox, roses, and bergamot.

Pond: The tadpoles in March are turning into frogs. I wonder if the barometric pressure affects the croaking of the frogs. They were loud before a thunderstorm but became quiet after the storm passed. Frogs are mating.

Frogs mating, three is a crowd

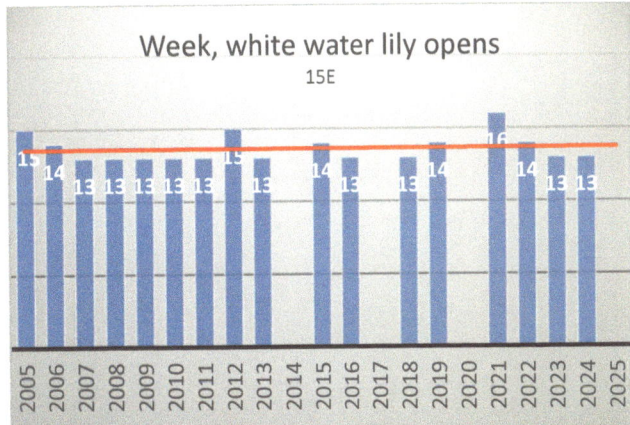
White water lilies open during the last week

Vegetables: This table shows what is happening in the vegetable garden in June.

Vegetables in June					
Week		1st	2nd	3rd	4th
Turn over		May 16th			
Asparagus	pick	x			
	3' fern				x
Beets	plant		x		
	up				x
Cauliflower	pick			x	
Broccoli	cut	x			
Chives	flower				x
Cucumbers	seed	x			
	flower				x
Lettuce	seed	x			
	pick			x	
	plant plants			x	
Pepper	in flower				x
Rhubarb	pick				x
String bean	plant			x	x
Tomato, cherry	blossom			x	
Malabar spinach	seed		x		
Zucchini	plant			x	
	up				x

June 21st – vegetable garden

<u>Time Out</u>: June 2017, 7:48 p.m., 75F. I decided to just take time and sit on the back patio and listen and observe. Something I do not do enough, just like not taking time to smell the roses. It is clear, the sun is setting between the upper spruces on the hill. A rabbit is eating the back lawn, a frog croaks from the edge of the pond. A robin sings its soft evening melody. Blue jays are yelling softly to each other, that is if you can image a soft yell. A female cardinal that lands on the feeder, sees me, and immediately flies off. Up the hill a neighbor's dog barks at something. The din of the continuous rush of the evening traffic on the nearby highway is always in the background. A house wren trills in the evening air. Another robin lands under the feeder to pluck a worm from between the empty sunflower husks. Cardinals are softly chirping. The air smells fresh.

At 7:55 p.m., a male cardinal lands on the feeder of sunflower seeds, flies off. The silence is interrupted by the roar of a motorcycle speeding down the highway. The upper branches of the pin oak sway in the evening breeze, sunbeams are filtering through the spaces of the oak branches. Again, interruption from a song blaring from a car. Another robin trills a cyclic call. I see a squirrel tail rapidly twitching from behind the dogwood tree by the pond. Suddenly the squirrel jumps straight up into the branches of the adjacent rhododendron, then drops down and runs off! Two doves land on the feeder, sup for a few minutes, then calmly fly off to roost. Many frogs are now striking a symphony. Downy woodpecker is at the peanut butter log. A red-bellied woodpecker calls from high in a spruce tree. I spot a tree swallow ballet over the grass on the hill. There are four goldfinches on the thistle feeder. Thankfully, there are no annoying bugs.

It is calm and peaceful at 8:13 p.m.; a robin now sings from its perch on the electric wire. A catbird comes for a second trip to peck at a suet cake. Dusk is approaching. A pair of house finch partake of safflower; the robin on the wire is still singing. Three chickadees, which are late eaters, are at the split peanut cage and peanut log. A female cardinal interrupts the chickadee at the split peanut cage, a male cardinal is on the ground below the cage feasting on the fallen scraps of peanuts. They fly off together to go under a nearby azalea, come out, and fly up the hill. Now I hear a flicker call. The leaves of the 50-

year-old skunk cabbage by the pond is still bright green. A rabbit bounces down the steps of the terraced gardens at 8:25 p.m.

At 8:35 p.m. the robins are coming to roost in the spruce trees. Some robins stop below the feeder to hopefully get worms before roosting. As it gets dark, more frogs are joining the symphony. They are not only by the pond but also spread out among the gardens. The first lightning bugs are rising from the lawn at 8:40 p.m. At 8:53 p.m., almost dark, big brown bats are flying over, robins are still coming to roost, which is causing the catbird to give a disturbing chirp. At 9:05 p.m. a thunderstorm is approaching.

It is inevitable the annoying bugs have arrived and attack. Time to go (rush) indoors. It has been a wonderful hour plus. I should do it more often. Try it.

Review of Data

Flora

<u>Insects</u>: Because the average monthly temperature is warmer, the white sulphur (cabbage butterfly) and eastern tiger butterflies are hatching a week earlier. Lightning bugs are rising from the lawn two weeks earlier.

<u>Perennials</u>: I have a period of 14 years of dates for June's perennial blooming and waning. Decided to investigate to determine if there is a trend in the change of those dates over the period.

The data shows the blooming dates from year to year look like a yo-yo. Some dates spread as much as two weeks from one year to the next. Why? The cause for the differences is in the amount water and temperature. The amount water is looked at in two ways, the monthly amount and the year-to-date amount. There can be a dry June, but the year-to-date can still be normal.

With water and temperature there are the average numbers which comes from years of data, which I will call normal (N). There are times when there are low (L), very low (VL), high (H) and very high (VH) readings. These readings come as combinations, such as a normal rainfall with a very high temperature or very low rain with a very low temperature or any other combination. During the period, normal rain and temperature have occurred together only five times. Reviewing the *monthly* amount of rain and temperature a trend did not develop. When the perennial blooms at the beginning of the month, then I referred to what weather occurred in the previous month. Using the *year-to-date* amount of rain and monthly temperature shows a trend of blooming five to 14 days earlier than 14 years ago.

June Weather Weather normal for month 4.3", for year 21.1"																
Occurrence	05	06	07	08	09	10	11	12	13	14	15	16	17	18	19	20
Rain in month	L	VH	N	L	VH	VL	N	N	H	H	VH	VL	N	L	H	
Rain year-to-date	VL	N	N	N	L	N	H	L	N	VL	N	VL	N	H	VH	
Average temperature during month	N	N	N	N	L	H	N	L	N	H	N	N	L	N	N	

When the ground at four inches has not warmed up to 50F or there is a drought condition the perennials come out later. In 2017 April and May were warm, but June was normal and April showers were higher than normal, many perennials, shrubs, and trees came out two to four weeks ahead of other years. When we have a lot more rain than normal, the perennials come out later because of the lack of sun. Most of the perennials showed the effect of the 2012 dip by blooming two weeks early; 2020 may be a repeat.

On the Average

Evening primrose blooms from the fourth week of May into the fourth week of June. They wane in two weeks. If June is wetter than normal, the blooms last three weeks but they still wane two weeks later.

Potentilla is out the second week; if the year is low on rain it may wait until the fourth week.

Loosestrife, Stella de Oro, and veronica are out the first to the fourth week. The veronica wane in eight to 12 days. The trend of these perennials from 2005 to 2019 is that they are blooming one to two weeks earlier.

Trumpet vine and tiger lilies bloom the third to the fourth week. Tiger lilies bloom the fourth week when the month has excessive rain.

Coneflower, daylily, hosta, milkweed, and yarrow have stayed steady to bloom the fourth week.

<u>Astilbe</u> bloomed the fourth week from 2005 to 2015 but bloomed the second week in 2019-2024

Butterfly weed blooms the fourth week; in 2017 they bloomed in the first part of the fourth week.

Coral bell was blooming the fourth week; in 2017 they bloomed the last week of May.

Rose campion was mostly blooming the first week; in 2017 and 2018 none came up.

Red hot poker bloomed the third week in 2005. In 2017 it bloomed the last part of the fourth week of May. It wanes in a week.

Globe thistle blooms the last part of the fourth week into the first week of July. The flower wanes two to three weeks later. After deadheading, they re-bloom. This the third old perennial that did not come up in 2018, all in the same garden.

Bergamot was blooming the third week, now it blooms the fourth week, opposite to the trend. Unlike three other perennials that did not appear in 2018, the bergamot flourished profusely

The sequence of blooming is not affected from year-to-date and monthly weather, but when they bloom does. The overall sequence is:

Blooming sequence of June's perennials from observed data	
Perennial	Blooming week In June
Evening primrose	1st
Loosestrife	2nd
Astilbe	3rd
Red hot poker	3rd
Rose campion	3rd
Stella D Oro	4th
Trumpet vine	4th
Tiger lily	4th
Daylily	4th

Arranging the sequence of the blooming weeks from the "trendlines," not the observed week, shows that the sequence of blooming basically remains the same from year to year. There is a different order than that generated from the observed data. *The first three are opening by one to two weeks earlier. The two in the middle of the month show no change in the opening week and the last three open a week later. Why? The first three open earlier because March and June are warmer.*

Sequence of perennials blooming in June and July using _trendline_ data				
Year	2005	2010	2015	2018
Week of June				
Red hot poker	2nd	2nd	1st	1st
Evening primrose	3rd	2nd	1st	1st
Rose campion	3rd	2nd	1st	1st
Loosestrife	3rd	3rd	2nd	2nd
Stella de Oro	3rd	3rd	3rd	3rd
Astilbe	4th	4th	4th	4th
Tiger lily	4th	4th	5th	5th
Trumpet vine	4th	4th	5th	6th
Bergamot	4th	4th	5th	6th
Daylily	5th	4th	4th	4th

How long do the flowers last before waning? With only having a smattering of data, coral bell lasts 20 days, evening primrose 14 to 18 days, rose campion 14 to 20 days, and Stella de Oro 10 to 16 days.

Shrubs: Mollis and mountain laurel blooming, with normal year-to-date rain and temperature, the Mollis azalea and mountain laurel showed no significant change over the period when they bloomed. Mollis bloomed nine days. Mountain laurel bloomed 13 days.

Depending on the species, the azaleas bloomed none to 20 days earlier from 2005 to 2019. The front lavender azalea showed no change. Others bloomed earlier; red-orange by the Chinese lantern changed 15 days, cardinal azalea changed 12 days, two big white in the back changed 20 days. Azalea, with normal year-to-date rain and temperature, red and red-orange bloom for 20 days; salmon, pink, and white bloom eight days. With high amounts of rain, add two to four more days to the blooming period.

With normal rain and temperature, the lilac bloomed six days; with very high rain the blooming period went to 12 days.

From flower to fruit, the blueberry takes 24 to 28 days.

Trees: Dogwood trees showed no significant change in blooming over the period. But to go to the other extreme, the date that the tulip poplar tree bloomed went from June 28th to May 20th!

Pond: Looking at the plant life shows that the water lilies open the first week.

Another significant change over the period is when the frogs are sunning or start croaking, but at what air temperature. Some years it is as low as 73F, most times it is between 76 to 85F. When they started croaking in late May, we had an early heat wave in the 90s. The trend shows that they have started to croak three weeks earlier over the period. With the warmer May's and June's, the graph shows that the frogs are croaking two to three week earlier. As the climate changes and it gets warmer, the frogs will croak earlier!

JUNE

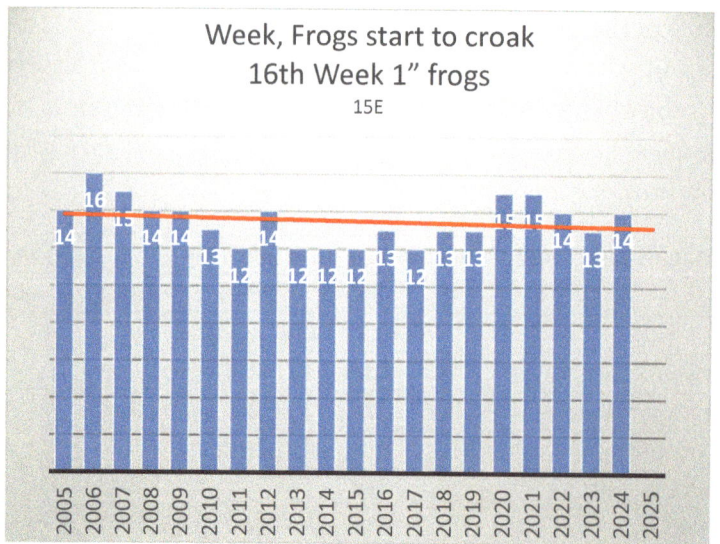
Temperature when frogs croak

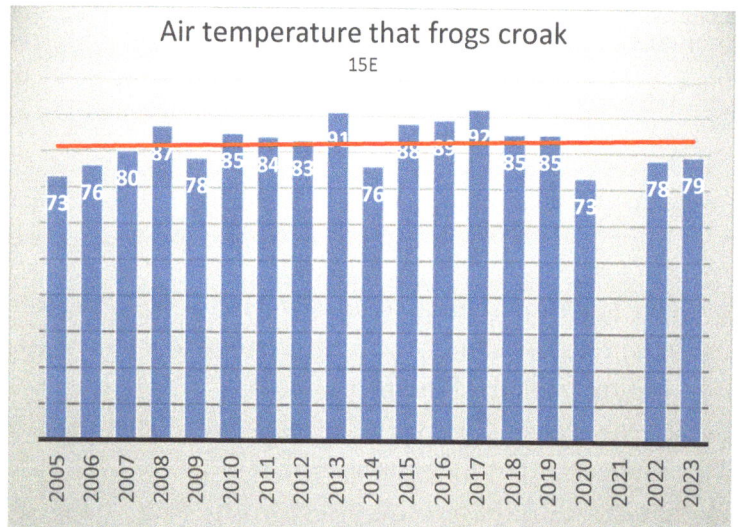
Week when frogs start to croak

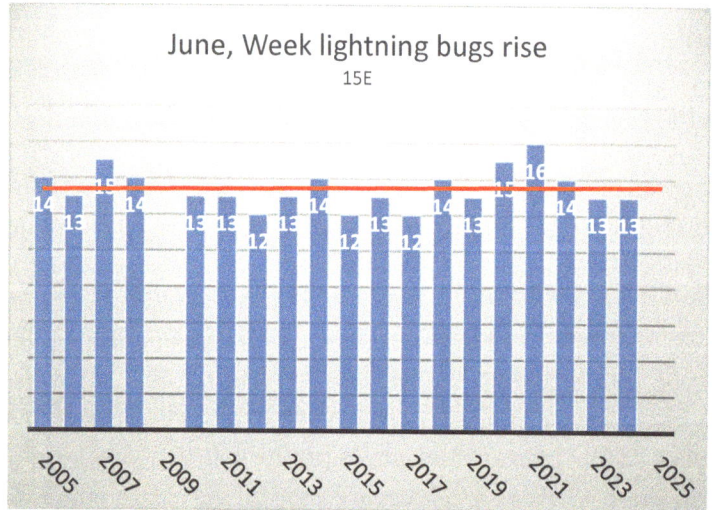
Week lightning bugs rising

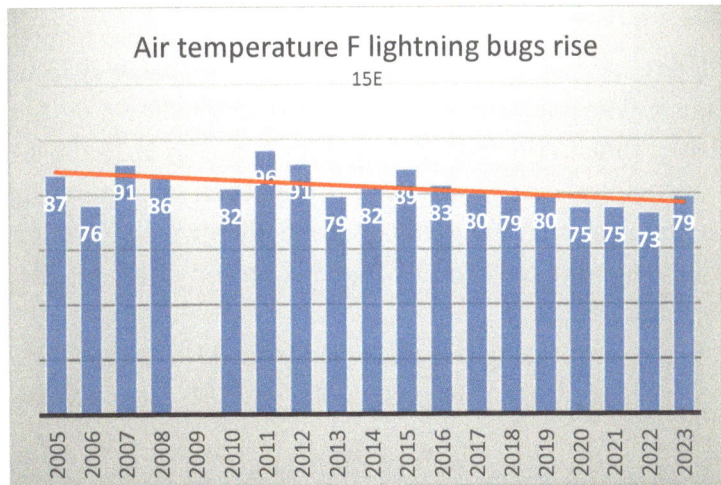
Rising temperature of lightning bugs

Fauna

Birds: With the charts of the flora, a trend can be projected, and the plus and minus peaks may be explained. The charts for the birds are more complex and harder to explain. This is because the transient species are arriving from southern areas such as Florida, the southern United States, and Central and South America of which I have no weather data. They migrate because their hormones say go north. The migration flight time is affected by the strength and direction of the winds, heavy rain, and fog. Just like a plane, any of these factors can cause a delay or layover of the migration. Reviewing 14 species, charts of arrival dates versus years were made. Some arrivals were in May, others in June. Only the rose-breasted grosbeak arrived consistently the same week for the period. Reviewing other charts, I choose a date that seemed to be a norm. In those charts are spike year arrival dates that are two to six weeks later than the norm. The spike years are prevalent in many of the species. I do not have the data for what occurred in 2009, 2010, 2011, and 2013.

Arrival dates of transient birds					
Species	Month	Normal arrival week	Winter resident	Spike years in June	Weeks from normal
Rose-breasted grosbeak	May	1st	Southeast	None	0
Hummingbird	June	1st	Central America	2009 May 16	+3
Male house wren	May	1st	Florida	2010	+4
Catbird	May	1st	Southeast	2006, 2009	+4, +3
Flicker	May	2nd	Southeast	2010	
Mallard duck	May	2nd	Southeast	2009, 2013, 2015, 2017	+3, +2
Goldfinch and Red-bellied woodpecker	May-June	June 5th to May 7th	Southeast	2011, 2011	-3, +4
Sharp-shinned and Broad-winged hawks	June-May June	none	Florida	2008, 2010, 2013	-4, +1
Grackle	May	1st	Southeast	2013	+3
Great blue heron	June-May	none	Southeast	2010, 2015	+2
Baltimore oriole	June	1st	Florida	2013, 2016	+2
Cowbird	June	1st	Southeast	2013, 2016	+2

Climate Change: In 2013 five of the species had early arrival spikes of two to three weeks; it happened in both May and June. There were no spikes in perennials or shrubs. In 2013 we had a normal year for both water and temperature. The goldfinch and great blue heron shown up four weeks early in arrival date, from the third week of June to the first week of May. Then in 2016 the goldfinch and red-bellied woodpecker started to overwinter. The great blue has now been overwintering in a local park for the past seven years; maybe it realizes that my goldfish are ready for picking when the pond is ice free. A sharpie has been overwintering and the broad-winged is arriving earlier. The other species have not shown a change in arrival dates.

The number of the species in the gang has increased from 11 to 16 for the period, but the number of migrating species dropped from 14 to three from 2005 to 2012, then increased to 10 by 2017.

Table of the Gang

Species/Quan. 2000	05	06	07	08	09	10	11	12	13	14	15	16	17	18	19
Blue jay R, G	2	2	1	2	3	3	3	2	1	2	4	4	2	2	4
young date						27					28		18	16	7
Cardinal R, G	3	4	2	3	2	2	2	2	2	3	4	2	7	2	3
young date	16	1				1			2		26				11
mange date		20		1					1						
Carolina wren R, G	1	1	2	1	1		1	1	1	1	1				
Catbird G	2		3	2	3	3	2	2	2	2	2	2	2	2	1
young date						27	4				28		9		
Chickadee R, G	2	2	3	2	1	1	1	2		2	2	1	2	2	3
young date	16			9											21
Chipper sparrow G		2		2	2	2		2					1		
Crow R	4		4	6	9										2
Downy wood R, G	4	4	2	3	1	2	1	2	2	2	2	2	2	1	2
youngster qua/date	1/13	5/2		1/14			1/18				1/20	3/2	16/2		
Goldfinch R, G	2	3	1	3	1	1	2	1		1	2			1	2
House finch R, G	4	3	6	4	2	1	2	4	21	2	6	2	6	2	6
young date		1	18					1	1		18	26	2	1	11
conjunctivitis											30				
House sparrow R, G	4	5	4	4	4	1		1	1	2	4	2	1	2	6
feed young date	10	21	18						10		27	5		1	
House wren G	2	1	2	2	2	6	2	1	1	2	2	1	1	1	1
young			18			17	2			19					
Hummingbird arrival			18	1	18	1		7	14	27	8			17	8
Mourning dove R, G	4	8	4	2	2	7	1	2	4	3	6	5	5	2	8
Red nuthatch arrival		27	24	13	2								4		
Robin	2	3	1	2	2	2	2	2	2	1	2	1	2	2	4
young date		19				1			10		29		19	2	27	2
Song sparrow R, G	2	3	1	1	1	1		1							
surrogate cowbird	25	27			2										
Titmouse R, G	3	2	3	2	2	1	2	2	3	1	1	2	1	1	1
young date				24			1								
White-breast nuthatch															1
Total gang 16 species	13	13	15	16	16	14	12	16	12	13	14	12	13	14	15
Gang population less Crows	37	41	37	35	28	32	19	26	20	29	39	19	30	22	43

TRENDS DUE TO CLIMATE CHANGE

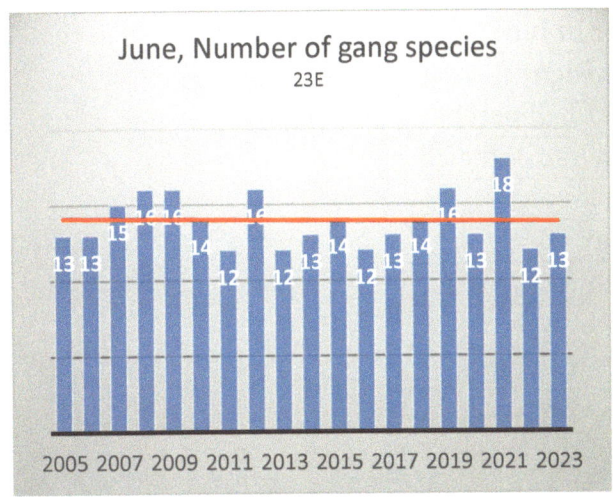
June, number of species in gang

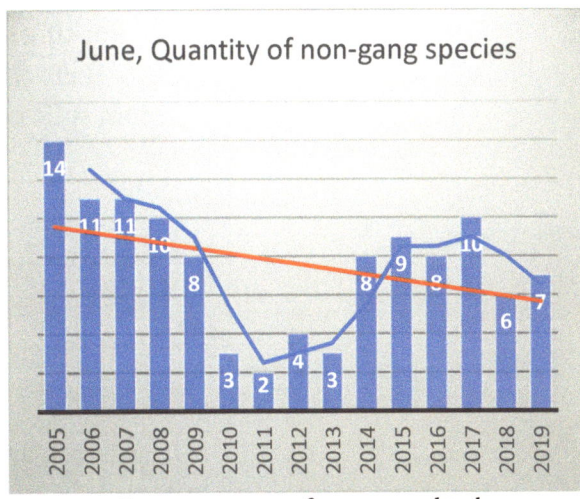
June, quantity of non-gang birds

Except for 2012, no crow, chipper sparrow, red-breasted nuthatch or song sparrow were observed after 2010! Crow demise is due to the West Nile virus. 2020, Crow pop recovers. Song sparrow demise is due to cowbird laying their eggs in the sparrow's nest.

Reduction in June's quantity reflects a loss of four species. The increase in 2015 is due to more mourning doves, house sparrows, and house finches. These species are around the yard or flying over but not necessarily migrating. Compared to May's migrating chart, they go in an opposite trend. This could be due to climate change warming in that birds are migrating sooner.

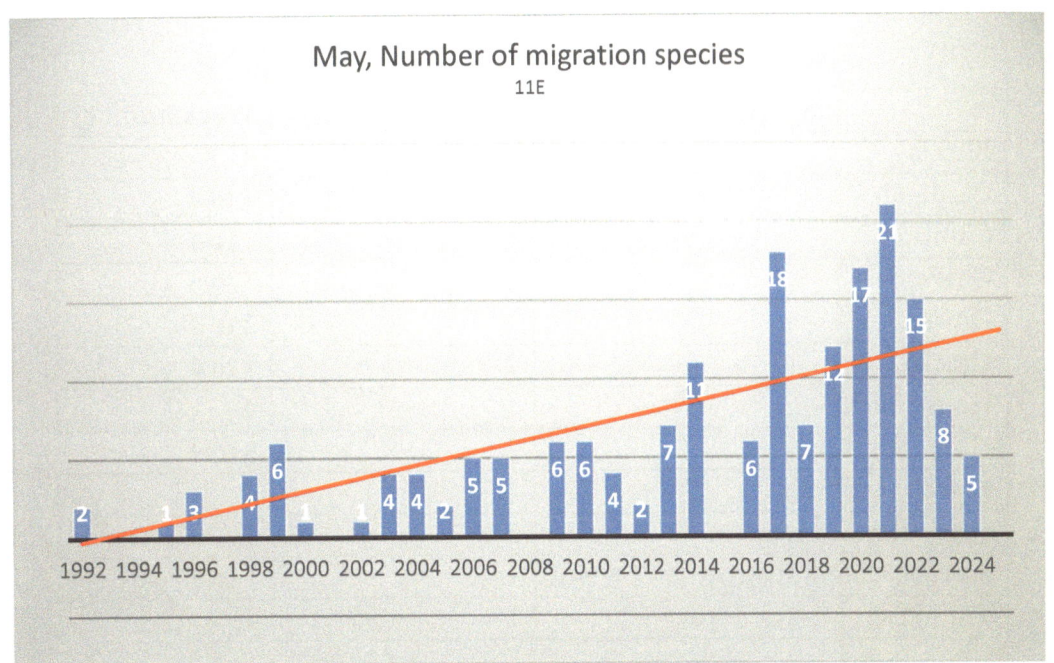
May, number of migrating bird species

Like other months there is a lot of pecking order going on between species. This is due to the need for food for the fledglings, especially for split peanuts.

JUNE

Pecking order in June	
Left chases Right	
From split peanut ball perch	
Dove	White nuthatch
Downy	Titmouse
Grackle	Jay
Jay returns	Grackle, wins
Jay	Downy
Jay	Pair Catbird
Titmouse	Red-breasted nut
In general, around the feeders	
2 Catbirds	3 Jays
Cardinal + Robin	Sharpie
Crow	Broad-winged
Grackle	Dove
Jay	Crow
Jay	Robin
Robin	Grackle - Jay
Song sparrow	Chipper

How many fledglings do species bring to the feeders and what week during the month?

When parents bring young to feeders in June		
Species	How many young	Week number
Blue jay	1	4
Cardinal	1 to 2	2 to 4
Catbird	1	1 to 4
Chickadee	1 to 2	3 to 4
Downy	1	2 to 4
Grackle	2	1 to 4
House finch	1	4
House sparrow	2	3 to 4
Mourning dove	1	4
Sharp-shinned hawk	2	4
Surrogate Song sparrow to Cowbird	1	1 to 3
Titmouse	1	4

So, ends the productive month of June.

Index of July

July Overall 164
Weather 164
First Week 165
Second Week 173
Third Week 178
Fourth Week 186
Review of Data 193
Fauna 193
Table of the Gang 194
Flora 199
Perennial Blooming Sequence 200
So, ends July 201

July

Week number for the year of 48 weeks					
Relative Month	Month/Week	First	Second	Third	Fourth
Previous	June	13	14	15	16
Present	**July**	**17**	**18**	**19**	**20**
Following	August	21	22	23	24

Weather

The average high temperature on the 1st is 82F, the low is 61F.
The average high temperature on the 31st is 84F and the low is 63F.
The extreme high of 105F occurred on the 3rd in 1966, the low of 46F was in 1943 and 1997.
The full moon is called Buck Moon or Thunder Moon.

Overall Summary of July

July is not only the hottest and most uncomfortable month for us humans, but also for the fauna and flora. Birds will pant, lay with their wings spread out, and take frequent baths. Some of the plants droop, others point their leaves up to catch moisture. Grass becomes crunchy under foot. When it gets hot and there is a lack of insect protein, parents of many species of birds bring fledglings to the area for water and feed. I continuously refill the peanut butter log. There seems to be an abundance of gnats, mosquitoes, and no-see-ums that the birds do not reduce. On the nice side, July produces lots of colorful perennials, also the earning of the first home-grown tomatoes happens at month's end.

Weather

The first week of July is usually the hottest of the year with temperatures going to 98 or 100F. A heat wave is three days of over 90F; one year we had 10 days of over 98F and high humidity. It is a high pressure over Bermuda that brings us the heat and humidity. You feel hotter when it is humid because there is less evaporation of perspiration. A thunderstorm can develop any time but mostly in the late afternoon. After a storm, the air smells sweet in the evening coolness.

When it rains at night the humidity is gone the next day, but it can still be 98F. Some years there was a drought all spring and into June. The ground had cracked the width of a pencil and was hard like concrete. Then it started to rain at the beginning of July and then back to dry conditions. In 2006 we had rain for 12 of the first 15 days of the month. One of the wettest Julys with 8.9 inches was 2010. It rained for 12 straight days. One day it was raining four inches per hour. That's a lot of rain. Another year we had rain every other day. By the 21st, due to the lack of sun, I only had one ripe tomato.

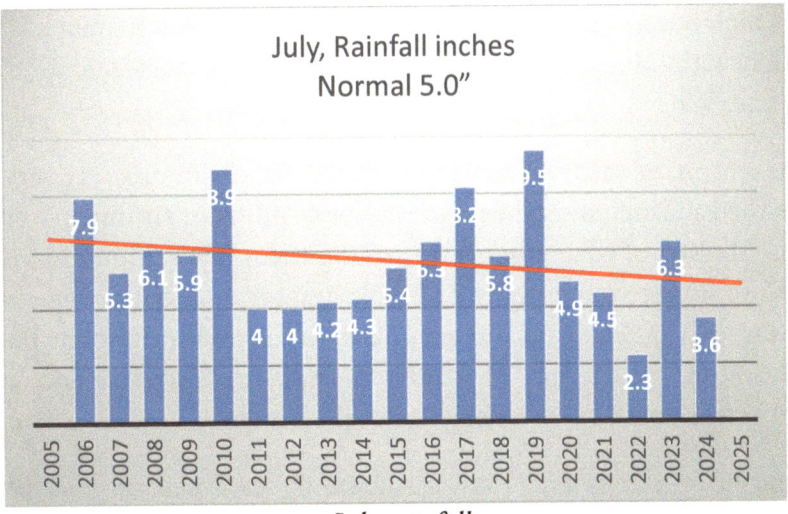

July rainfall

On the downside, these conditions produce a fungus that attacks the shaded back lawn and if not treated the whole lawn will die in a few weeks.

On the upside, the heat produces gigantic billowing, white cumulus clouds with thermal heat columns that allow the hawks and turkey vultures to gracefully soar in a kettle for a long distance.

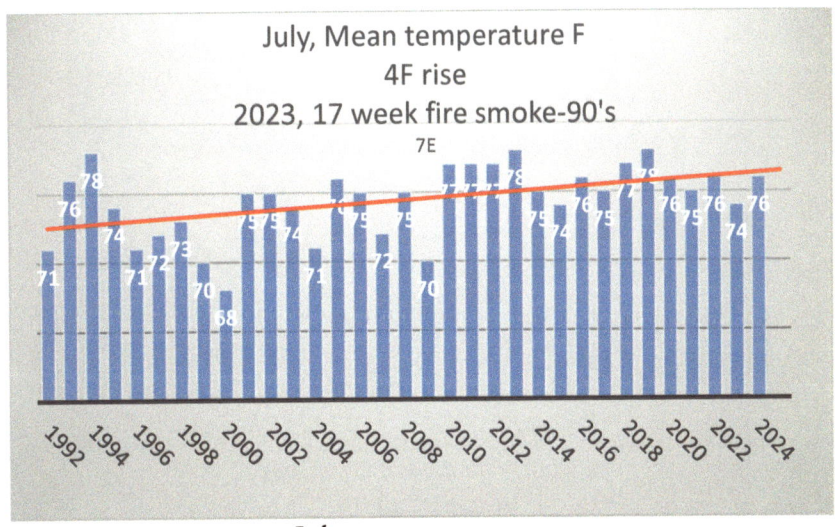

July, mean temperature

First Week

There is a lot to observe this month, but I do not know what it all means. Summer evening, 8:50 p.m., lightning bugs rise from the lawn. The evening star, Venus, has been shining brightly in the western sky for the past seven nights. The bats are crisscrossing across the sky at 8:55 p.m. At my head level, the mosquitoes are biting furiously, and the frogs have struck up a symphony. A northwest breeze develops blowing out the humidity. At 10:30 p.m. the lightning bugs are as high as the top of the oak. Distant flashes of lightning can be seen in the west. At midnight the storm was upon us, sharp flashes with a crack of thunder only seconds after. Then came the downpour in three or four pockets. The thunder was in the distant east as I fell asleep.

Weather: With the high humidity, I should spray apples, raspberries, hemlock hedges, roses, string beans, and asparagus with fungicide and horticultural oil. With the humid air, the boxwood gives a fragrance of cat scat! Whew!

Animals: For some reason, a squirrel is chasing a rabbit. The chipmunks are very frisky; they chase each other all over the yard including the patio. When standing or sitting on a chair, they may run right over my feet, oblivious of what is going on around them. Two chipmunks nuzzle each other's noses, playing Eskimo. Three young squirrels are playing together in the dogwood; one misjudges and falls out of the tree. It shakes itself and goes back up to continue to play. I observed a squirrel eating a green spruce cone, new to me. At 10:30 p.m. while driving home and near to my house, a red fox runs safely in front of the car. A delightful sight at night.

Birds: At 6:00 a.m. the sun is up and I am fortunate to see a pair of great blue heron go west squawking at each other in a morning conversation. A song sparrow, house wren, and a Carolina wren all enlighten me with a bright morning song. The birds have enjoyed most of the blueberries, none are left for my cereal. Birds get mites in the hot summer. House sparrows and wrens take a dirt bath to get rid of the mites. In the hot, dry weather cardinals get mange and lose their head feathers; they now look like miniature vultures.

Mange on a young cardinal

Ten species bring fledglings to the feeders during the first week of July. They enjoy the peanut butter. Young sharpies are also near the feeders eyeing the fledglings. Teenage cardinals are turning from beige to red. Two young hen doves are at the feeder. A bull dove lands and chases the hens around the platform. The hens are smart enough not to stay and take off fast. A downy woodpecker is eating the young punk on the cattails. Youngsters, either birds or animals, unknowingly get into trouble or upset others. I look out to see a freckled-breasted robin hopping down the steps as fast as it could. Right behind it was a chipmunk nipping at its heels. I wonder what caused that? A starling chases a blue jay from the split peanuts. Two robins chase a grackle from the feeder, but the grackle puts up a scrap before leaving. When a not so subtle young sharpie lands on top of the feeder all the small birds scatter. I open the window, it flies off. The small birds are all back in a few minutes, but within an hour the hawk had a dove. Two male chickadees go at each other fighting for a maiden.

Because of the shape of their beak, cardinals cannot pick up seed by going straight down on the seed. They must turn their head sideways, like lying down, to get the seed. After replenishing the peanut butter log, catbird, chickadee, and house sparrow are at it within minutes.

Hey, where is the smooth stuff

A male downy sees another male landing nearby. It lowers its body flat, thrusts its head down with beak pointing straight out, and charges at the intruder who flies off. Friendly kin! This is the week when a female song sparrow acts as a surrogate mother to a fledgling cowbird. A male and female house finch have conjunctivitis. I spot a squirrel that looks like it has the same disease!

House finch with conjunctivitis.
In the reflection you can see the still good eye.

Late afternoon is when hummingbirds are at the nectar feeders and bergamot. Sphinx moths are on the bergamot and summer phlox. From a distance these moths look like a hummingbird.

Hummer at the feeder

Sphinx moth on the bergamot

After a rain on a hot afternoon I look out to see a robin cock its head then quickly pull a worm from the lawn. Unfortunately, sometimes a young robin gets too close to the edge of the pond, falls in, and is unable to get out. A young aggressive grackle on the lower level of the feeder points its beak straight up to a jay above it. It tries to hit the jay. The jay steps aside and goes after the grackle. The grackle promptly leaves. Just like June, the pecking order goes on. A blue jay meets its match when the larger grackle chases it from the feeder. The jay brings back two more brothers, the grackle leaves and comes back with brothers, who chase all the jays. A downy chases a male house sparrow from the split peanut feeder swinging perch.

Pecking Order in July	
Left chases Right	
Grackle	Jay
Young starling	Jay, jay wins
Jay	Hairy woodpecker
Titmouse	Young Grackle

Then there is the raising of the wing, which is like saying, "Do not tread on me." When another bird or animal gets too close, the defender raises one wing high.

Dove with raised wing to chipmunk

Dove with raised wing to a grackle

Jay with raised wing   *Jay and cardinal sizing each other up*

Raising a Wing in July	
Left side raises wing toward the right side	
Dove	Grackle
Dove	Jay
White-breasted nuthatch	Downy
Jay	Another jay
Titmouse	Young Grackle
Dove	Chipmunk
Grackle + Dove	Jay
Downy	House finch

On a clear, hot, day a broad-winged hawk soars up high.

Hawk soaring

    In the late afternoon I look up and see four goldfinches gliding in a cyclic, wavy path and hear them calling T-hee, T-hee, T-hee. Goldfinch lands on a young globe thistle for the unripen seed; the stem droops over from the weight of the bird.

The house wren sings an evening lullaby at 7:30 p.m. Fourth of July celebration brings fireworks. Roosting robins do not like the fireworks. They give off a warning alarm after each firework is fired. Robins and doves are coming to roost in the evergreens.

A starling helicopters up to the suet

Of 21 species in the yard, 18 had mates. A dove eats 210 safflower seeds in three-and-a-half minutes (pig). I had a hard time keeping up with the count.

Downy, mom feeding junior

Cousin hairy

First, six robins, then 12 are yelling in the high grass; a crow is going after a young robin. The brotherhood of robins came to harass the crow. They succeed in chasing the crow. The young robin had feathers plucked from its body but appeared to be okay. To go one further, I look to the sky to see four grackles chase a crow that had a baby grackle in its beak.

Two young doves are having a wing flapping contest at the feeder. A jay lands on the perch and gets the split peanuts as the perch swings back and forth.

Fruits: Blueberries are picked clean by the birds the first through the third week. Again, another surprise, I had put a net over the blueberries to keep the birds away but see a robin fly under the net to get the berries. Raspberries are also ripe, which again, the birds love.

Insects: Yellow jackets take over a house wren's house and hummer feeder. Swarms of bugs dance in the evening sun beaming under the branches of the oak tree

Yellow jackets at the hummer feeder

Perennials: Evening primrose and rose campion are done, hosta is in bud, bergamot and phlox are three feet high. The red trumpet vine is out. The yellow trumpet vine follows the red vine in opening. The hummers seem to prefer the red.

Red trumpet

Yellow trumpet

This is the week to trim the stalks of the double blooming iris behind the pond. The butterfly bush is out full. Elephant plant that wintered in the garage is getting new leaves. The japonicas have lots of new growth. The first red daylily is open on the third terrace.

Some spring pansies are still blooming next to the petunia

Same plant with different name: German mint, bergamot, monarda

Bergamot opens the last week June to the first week of July and wanes on the fourth week. When we have the 3H's (hazy-hot-humid), powdery mildew develops on its leaves.

French marigold

Impatiens

In 2012, fungus killed all the impatiens in one day! In 2015, a fungus killed all the marigolds.

<u>Shrubs</u>: The wild honeysuckle bush has red berries. The butterfly bushes are out full attracting lots of sulphur butterflies. Last year's trumpet vines are poking through the lawn. If I don't get the deep roots, they will reappear the next year. White daisies, miniature cosmos, and white balloon flowers are all opening. Now is the time to remove the scape (stalk) from garlic plants and sauté them.

<u>Vegetables</u>: Young beet plants are struggling. I had put up a two-foot high chicken wire fence around the vegetable garden to keep out the vermin. To my surprise I saw a chipmunk climb right over the fence. Not only does it go over the fence, but it also bit through a cucumber vine! Cucumbers are three feet high on the trellis, tomatoes have small fruit, dill is three feet tall, cabbages are forming. String beans are ready to pick. Brussels sprouts and pole bean stalks are recovering after being a meal for a groundhog.

Second Week

Early morning, sun is up, I open the door, the air smells sweet. A new batch of pole beans that were planted six days ago are coming up. With the 3H's, powdery mildew is showing on the phlox and bergamot. Time to spray the lawn, fruit trees, and flowers with a fungicide.

<u>Weather</u>: These are the "dog days of summer" called that way because the constellation Dog Star or Sirius rises with the sun purposively adding more heat to the atmosphere. It is 95F, a strong thunderstorm builds up and releases rain for 20 minutes. At the end, the temperature drops 20 degrees. It takes a half inch of rain to penetrate a full canopy to get you wet under a tree. Another extreme in July is after a great deal of rain it gets hot and humid, then no rain. The ground bakes into cakes and cracks as we go into a drought. Leaves are dropping off the west end dogwood. Still with all the hot weather some shrubs need a second trimming.

On the afternoon of the 9th in 2015 a tornado warning goes out. The local TV station cancels all programs so the weather people can give a minute by minute, block by block, travel path of the tornado. This goes on for almost two hours. It does a lot of damage to homes and crops.

<u>Animals</u>: It's 100F, a rabbit digs a shallow hole under the feeder, lies in it and pants.

A chipmunk is up in the apple tree; as bad as the squirrels. The chipmunks chew into the plastic bag of lawn fungicide. In need of food, a squirrel chews a spruce tree cone. A bunny stops eating clover, leans back, raises its leg to scratch its right ear, stops and repeats to scratch the other ear, then goes back to eating the clover.

Sitting on the patio I noticed movement in the upper branches of the dogwood tree next to the pond. It was bigger than a squirrel and brown in color. I got up to take a closer look. To my astonishment it was a young groundhog. I do not know what it was doing or wanted from the tree. I got the garden hose and gave it a bath until it came down.

Groundhog

Voles are scampering around the rock walls. Then I see a red fox, probably going after the voles. Bats are out at 8:45 p.m.

Birds: A hummingbird goes to the pink phlox, geranium, German mint, and hosta flowers. To my surprise a catbird and a cardinal take a whole peanut. An Eastern phoebe may pass through! They also passed through in March.

The birds have a hard time with the weather. Birds do not perspire. The house wren and Carolina wren look bedraggled; they take baths if available. They let their tongue stick out to evaporate internal moisture.

100F – hot Carolina wren

It is so hot that when it does rain a male dove raises one wing up into the rain and then switches to the other. Another dove lies under the main feeder's exposed soil with its wings spread out.

95F heat – dove with wings spread out

A group of eight grackles, including two fledglings, land on the feeder. The leader keeps puffing up its breast and straightens its body to remind the others who is the boss. He chases a jay from landing on the feeder.

The leading grackle puffs out and stands straight

Birdhouse housekeeping for a new coming brood goes on as a house wren empties and refills a house sparrow's house. At least for two years I have observed a titmouse chase a house finch from the safflower feeder. Then a reverse, a house sparrow chases the titmouse from the feeder. The pecking order is still in place. A jay comes near a robin's nest, the robins chase it away; the jay hits the kitchen window during the flight.

<u>Insects</u>: Locust are calling in the mid-day heat. From late July through August they emerge from a half-inch diameter hole. The hole has no residue or piling of dirt around it. I do not know if the dirt is digested through their body as they dig to the surface. They then crawl two to 10 feet up a tree trunk, where the blackish-brown two-and-a-half-inch long locust breaks out of its cocoon shell. The shell stays adhered to the tree. With a three inch wingspan, the locust flies to treetops to produce a

high pitch buzz or pulsating sound from their abdomen that we hear during the heat of the day. The racket of their sound may go on for two weeks. A house sparrow flies after and gets a locust.

Locust

Fritillary butterfly

It's noon and a fritillary butterfly (cream yellow) glides over the pond. A viceroy butterfly, which looks very similar to a monarch, is on the coneflower. On a clear day the bees do not arrive at the flowers until 8:00 a.m.

Fruit: The apples are now large enough to attract squirrels; they use the old adage, an apple a day will keep me coming after them. As soon as a blush appears on the skin, the house finches peck at the blush and ruin the apple. I hang CD discs among the branches to try to scare them away. I put protective bags around some of the apples to keep birds and insects from the fruit. The brown baffle on the trunk is to keep squirrels from getting into the tree. I must brace up branches when the Empire apple has an abundant crop. In July the young apples are the size of a peach. I cull out the apples from five on a cluster to two or three, if not I will have a lot of small apples.

CD disks to scare the birds

Lawn: With the 3H's, mushroom fairy rings form in the second and third week on the hill and under the hemlock branches. The broadleaf weeds in the lawn are out full.

Perennials: Caladium tubers planted in April finally emerge. Frankly, I had forgotten about them. Within a week they are four inches tall. The white-green leaves grow to 12 inches, the red-green leaves stay at six inches. Trumpet vine sprouts are coming up in the lawn. Their taproot is very deep, and they are hard to eliminate. During this week the tiger and daylilies are opening in their array of reds, yellows, oranges, and whites.

Daylily blooms first and second weeks of July 2024, blooms third week of June!

Coneflower blooms second through fourth weeks

Phlox bloom second through fourth week

On a hot afternoon with a slight breeze the many parachutes of wild thistle are floating along in the trails of the air. It looks delightful now, but they will be next year's weeds.

Pond: Punk on the cattail has turned dark brown. White lily is out the second or third week. I must watch the water level in the cattail tote; these plants consume gallons in one day. When we have drought conditions the pond level can drop three to five inches. Replenishing with a hose does not help if there is a leak in the lower level of the pond. The pond level drops with the drop in the adjacent ground water level. I almost was going to empty the pond to find the leak, which I really did not want to do. I knew there was at least a foot of muck on the bottom, so I put on boots, got in, and stamped all around for 15 minutes. I refilled the pond and waited; the level held. A year later it still was holding.

Shrubs: Trim the yellow bell in the front for the second time in the season.

Vegetables: Put 10-10-10 on the annuals and vegetables, cow manure on the strawberries and rhubarb.

Third Week

Weather: It is very humid.

Animals: A squirrel with mange around both eyes and front legs goes for a drink. Its hearing and sight are poor.

This time of the year there may be an extra abundance of rabbits in the neighborhood, but there are hawks to keep them in check. To my surprise, the rabbits nip off the yellow daylily flowers and a chipmunk is going for the globe thistle flower.

Buck rabbit in the morning sun

<u>Birds</u>: When the heat or index is over 95F, there is very little bird activity. In the bird world, a downy is drilling out a nest in an old maple stump. Another downy lands on the cattail punk and pecks at it; it is gathering nesting material. Unusual, a yellow warbler passes through. A chickadee chases a red-breasted nuthatch from the split peanut cage.

Chickadees on split peanut cage

Goldfinch on coneflower

A house wren is singing quite a song. Goldfinches are eating the top of the coneflower.

At 5:00 p.m. a young sharpie flies into the area. I chase it away. Just before I go back inside a catbird is meowing at me. It is thanking me for chasing the hawk. A young grackle is on the lower platform of the feeder. It detects a young jay above him on the next level. Aggravated for some reason, it sticks its beak straight out and flies directly at the jay. The jay squawks, backs up, but does not leave, the grackle does.

Sometimes the robins cannot penetrate the hard, dry soil to try and get a worm, so they will go to the feeders to get suet or peanut butter. A catbird has a sweet beak; it gets nectar from the hummingbird feeder and sugar from an orange.

Catbird likes oranges

There are 10 species at the feeders in the morning. Male teen cardinals are turning beige to true red. A male cardinal is upset with a female and hisses at it. That's right, they hiss like a mad cat. I also observed this in the winter during feeding time.

I find a few feathers of a dove on the lawn. A short time later I spot a Cooper's hawk in an evergreen. The jay is the king at the feeders and rules the roost; except one day the jay was at the feeder when surprise, a Cooper's hawk sweeps in and took the jay! A young robin is creeping and hopping on the lawn looking for mom and pop; he better watch out because I spotted a sharpie earlier.

Unusual observations, a house sparrow is upside down at the suet house. It is hot, but there is civility among the species. A robin and catbird are lining up at the birdbath. First the robin bathes, then the catbird. Remember both are the same size and temperament.

Fruit: The apple trees are sprouting new growth.

Insects: In the morning when I go out onto the patio, I walk into spider webs.

Garden spider's web

Monarchs and black and yellow swallowtail butterflies go through. There are lots of bumblebees on the globe thistle and loosestrife.

Bumblebee on purple loosestrife

Crickets are calling at 7:00 p.m. They chirp faster as the temperature increases. The heat index could be 107F. When it is hot the katydids and cicada call, which is a part of summer.

Katydid, first through fourth weeks

Cicada sing third through fourth weeks

<u>Lawn</u>: Purslane, yellow wood sorrel, and white clover that were thriving in June is still going strong. Continue to put down fungicide on the lawn, flowers, and vegetables. Time to fortify the flora with 10-10-10 fertilizer.

<u>Perennials</u>: Foxglove did not do well in my soil, they only lasted two years. Daises are out.

Daisies

Deadhead the bergamot. The hibiscus is out and the elephant ear plants are flapping in the breeze. The lungwort looks healthy; it will dry out in the hot weather.

Lungwort

Elephant ear plant

This is the week that male spikes form on the cattail stalks. The black-eyed Susan comes out to add lots of color to the garden.

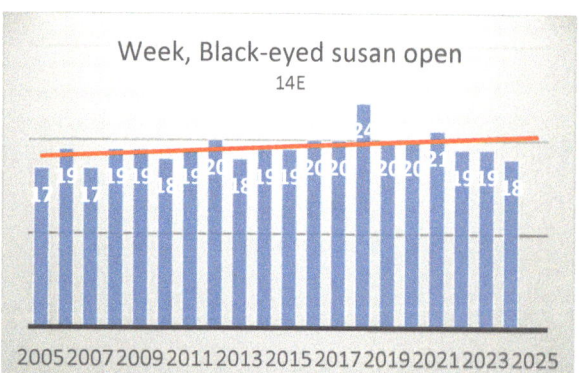

Black-eyed Susan blooms third and fourth weeks

Honeybee on globe thistle

Foxglove

Pond: The frogs like the hot weather and croak in a loud symphony all night. Many of the tadpoles are now small frogs. I must be careful not to step on them in the lawn. They are the size of a dime or nickel. Blue and brown dragonflies hover over the pond. More often you see them in September and October. Pickerel weed's light blue flowers are out the third week.

I put in new waterfalls. I must make sure that the edge of the spillway is level, has a sharp edge, and has no projections like a nail or screwhead; if not, the falls will not fall smoothly nor flow straight across the spillway.

Mexican face is the first falls

Lower falls

Splashing water puts air into the pond

Around they go

JULY

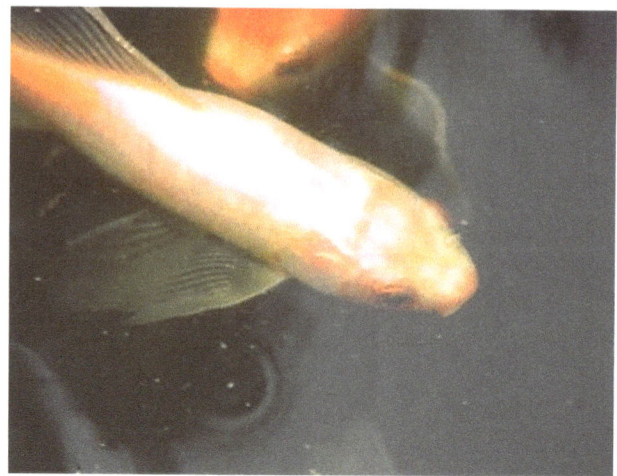
Had fat, white fish 14 years; great blue got it

Getting noon day sun

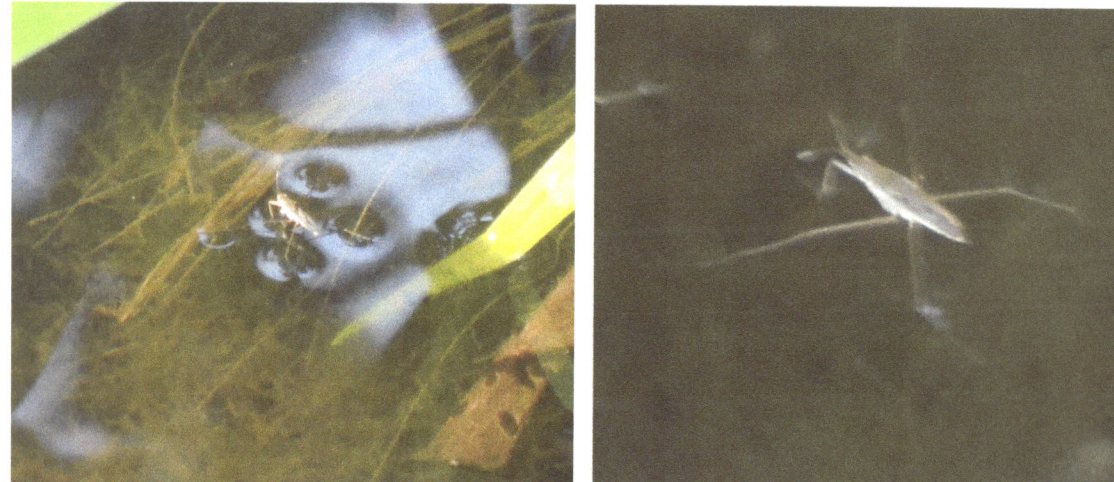
Water strider, some call it the Jesus bug because it can walk on water

The high temperatures produce an algae bloom; add a quarter cup of algae preventer. Pink water lilies are out.

Reflection in many ways

To me an event that makes summer is seeing the dragonflies and damselflies hover over the pond. They hover in position over the water with two sets of wings and in a second can be in another position. The bodies can be blue, green, and brown with big multi-lens eyes. They seem to say, "Don't tread on me." They will even go after a hummingbird that is at the pickerel weed flower.

Shrubs: Purple butterfly bush is out. Lots of new growth on the azalea bushes. Rhodie leaves pointing up at 60 degrees. The rhododendron that I trimmed a couple of weeks ago is also getting new growth. Time to trim the front bushes.

Trees: Every three to five years the oak trees develop a large abundance of acorns (2014 and 2019), called a mast. This is the result of the weather conditions from a couple of years before. The mast develops the third or fourth week. It is a challenge to pick up the acorns. I save them in metal buckets to feed the animals during the winter.

Mast crop on the black oak

<u>Vegetables</u>: The pole beans are growing at six to eight inches per day. The heat is not all bad. The tomatoes, zucchini, and cucumbers love the heat and may ripen the first week. Usually 75 percent of the tomatoes are ready to be picked by the 20th. The lettuce seed I planted 10 days ago is up. Japanese beetles are on the rhubarb. Asparagus fern is three to four feet tall. Be careful when you cut it back, this fern has thorns. Swiss chard is ready to be picked. Older black seeded Simpson lettuce has sprouted stalks that have light blue flowers that the bees like.

Fourth Week

Summer glory

<u>Weather</u>: Dusk starts at 8:45 p.m., The pot of the big dipper is pointing straight down. The end of the month can be miserable with 60 to 70 percent humidity, temperature 73F, dew point also 73F. When the air temperature and dew point are the same, it is a sure sign for heavy rain and thunder. With the high temperature and humidity, the locust is calling at the height of the day. Many days it feels like 100F. At 5:30 p.m. it is the muggiest, when all of a sudden we get a 10-minute downpour. Ah! Refreshing.

After a short thunderstorm, a vivid double rainbow is seen in the eastern sky in front of billowing cumulus clouds.

The last week can be just hot, 95 to101F, lots of thunderstorms. One day it got to 104F.

Afternoon in July

<u>Animals</u>: With my window open for fresh air I am awoken at 3:00 a.m. by the strong fragrance of a passing skunk; so much for fresh air. A squirrel takes dead grass up tree for a nest. There are piles of small oak and apple twigs on the lawn from misplaced positioning in the nest.

I see a piece of bread at the bottom of the feeder move. I look closely and see it is a vole trying to take it to its nest.

Voles, red fox, and skunks are around the fourth week. I wonder who is going after whom? Still skunk in2024.

The fourth week of August is when I usually see a pair of rabbits munching on the lawn at dusk. That week I would spot one or two bunnies. Taking a walk at night or at twilight, I like to see the white of the cottontail flash across the lawn as I go by.

A chipmunk jumps straight up 18 inches to nip the flower off a red daylily! Two young squirrels are on the chase.

<u>Birds</u>: The number of species for breakfast changes each year depending on the weather. With a drought, high temperatures result in low availability of insects and seeds, increasing the species from seven to 13.

Carolina wren starts singing for a mate. Not to be out done, the song sparrow is also singing. There is rivalry between two male downy woodpeckers. One gets down low on the feeder with its beak pointed straight ahead and makes a beeline toward the other which promptly flies off. There are two fledgling robins hopping and chirping around the lawn.

In the bird world, there are 13 to 15 species in the morning. A hungry jay is eating impatient roots! A chickadee and goldfinch are at the thistle tube; they look at each other to see who is going to go first, the chick wins. House sparrows are active at houses again.

For 15 minutes, a pair of robins chase a sharpie away from their nest. I hear, but do not see, a flicker. A catbird is chasing a hummer! A hummingbird chases yellow jackets from the hummingbird feeder. There is young blue jay, downy, house sparrow, robin, cardinal, and house wren on the feeders.

Female jay with fledgling

One evening the parade of visitors to the feeder was interesting. Two blue jays were peacefully eating when a bull dove lands, raises its wing, causing the jays to leave. The dove was replaced by house finches, which were replaced by the more aggressive house sparrow. Then a downy took over. When it finished a pair of chickadees came in, then a female hairy woodpecker had dinner, and finally a pair of titmice finished up. All of this in 12 minutes.

The parent house wrens are still supplying insects every three to five minutes to three fledgling whose beaks are wide open for more-more-more.

A pair of adult robins chases a sharpie for 15 minutes as it was going for their fledgling on the ground. A titmouse and downy are on the swing perch going for the split peanuts at the same time.

On the 24th of 2017, a flock of 30 robins flew over going northeast! Something is out of sync; this is the wrong time of the year and the wrong direction.

A jay is eating the roots of the New Guinea impatiens! A woodpecker is making a cavity in a branch of the poplar. In the evening, I spot six chimney swifts circling overhead. Catbirds like blackberries, watermelon, apricots, and jelly.

Insects: Four white sulphur butterflies flutter erratically from flower to flower. A sphinx moth is getting pollen from the pink phlox. Some years this is the week that the hordes of Japanese beetles strike everything. The traps get full fast. Locate the traps away from the flowers and vegetables so the beetles are not attracted to them.

When weeding I may see a scarlet-and-green leafhopper; it is only four-tenth of an inch long. This little creature injects saliva into the stalks of both weeds and flowers which causes the plant to bleed from the stalk, eventually killing the plant.

At 7:15 p.m. there is a swirl of small bugs in a sunray coming through the branches of the evergreen on the hill. The ray reflects off the top of the hemlock hedge and through the leaves of the apple tree.

Almost dark, 8:16 p.m., the crickets start their symphony as the locust shut down.

Noon, a monarch flies through, a vole on the wall pauses to look at the butterfly.

While gardening, you get a spider bite that becomes a continuous itching bump that takes two weeks to subside. Wear long sleeve shirts and long pants.

Early in the morning I look at the top of a boxwood. It looked like a thin sheet of cotton candy was draped over the top. Looking closer it was a spider's web with an inviting welcoming black hole.

Spider's blanket web on boxwood

I spot green snowy tree crickets as long as a dime, and a cricket with a brown body with wings that have a white edge. As it gets dark, the crickets start to chirp. It is the male that makes the music by rubbing its wings together. The snowy tree cricket makes most of the evening music. The larger black house cricket is the one that tries to get inside in the fall. A yellow and a black swallowtail butterfly pass through in mid-afternoon. They were first seen in June.

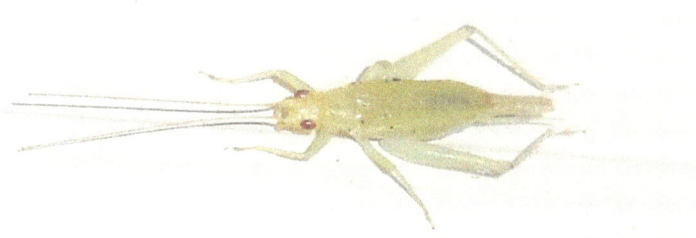

Snowy tree cricket eats aphids

There are three other insects out now that can drive me indoors. First, the gnats are out in force. Second, the no-see-ums, just that, are all around your exposed body. Third, as the clock strikes 9:00 p.m. the female mosquitoes attack. The only way I know I can attack back is with sprays for the mosquito.

Locust starts to call as the temperature rises. They will call until late in the evening.

<u>Lawn</u>: The lawn has brown spots from fungus. Remove dead grass from back lawn. Now is the week to spray for clover and put down crabgrass killer. Put vinegar on the weeds. Crabgrass, white clover, purslane, and black medic are all out the beginning of the fourth week. Time to put down weed preventer.

With the 3H's, fungus will form the third week. Mushrooms will come up the fourth week.

<u>Perennials</u>: Put 10-10-10 around the flowers and apple trees the third week.

The annual dahlia have opened.

Dahlia

Naked lady develops very fast. Naked because it comes up in March-April with lush foliage that dwindles and disappears. Then the stalks appear in the last week of July, quickly grow to two feet high in six days, blooms without any foliage, hence naked. Then wanes in a week.

July 28th – naked lady   *July 29th – naked lady*

TRENDS DUE TO CLIMATE CHANGE

August 2nd – naked lady

The cardinal flower (Lobelia) has bright red flowers. There is powdery mildew on perennials. White and pink phlox are out. Cut back yellow coreopsis and Montauk daisy.

Globe thistle, plant died in 2017

Yellow coreopsis

JULY

White phlox

Verbena on the hill

Convolvulus blue from seeds

Miniature snapdragon and herbs

Pond: I spray the pond with a liquid film that kills mosquito eggs. Arrowhead white flower is out the third to fourth week. The pink lily comes out the fourth week. This week the pond iris gets thinned out.

It is hard to pin down when brown and blue dragonflies will arrive. They seem to mostly hover over the pond the second and third week when the air temperatures are in the high 80's to 90's.

Arrowhead flower

TRENDS DUE TO CLIMATE CHANGE

Purple iris to be thinned out

<u>Shrubs</u>: This week spray horticultural oil on the hemlocks and trim the shrubs. Verbena on the hill has been blooming for the last two weeks. Trim arborvitae and rhododendron the fourth week. Put aluminum sulfate on acid-loving plants the first week.

<u>Vegetables</u>: I'm picking cucumbers, pole beans, lettuce, and large tomatoes.

Zucchini and cherry tomatoes are ready to be picked. Trim off energy sucking spurs from the graft root at the base of the Empire apple tree. Spray the apples for skin fungus. Pick large tomatoes and oval cherry tomatoes off the trellis. Harvest four cups of rhubarb.

The cherry tomato vine has turned into a monster; it grew onto the roof. Picking tomato, cucumber, zucchini, broccoli, and oriental cabbage. The yellow hot poker flowers are done by the fourth week. Globe thistle flowers are deadheaded, new buds will form in two weeks.

Review of Data

On an Average

Fauna

<u>Animals</u>: The first week, young squirrels are making a pest of themselves as curiosity gets them in trouble with me. By the third week they are on the chase to get a mate and the fourth week they are climbing up the oaks and spruce with nesting material. The third week is when they may get mange.

JULY

Table of the Gang

Species/Quan.	\multicolumn{15}{c}{Species in the gang for July Day/Quan.}														
	05	06	07	08	09	10	11	12	13	14	15	16	17	18	19
Blue jay R, G	2	1/2 31/7	7/1	23/2	27/2 6/1	2 3/1	16/2 20/3	18/2 22/1	1	11/2 11/2 29/2	2 26/2	2 21/6	2	1 4/7	4
Cardinal R, G young molt mange	2 11/1 16,26	2 4/2	2 10/28	2 1/30	2 1/30 18	2 3/1 6	1	2 7-1 8-23	2 12/1 8-30	2 1-18 13 30	2 3/2 31	2	2 1	2 21	4
Carolina wren R, G young	2 3/1	2	1	1	1	2 29/1	1	2 8/1	8/2 13/1	30/2 31/1	29/2	1	27/2		1
Catbird G young	2	2	2	1	1	2	1	2	2	2 21/1	2 31/1	2	2 1/2	1	1
Chickadee R, G young	2 24/2	2 28/2	2 31/1	4	2 18/1	2 1/1	2	2 7/1	2	2 13/3 29/1	4 15/2 30/2	2 1	2 1	2	1
Chipper sparrow G young surrogate cow	2 29/1	1 29/1	27/2 12	3 21	0	1	2 9/1	1	16/1	30/2	0	0	0	0	0
Downy wood R, G young	2 1	2 6/2	2 12/2	2 4/1	1	2 21/1	1	2	1	2 1/2	2 20/2	2 12/1	1	0	1
Goldfinch R, G young	2/4	2/4 2/1	1	2/1	2	2	1	2	1	2	1	1	2		2
House finch R, G young conjunctivitis	2 26/2 28	2 28/2	3 12/2 12-18	4	2 30/1	4 29/1	5 9/1	4	2	4 30/2 7-14	6 17/1	2 1	6 12/1	3	5
House sparrow R, G young	2 5/1	4 6/2 17/1	3 9/1 25/1	4	2	4	0	0	1	2 9/1	2	2	4	1	12
House wren G young	3 19/1	31/1	2	15/1	2	2	2	2 25/1	2 4/1	2	2	2	1	1	4
Hummingbird	28/2	3/1	1/1	14/1	none	14/1	24/1	7/1	15/1	15/1 23/2	26/1	27/1	none	6/1	1
Mourning dove R, G young	6 16/1	8	2	4	5 6/1	4	3	6	1	3 6/2 17/1	4 13/2	5 26/1	6	2	5
Red nuthatch		1	12/1	18/1	3/1	1/1									
Robin young	3	2 19/2	1	3 3/3	3 4/1	1	2 22/1	1	1	2 1/1	2 4/1	0	2 20/2	2 1/2	2
Song sparrow R, G surrogate	2 22	6/2 2/1	2 15	2 30	1/30	1/13	1/16	1/31	0	1	1	0	0	0	0
Titmouse R, G young	2 20/3	2 1/1	2 1/1	2 1/1	2 18/1	2 29/1	1	2	2 10/1	2	24/1	2	2	1	1
Total population	43	41	38	43	36	34	33	32	29	49	47	29	36	19	43
Total species	15	16	16	16	15	16	14	14	15	15	14	12	14	10	15

All month long, the big brown bats are doing acrobatics at 8:30 p.m. Their July quantity has dwindled down from six in 2005 to none from 2010 through 2018, in 2024 there were two.

Number of species in the gangs decreased from 16 to 10 for the period of 2005 to 2018 then goes back up to 15 in 2019.

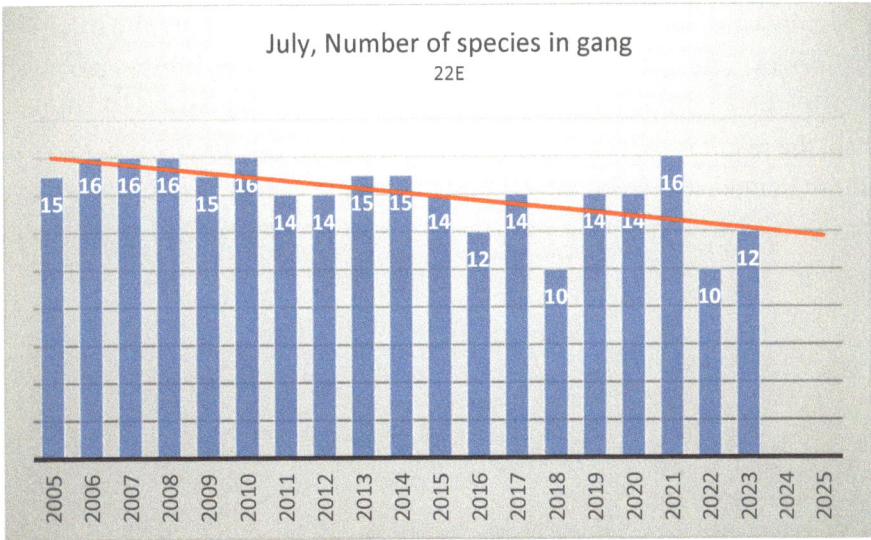

Number of species in the gang

Then there was a real catch 22 that has taken years to come to fruition. The female cowbird lays her eggs in a song sparrow or chipping sparrow's nest. The female sparrow hatches and raises the larger cowbird chick rather than her own chicks. I then see the sparrows surrogating the cowbird chicks during the third week. The result after time is there are no sparrows. As a result, there are no nests for the cowbird to lay her eggs and then no cowbirds. This showed up in the data, no song or chipper sparrows, as well as no cowbirds in 2016, 2017, and 2018. Others that are not arriving are the red-breasted nuthatch, goldfinch, downy woodpecker, and Carolina wren (all four of the species were observed during the winter of 2019). Some of the species did show up in August.

So, what has happened to the total population of the gang since 2005? It has dropped substantially, from 43 to 19! Due to house sparrow explosion, the population was back to 43 in 2019.

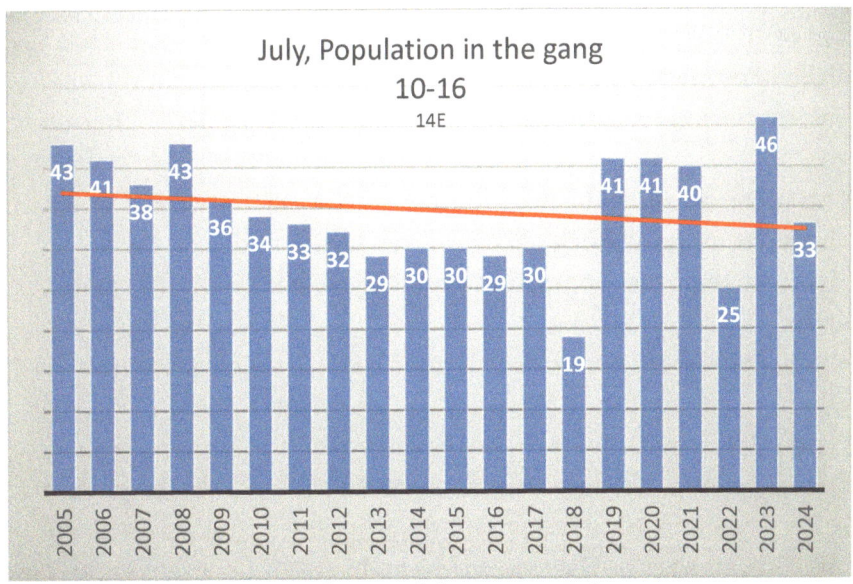

Population of the gang in July

Forty-three is quite a population for one yard to sustain, considering the different wants of food for the different species. The rise in 2014-2015 is from house finch, house sparrow, and mourning dove. Then in 2016 their populations drop again. Adjusting the data of 2014 to what I think it should have been, the result is that the trendline is steeper. In July of 2014-2015, many gang species that usually did not fledge went from none to one or one to two or three. "It was a very good year." It is concerning to observe both a drop in the number of species and the population.

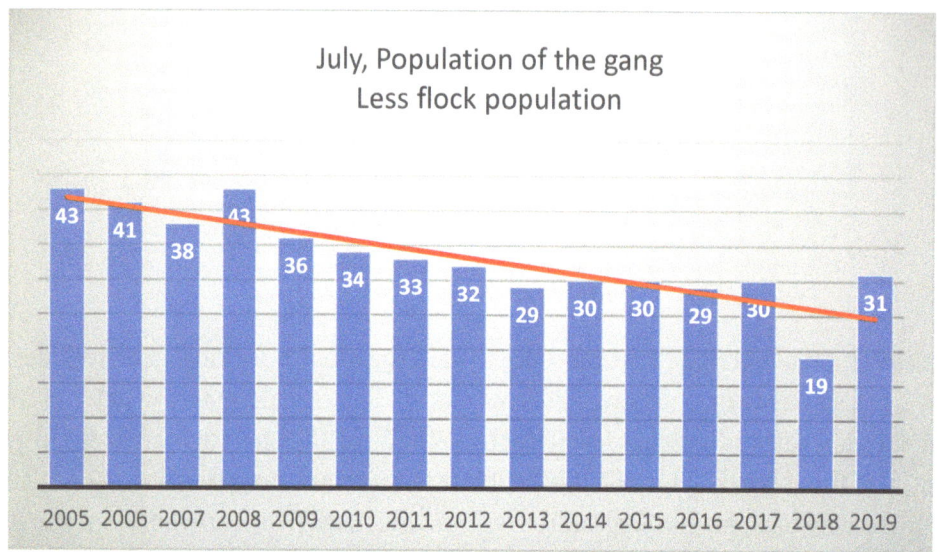

Gang population with adjustment of flock population in 2014, 2015, 2017, and 2019 shows a definite downward trend

The quantity of fledglings that a species has can depend on the weather. July 2011 was a very hot month, in the high 90's and into the 100s. On the 22nd, the temperature at 3:45 p.m. went to a record 104F. Seven of the 14 gang species had only one or no fledglings compared to an average of two.

In 2008 six out of 16 species had one or more fledglings than normal. That year we had a heat wave, 3H's, and above normal rain for the month and year. The vegetation was lush.

Year after year, the various species bring their fledglings to feeders either the first or last week of the month.

The population of the blue jays during July is normally one to two but in 2006, 2017, and 2018 out of nowhere a loud squawking flock of six to seven jays would be at the feeder. They would stay two to three days and then the population would go back to the normal one to two. It was interesting to watch.

Chimney swifts are observed in the late afternoon to dusk doing acrobatics for the evening meal in the last week.

In the 1980s and 1990s the trumpet vines were more prolific, growing high into the maple trees. Their southern facing red and yellow trumpets would shine in the bright sun. Hummingbirds would come for nectar. Six to 10 males would do a figure eight dance in front of the flowers to court the females. As time went by there were less and less trumpet flowers and less hummers. By 2000 the hummers would go to the red perennials and hummer feeders. The normal observation went down to one hummer, either a male or female. Now and then a pair would be gathering nectar. Until now, I only have one hummer coming to the feeder. However, when another one does arrive there is a rivalry of who will rule the feeder. They go at each other for 10 to 15 minutes, until one leaves. During this engagement they will rest near each other on a branch or wire, study each other, and then attack. The newcomer usually loses.

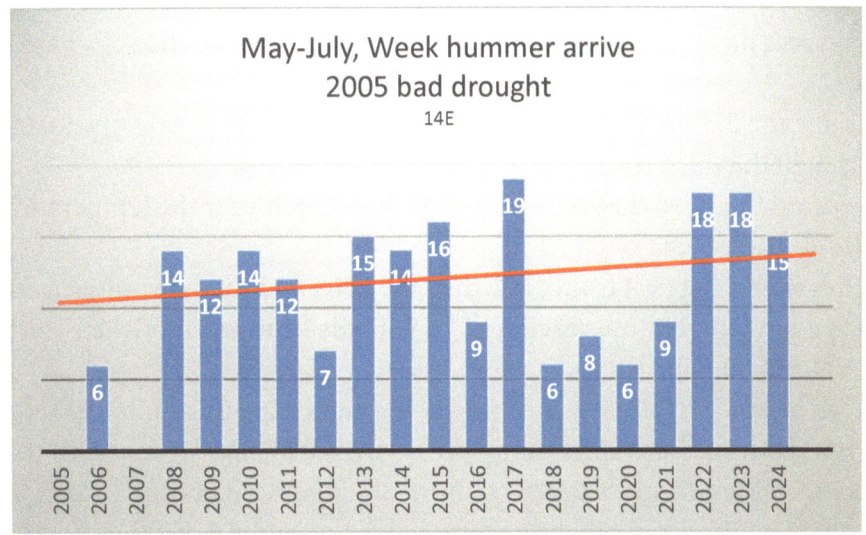

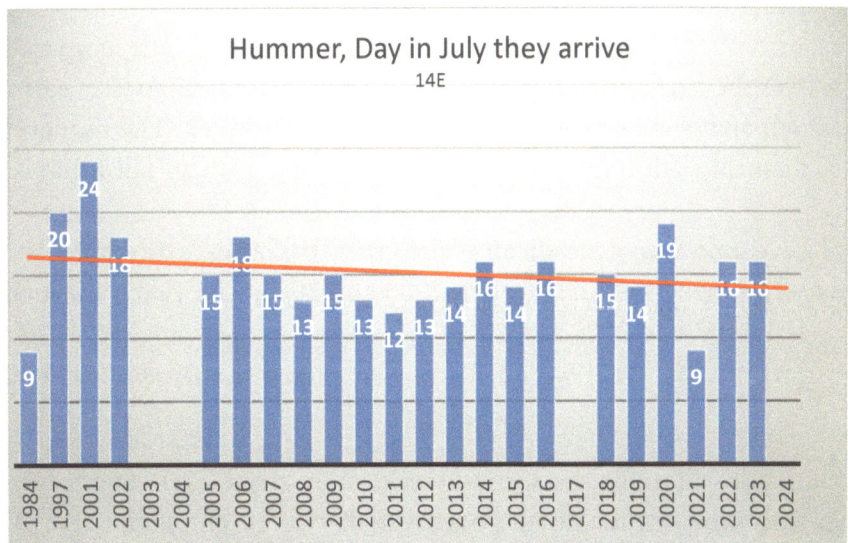

Hummer arrival day goes all over the month

Hawks migrate through in spring. Some years I would still observe a broad-winged in the first week of August. I used to see a Cooper's hawk into the fourth week; none seen in August after 2009. A sharp-shinned may still be around the third to fourth week. The red-tailed hawk is usually not in the area, but I observed one on the 20th in 2016.

Insects: Many of the observed insects would appear like clockwork every year, except in 2012, 2013, 2016, and 2017 there was an absence of most of the species. In 2012 and 2013 there was a lower amount of rain for the month; 2016 and 2017 had higher rain for the month. (All four years had the average expected temperature for the month.) Like the bats, from 2009 the insect population dropped to almost non-existent. Japanese beetles disappeared after 2009. The lack of insects is contributing to the decrease in bird population.

Lightning bugs arise the first week at 8:45 p.m. from 2005 to 2015, but none were seen in 2016 and 2017. They were again seen in 2019.

Four to five bumblebees are pollinating during the first three weeks.

Very few honeybees have been observed during the entire period; in 2018 there seemed to be a little come back.

When the air had the 3H's and the afternoon temperature is 85 to 90F, the locust can be heard from the first through the third week.

Cicadas were heard the third through the fourth week. Each year the temperature was between 87 to 94F and the air had the 3H's.

Japanese beetles were a big pest from 2005 through 2009 and then dwindled to almost nothing.

When the afternoon temperature reaches 88 to 90F, the little green crickets start their calls the third week at 9:30 to 10:00 p.m.

Spider webs can be seen in the early morning at the end of the third into the fourth week.

My observation about the butterflies in my yard is they are active in July with the temperatures between 79 and 96F. Different species seem to have different temperature ranges that they like to take flight. They have a very erratic flight pattern, up-down-around, yet they do get from "a to b." They are fun to watch, but hard to get a good picture.

White sulphur, or cabbage butterfly, pass through the first week with temperatures 87 to 98F. Yellow sulphur arrives the first and fourth week with temperatures 87 to 91F; eastern tiger arrives at 79 to 96F. Monarch arrive at 84 to 92F, the third and fourth week. The monarch will continue to flit through on their migration into September and during a warm fall I may still see a few in October.

Some species I have only seen sporadically. The great fritillaries, azure blue, and silver spotted skipper were only seen during the first week in 2005 and 2006. The viceroy, which is a look alike of the monarch, was observed in 2007 and 2008.

Similar to the other insects, there was an absence of some butterfly species in 2012, 2013, 2023 and 2024.

Pond: Frogs croak loudly all night the second week. They strike up a symphony in July that is louder and longer than June. Some years it goes on all day and night. Some years, there are lots of frogs and other times none. Small frogs, a quarter of the full grown, can be seen the second week. Frogs seem to like an air temperature over 83F; the hotter, the louder the symphony. This is could be misleading because I do not have the water temperature for comparison. 2024, garter snakes ate the frogs.

TRENDS DUE TO CLIMATE CHANGE

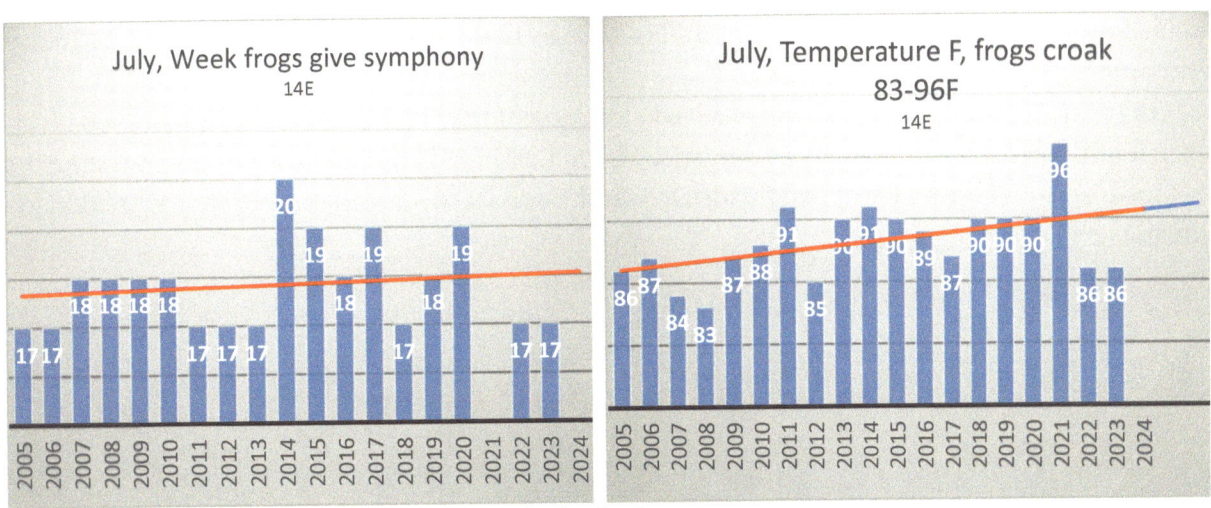

Weeks for frog's activity

Flora

<u>Perennials</u>: There are not many perennials in my garden that begin blooming in July. Most start to bloom in the last week of June and carry over into July. The temperature data shows a rise of six degrees from 1992 to 2024.

Phlox show a trend of blooming earlier by a week. It blooms at the beginning of the third week but with a very wet June it blooms two weeks earlier; with a wet and hot July, it blooms one to two weeks later.

Black-eyed Susan blooms all of July into August. White daisies come out the first to second week. Coneflower blooms the first week of July. Blooming period is 18 days before deadheading.

Trumpet vine shows no change in the blooming date. The yellow trumpet vine blooms 10 days after the red trumpet vine.

Hosta bloom the first week of July. The blooming period is 25 to 40 days.

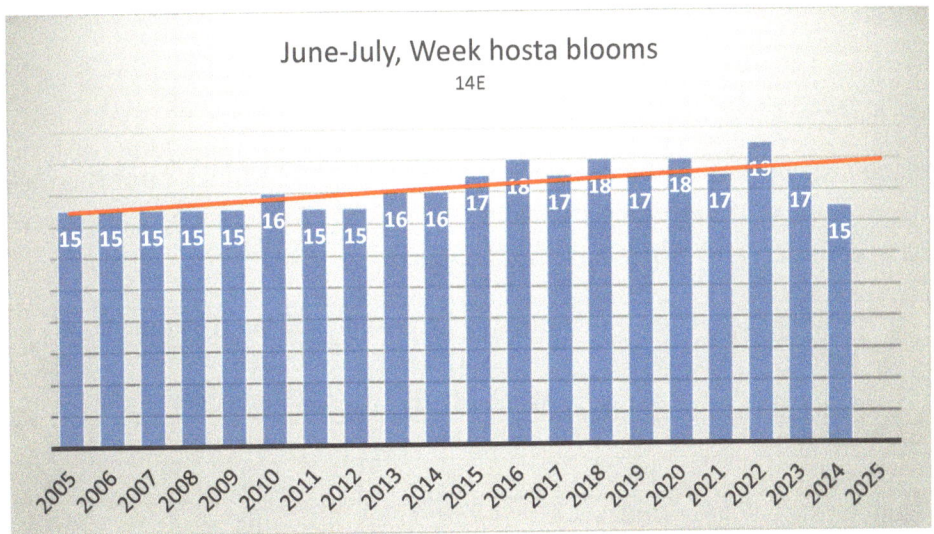

Blooming week for Hosta, two weeks spread

199

Yellow poker (ice plant) blooming period is eight to 12 days, loosestrife's period is 23 to 25 days and orange daylily are 18 to 22 days.

One year, I sowed a bag of wildflower seed at the top of the hill. Many flowers blossomed nicely, except unknowingly it contained wild thistle, which blossomed and opened its parachutes the first to fourth week. The next year I had thistle all over. Lesson learned, read the label for content before sowing.

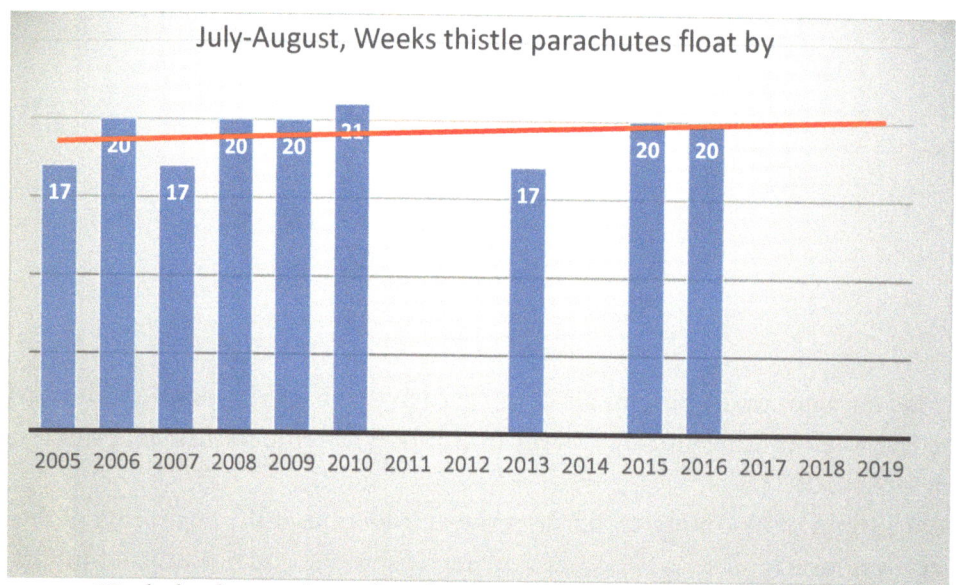

Week thistle parachutes float by, four-week spread; none 2017-2023

Blooming sequence of July's perennials from observed data	
Perennial	Blooming Week
Globe thistle	1st
Hosta	1st
Coneflower	2nd
Pink phlox	3rd
Black-eyed Susan	3rd

Arranging the sequence of the blooming weeks from the "trendlines," not the observed week, shows that the sequence of blooming basically remains the same from year to year. Some are blooming two weeks later, and others have not changed.

Sequence of perennials blooming in July-August using trendline data				
Year	2005	2010	2015	2018
Week of July				
Globe thistle	1st	1st	2nd	3rd
Hosta	1st	1st	2nd	3rd
Coneflower	2nd	2nd	2nd	2nd
Black-eyed Susan	2nd	3rd	3rd	4th

Pond: I wish I had recorded the water temperature over the years, since all the growth, sights, and sounds of the flora, frogs, and insects depends on its warmth.

Overall, at the End of the Month

Looking at all the data of the fauna from 2005 to 2018, I saw that with the many species of the birds, in 2011 there was a reduction in the quantity. Then I noticed that the same thing happened with the insects and butterflies. Looking into the weather data, I find that July of 2011 was very wet, causing the flora to rot or not bloom from lack of sun, which led to fewer insects for the birds to eat, who in turn had less fledglings. There were other years that were very hot, resulting in the flora not producing, resulting in less quantity of birds.

So, ends the hot 3H's of July.

Index of August

Weather	203
August Overall	204
First Week	205
Second Week	208
Third Week	213
Fourth Week	219
Review of Data	219
Fauna	219
Animals	219
Insects	219
Birds	220
Table of the Gang	221
Flora	223
Perennials	223
Trees	223
Pond	223
Overall	223
So, ends August	223

August

Week number for the year of 48 weeks					
Relative Month	Month/Week	First	Second	Third	Fourth
Previous	July	17	18	19	20
Present	**August**	**21**	**22**	**23**	**24**
Following	September	25	26	27	28

The average high temperature on the 1st is 81F and the low is 62F.
The average high temperature on the 31st is 74F and the low is 53F.
The extreme high was 100F on the 2nd in 1934, and a low of 41F on the 30th of 1934 and the 31st in 1986.
The full moon is called Sturgeon, Red or Green Corn Moon.

Weather

The downpours can be very heavy; in 1933 there was one of 4.7 inches. The monthly rains can be three to six inches or there can be a drought. In 1998 there was no rain for seven weeks with 103F temperatures. Since 2005 there have been eight dry Augusts and three overly wet. Temperatures go to the high 90's to 100's. The last week is when the Perseids meteor showers occur. For years I would go out at 2:00 to 3:00 a.m. to observe the show to no avail. I gave up, better to just stay in bed.

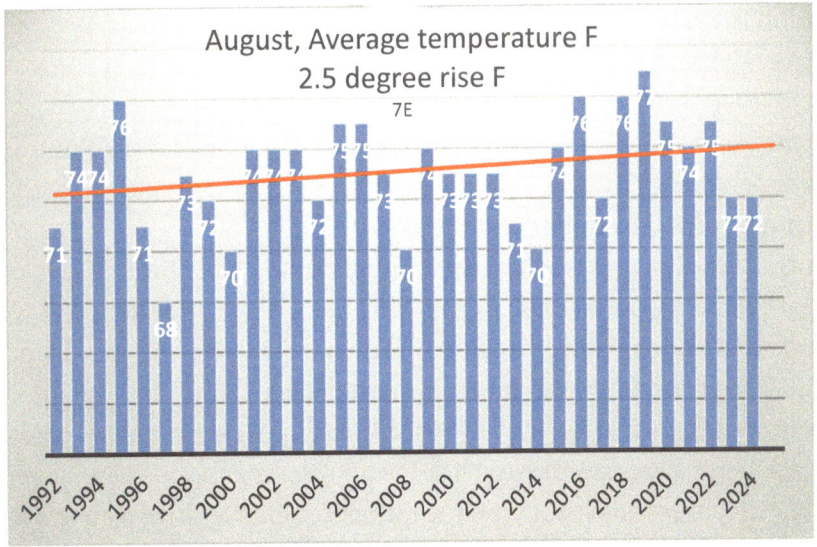

August mean temperature

As a reference following is the chart for July.

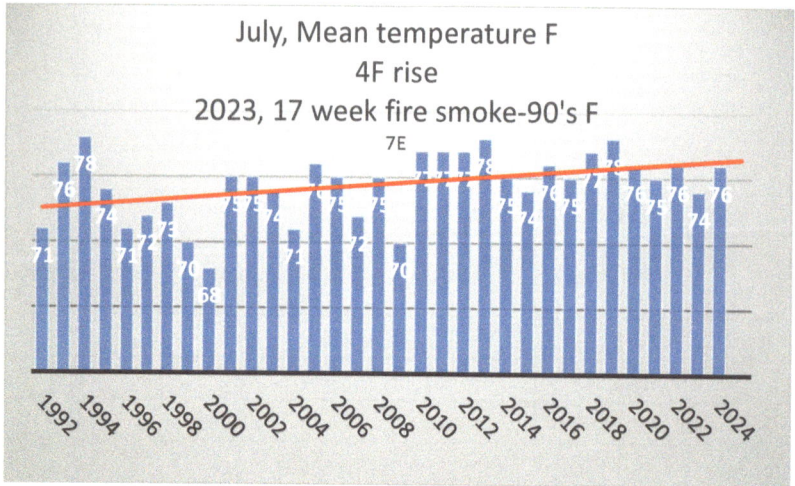
July mean temperature

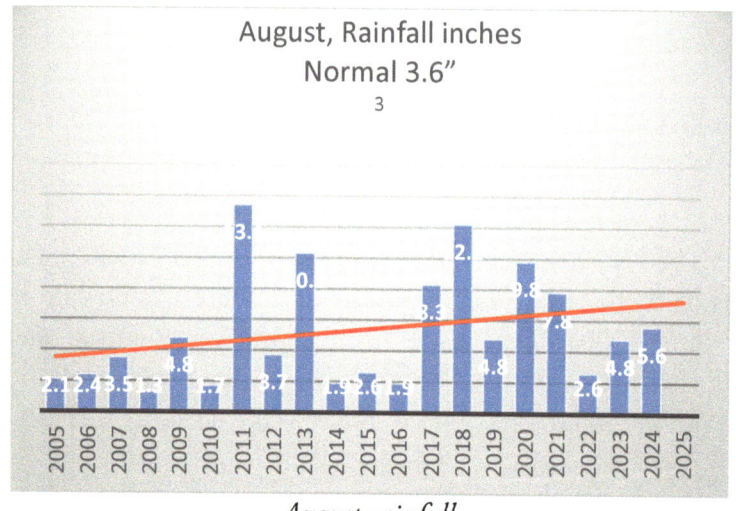
August rainfall

Overall Summary of August

    This month is the real beginning of the hurricane season. "Dog Days of Summer" (July 22nd to August 22nd) is a superstition that was a period that dogs went mad, also the constellation of a dog rose in the sky. Another 3H's (hazy-hot-humid) month.

    Mushrooms form in the lawn. It is so hot for so long that the fish die from lack of oxygen (96+F). A young rabbit (kit) is rolling on its back to get a dirt bath. The rhododendron leaves are pointed up at 45 degrees. The west dogwood leaves turn brown-green-red and are dropping. Locust and katydid are calling. The ground turns to hard clay and cracks.

    When the weather gets excessively hot, we start picking large tomatoes midway through the month. With the excessive heat the cabbage heads and tomatoes split open. When we have a very wet-cool month the tomatoes may not ripen until September and then they have a bland taste. High temperatures turn the lawns brown and there may be watering restrictions. It stresses the annuals to the point of drooping at mid-day and recovering at night. The apples, peaches, and berries will be smaller than normal.

On the positive side, the high temperatures ripen the tomatoes early providing that there is rain. If there is no rain the lawn gets like straw. A continuous stretch of hot, dry weather produces fewer wild seeds and bugs. It is not only stress on the birds but also the bats. The bats depend on the night bugs, which are far and few between during a dry, hot spell.

August 16, 1944, the temperature reached a record low-high of 59F for the day. A year later, on the same day, the temperature went to a new record low of 49F! What a year will do.

Examples of Yo-Yo Weather

In 2013 it was very wet and cool. On the 11th there was heavy rain, on the 29th there were heavy thunderstorms and it was very humid. The temperature did not go over 90F for the whole month. The lawn was lush and had to be cut twice a week.

In 2014 August was a very dry, cool month. The hottest August on record is 2019. The ground cracked; high 80's and 90's all month. It was humid and hard to work outside.

In 2017 we had a very unsettled summer, from no rain to sudden downpours.

In 2018 there was a heat wave from the 1st to the 7th with dew points over 70F and feel-like temperatures of 105F. The second heat wave was the 14th to the 17th. Dry, mold high, then rain. The third heat wave (94-95F) was the 28th to the 31st. Tomatoes cracked, tomato and cucumber vines turned brown, back lawn had fungus, few fledglings, crickets didn't chirp, locust did not sing, few lightning bugs. August 2018 was the fourth wettest in history.

First Week

Animals: Rabbits are eating anything in the rudbeckia family like black-eyed Susan.

Birds: Between 2005 and 2012, I would observe various bird species the first week of August: magnolia, myrtle and yellow warbler, Savannah sparrow, redpoll, red-winged blackbird, and turkey vulture. Until 2010 the red-breasted nuthatch would be a summer resident. They have not been observed in August since 2010.

Fruit: Spray the apples for black spot during the first week.

Insects: With the 3H's, the crickets seem louder than usual. Using a flashlight, I found four types of crickets, including black, brown, and green colors. Even with the flashlight off they sense my approach, maybe from the vibration of my steps. They go quiet when I am 10 feet away. I keep on walking to where I heard them last, then turn on the light and look over and under the leaves. Each species seems to have a different song, but it is difficult to verify.

The cricket symphony usually begins the first week and continues into the fourth. Locust come out the first week. Their song continues into the third.

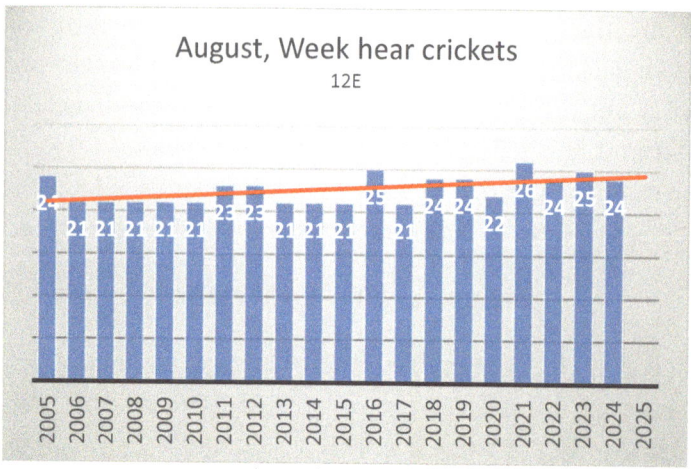
Week crickets start to sing

We had very hot summers in 2005, 2016, and 2018. The second hottest August on record was 2016. The summer had 33 days of 90F or above, it was the 14 driest with five heat waves. Bottom line, crickets do not seem to like to chirp in very hot weather, so they put it off until September. The hotter it gets the faster they "crick."

Ticks are prevalent this time of the year. Black and yellow swallowtail butterflies sail through visiting the black-eyed Susan.

When we have an infestation of Japanese beetles, the birch tree leaves look like lace. The traps get full fast and stink if left in the sun. The traps should be put a good distance from the plants to be saved, if not, it is like inviting the beetles to dinner. The beetle population has dropped significantly over the years because people spread their lawn with insecticide or put down milky spore in the spring to kill the grubs. Unlike chemicals, milky spore is a live thing that gets in the intestinal tract of the grub and kills it. Once established the effectiveness of milky spore can last for 10 to 15 years. Putting down the milky spore powder is another thing. The instructions say to lay a teaspoon down along a line every four feet and place the lines four feet apart. I quickly realized that the powder disappeared as I put it on the lawn, so I did not know where I put it down. So, I got four twelve-foot-long pieces of one-inch lath and made a grid that looked like a large tic-tac-toe. Now I had 12 points, four in the middle and eight at the end of the laths. I laid it down, put powder at the points, picked up the grid and moved 12 over and up and down the lawn. The neighbors really scratched their heads. It worked fine and got rid of the grubs.

<u>Lawn</u>: If I do not put down fungicide on the lawn, brown spots will start to appear and spread rapidly. This is the week that red clover appears in abundance in the sunny areas and then spreads. When the ground is not too hard the weeds still grow and need to be pulled. When the ground is dry, applying chemicals in July or August does not do much good against the clover, they need water to absorb the herbicide. I pick the sorrel by hand; it has one strong root.

With the high humidity and maybe rain, mushrooms appear in the lawn, sometimes they form a fairy ring. The mushroom spores can be dormant for years and when the conditions are right, they reappear.

Perennials: Time to transplant iris. Transplanting a butterfly weed usually is unsuccessful because of the deep taproot; it is better to buy a plant from a nursery.

Pond: A white dragonfly may hover over the pond. Punks are forming on the cattails.

Shrubs: Hydrangea is blooming at the end of the month; the pink will fade to a dull brown.

Pink hydrangea

Trim the holly and yellow bell this week and maybe again in September.

Trees: When it gets dry, the west dogwood leaves turn brown-green or dark maroon and start to come down, by the fourth week you would think that fall has arrived. More than once I thought I had lost the tree. When the leaves drop early, the tree does not develop green berries until September. Then there is no snack for migrating birds. Poplar trees also lose leaves when it is very dry. Silver maples will prematurely turn brown.

Vegetables: Finally pick broccoli grown from seed.

This Week in the Bird World

Catbird, flicker, house wren, and hummingbird sometimes are absent for two or three weeks in August and then reappear the fourth week. Sometimes I wish that the catbird would not come back until the middle of September because while it is here it pecks at the ripe tomatoes! Thought there was a different species in the yard until I realized it was a catbird with no tail. It flew a little unbalanced.

House sparrows are not pretty birds. They are aggressive at the feeder and multiply faster than rabbits, but they do go after and get Japanese beetles and moths. I heard a lot of chatter at the side of the house. There were 20 sparrows enjoying my tomatoes! Another time a sparrow was enjoying suet!

Even though the daily high is 90F plus for another heat wave, a sign of the coming fall is to see 10 to 20 local Canada geese fly over just after sunrise. A male Carolina wren is singing loudly calling for a mate. This lasts only for a week or so, and then it goes is quiet again.

House finches eat the pulp from the sliced orange. The catbird goes further, it likes the grape jelly that I put on the orange slice as well as watermelon. In this case they both have a sweet beak.

During the summer robin couples are independent (anti-social), but this week there is a gathering of the clan. In the morning, there may be a flock of 10 to 12 on the front lawn ready to go south. Among the flock there are teenagers with speckled breasts.

The pecking order is in effect. A cardinal chases a house sparrow, then the house sparrow chases a song sparrow. House finch chases a chickadee. Catbird chases a hummer and house finch. Notice that it is usually the larger chasing a smaller but not always. A group of six house finches prevent a young cardinal from landing on the feeder. Or an adult house finch chases a young cardinal as to say learn your manners. Pecking is not necessarily just among birds. I saw a chipmunk chase an ovenbird from the feeder area and a squirrel chase a chipmunk.

Throughout August a lot of species bring their fledgling to the area (cardinal, Carolina wren, titmouse, robin, cowbird, and chickadee). The blue jays brought three fledglings! A female dove has pieces of grass in its beak to start a new nest.

There were some split peanuts lying next to the pan of water. A grackle lands next to the nuts, picks one up and dunks it into the water and consumes it. Maybe it is the same one I saw in May. Two weeks later I observed the same thing.

Broad-winged lands at the top of the oak. Within minutes six crows arrive and harass it until it flies off. I witnessed the same thing in February.

Second Week

Weather: This is the week that there is an annual meteor show at 4:00 a.m. Many years I did go out to observe it, but I have never been successful in seeing any. Afternoon, with the humidity and high temperature, it feels like 105F, there are no birds in sight. That night a thunderstorm arrives, cooling things down.

Animals: The chipmunks and squirrels are eating the Empire apples. Squirrels are gathering large walnuts. Some nuts are as large as their head. This is first nut to be harvested for the season. Two squirrel kits are frolicking together; they chase each other, and then start to climb a tree, just to turn around and go back down. Then they grapple each other and become a ball of fur, rolling and tossing in the grass. The adults ignore them and proceed to eat the dogwood berries. Meanwhile there are females high in the oak tree causing a rain of small branches that they did not use to build their nest.

Rabbits are basically nocturnal; they come out to feed in the middle of the lawn around 4:30 p.m. They are very vulnerable to the red-tailed or broad-winged hawk that is also out for dinner. The rabbits like to get in the flower garden to eat tender coneflower shoots and petunia. Just as bad, the chipmunks are eating the Swiss chard.

Hawk with a rabbit that came out too early in the evening

The jays bring two to three fledglings to the area. A brazen grackle takes a whole peanut from right in front of a surprised jay. Cardinal and catbird are eating blackberries. Hummer goes from a zinnia to a butterfly bush to the trumpet vines.

A skunk visited the base of the feeder at night; the fragrance woke me at 2:00 a.m.

A squirrel has mange, which come from a parasitic mite. It looks funny because it has lost most of the fur on its head.

Birds: Jays and cardinals are bald from mange. A jay imitates a hawk calling; the call is loud enough to chase the small birds from the feeders. Then the jay flies down to feed. Female house sparrow brings three fledglings to the feeders. A group of goldfinch glide through in their low roller coaster cyclic path. They repeat their tweeting call along the way. A catbird is hungry and is eating safflower. Chickadees, cardinals, house finches, jays, and song sparrows are all molting.

Blue jay with mange

A Carolina wren and a downy are both on the peanut butter log looking at each other. Later a chickadee and catbird come for peanut butter. Because of the heat of the day, peanut butter oil is dripping from the holes in the drilled-out log. I put a pan on the lawn to collect the oil.

Peaking order is still active; a downy chase a chickadee from the split peanut feeder. Early in the morning a flock of 20 crows go north. A flock of starlings gargle. They make many noises; the gargle sounds just like when you gargle with a mouthwash.

A dove raises and spreads its wings to cool off in the afternoon heat. Rare to see, but a delightful sight, myrtle or yellow warbler rest here on their migrant journey. They may stay a couple of days. I'll also be looking for migrating tree swallows. Some years I see chimney swifts circling and crying in the evening sky catching hundreds of insects.

A titmouse is aggressively going after another male, which is its own reflection in a car door mirror. A bully catbird chases a female house finch and hummingbird from the feeder. Young hummers will come to the feeder into late fall.

Fruit: Spray the apples, annuals, and lawn for fungus. The apple skin gets brown stripes from the fungus. It does not look nice, but after you peel the apple, it is okay to use.

Winesap apple

Insects: Gnats are out. Mosquitoes are bad; an article says that they are attracted by to carbon dioxide. I guess we are not supposed to exhale!

Yellow jackets and bumblebees crowd out the hummingbirds to get nectar from the hummer feeder.

Paper wasps get water from the birdbath to build their nest. When it is humid, a musky walnut fragrance surrounds the nest.

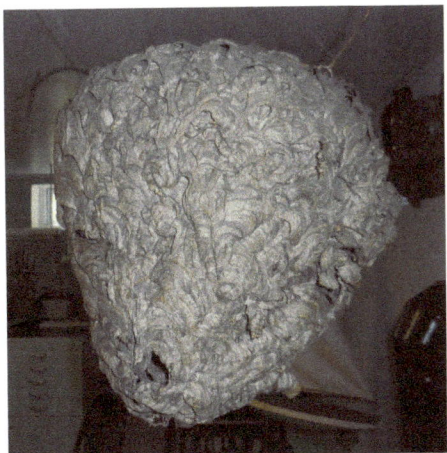

Paper wasp nest from hemlock hedge, it has streaks of orange and blue throughout

Two White Sulphur butterflies dance in the sun. Sulphur are quick, erratic in flight, not landing on anything for a long period. When two meet they have a quick dance of going up and down, then quickly part and go their way, only to return a few minutes later to do the dance again. They land on cabbage family plants to lay eggs. If it is warm, this dancing ceremony continues through September.

White Sulphur *Yellow Sulphur*

At 6:30 a.m. I find a big garden spider web hanging from the front dogwood. The spider will retrieve the webbing by noon. One morning with the dew still glistening on surfaces I noticed a spider web strand going from the top of the 20-foot flagpole to the ground, almost 30 feet long. How was that done?

Gnats are out in force during the day and mosquitoes in the evening.

Lawn: Purslane and sorrel are coming out stronger than in July, mostly in the edge of the flower gardens. With the 3H's, a fungus can form on the lawn overnight resulting in the whole lawn dying quickly. I learned to put down fungicide at the beginning of June and the end of July. When the grass dies, it takes three to four hours just to rake up the dead grass of the back lawn. Ugh!

Perennials: Bergamot that were deadheaded in the third week of July are getting new flowers.

Pond: Cattail punks should be forming. The water plant foliage gets so dense that I cut it back for the fish to receive sunlight and also to have adequate water surface to apply a treatment to kill mosquito larva. A blue damselfly is hovering over the pond. A white waterlily is opening in the morning sun. This is the week when the skunk cabbage leaves go brown.

Shrubs: Butterfly bush blooms the second week of August; in 2013 it bloomed the third week of September, in 2019 it bloomed the fourth week of August!

Vegetables: Time to pick pole beans and cucumbers.

Pole beans

Pepper plants have buds, starting to pick cherry and large tomatoes. Pick cucumbers, string beans, beets, and zucchini. Pulled carrots grown in a pot, the center ones were 12 inches long, the ones around the edge only six to eight. If August has been cool (in the 70's) it may take until the 28th for the tomatoes to finally turn red. Cucumber vines can die overnight from a beetle or fungus.

A groundhog can eat a whole vegetable garden in an hour. The vegetables are eaten to the ground at the beginning of the feast, and then as its belly fills less of the vegetables are eaten, until at the end of this 10-foot-long feast only the tops get chewed!

Third Week

<u>Weather</u>: On a hot and humid afternoon, 4:30, a thunderstorm is coming. The storm arrives and dumps three quarters of an inch of rain; the yellow jackets and birds take shelter, so do I. At 6:00 p.m. the storm is over and the sun comes out. Small birds resume eating, jackets called it quits for the day.

<u>Birds</u>: Hummer goes to a dahlia. Flock of 50 grackles go overhead, then a flock of Mallard ducks. Some years I am lucky enough to see and hear the wonderful melody of a passing brown thrasher. A flicker arrives and will stay a week or so. While it is here, it calls out loudly a little after sunrise. A hairy woodpecker will share the suet feed with the downy. The house wren will be leaving soon to go south. A male cardinal feeds its young. Cowbirds gather to go south. A female dove has pieces of grass in her beak to make a loose nest in the boughs of an evergreen.

I go back outside with a book and sit on the patio facing the main feeder. There are two horizontal wires 12 feet apart at an eight-foot height that cross the yard. One holds the small feeders and the other is the power line to the main feeder. I look up and see a hummer sitting proudly on the power line. I usually do not see a hummer sitting still for any length of time. This one is sitting for 15 minutes. I look up again and see another hummer on the other line. The two are facing each other. Ten minutes later they are still looking at each other. The one on the power line takes off, flies around, and comes back to the same position on the power line. A few minutes later it takes off and lands 10 feet away from the other on the feeder line. They look at each other. The one that was on the feeder line now flies to the hummer feeder, gets a drink and lands on the power line for a short time then flies away. It now is dusk. The one that is on the feeder line gets a drink then lands on a branch stub. This is longest I have seen hummers stay stationary.

Hummer on the branch

At 9:00 a.m. 10 to 20 local geese rise from a nearby field and go north or west; 7:30 p.m. a "V" of 30 go over. They will be residents until next spring. The population may increase to 60. The geese are not the only large birds, I look up to see a broad-winged hawk and turkey vultures soaring on high. A female cardinal waits to the side of the feeder for a jay to take eggshells and a whole peanut.

Dusk, 7:45 p.m., I look to the western sky to see erratic low flying bats.

<u>Insects</u>: Do not see any lightning bugs this week; guess mating season is over. Dark, 8:15 p.m., the crickets are calling.

A white dragonfly hovers over the pond. Over the years the canopy has over-shadowed the pond and as a result I no longer see damsel or dragonflies. When there is a hot, late afternoon sun, I walk to the top of the hill which still has a wide-open space. I watch in awe as dozens of dragonflies fly radically in the sun with their two sets of transparent wings; an envy that the Air Force wishes they could duplicate.

<u>Lawn</u>: Before 2005 there were lots of Japanese beetles which produced grubs, which ate and weakened the lawn. Then homeowners battled both beetles and grubs. By 2005, there were only a few beetles. By 2009 there were none and the lawns looked good in August. In 2016 and 2017 lawns were lush. It is not only the insects, the 3H's of August cause the fungi to thrive. Diligent applications of fungicide are required in the third week. If not, brown spots quickly appear, followed by hours of raking up the piles of dead grass. Clover and crabgrass can also be a problem. The plights of a homeowner.

<u>Perennials</u>: Globe thistle and loosestrife that I deadheaded two weeks ago are forming new buds. When it is hot the plants are stagnating and will not absorb fertilizer.

Black-eyed Susan may be starting to wane. Goldfinch will land on the flower to pick out the seeds. Their gold camouflages them into the fading gold of the Susan. Sedum gets green blossoms.

Fall sedum, green phase

<u>Trees</u>: Squirrels are eating the green dogwood berries. Some drought years there are no berries on the trees. In 2005 the berries turned red the third week of September; in 2017 they changed the fourth week of August, a six-week spread.

Fourth Week

<u>Weather</u>: One year after a morning rainstorm, I looked north to see an inverted rainbow. It was less than a quarter of a circle. I went to my *National Geographic* weather book to look up the phenomenon.

<u>Animals</u>: Chipmunks makes a chip-chip-chip warning sound when threatened; sometimes the call goes on for five to 10 minutes. Usually, the threat was unwarranted and due to another chipmunk being in its area. Put whole peanuts under the feeder for squirrels and blue jays. Instead a chipmunk comes and chases an ovenbird off the feeder and puts a whole nut in its jowls and another in its mouth. It stops halfway along the lower garden, eats the one in its mouth, then goes back and puts another one in its jowl plus another in its mouth. Again, goes halfway along the garden, stops, and eats the one in its mouth, then runs off to the nest with the two in its jowls.

I'm watching a chipmunk maneuver around obstacles I had put on a birdfeeder. I observed that the chipmunk favors a right versus left side paw just like we use a right or left hand. When it used its right paw, it could manage to maneuver around the obstacles, but when I switch the obstacle to the opposite side position and it had to use its left paw it had lots of difficulty.

Early morning as I get the paper, I watch two skunks frolicking around on the lawn. Why not?

<u>Birds</u>: Flocks of 50 to 60 cowbirds come through. A flock of 200 crows go west. A flock of 10 young grackles come through. They vaporize any seed that was on the feeder. Flocks of 100 geese fly over going south.

Barn swallows go over in the evening. However, if it has been very dry there will be very few bugs for the swallows. A nearby garden center has had barn swallows that have been continuously nesting for years.

Adult barn swallow

Barn swallow nest

Jays still have mange. A father cardinal is fledging four to five youngsters. A young cardinal goes after a chipping sparrow on the split peanut perch. Cardinals look nice but they are aggressive toward their own species as well as others. A cardinal will wait for a jay to take whole peanuts and leaves before its proceeds to acquire sunflower seeds on the platform. Adult doves are feeding their young. An ovenbird passes through.

Ovenbird stays for three days

    Song sparrow brings two young to the area. This specie is a ground feeder, so I don't usually see them on the feeders. Male and female house finches get conjunctivitis. Young male house finch breast feathers are turning red. Catbirds have a sweet beak, they like the pith of an orange, watermelon, and grape jelly. Hummers are still here, but not for long. The catbird and house wren have gone south. Goldfinches are changing to molted green. The bed and breakfast birds are the purple finch, tree swallow, ducks, and flicker. Individual robins are gathering in flocks of 60 for the fall migration. A red-bellied woodpecker has returned for a fall stay, it will move on comes winter. However, with the changing climate, the habit of the red-bellied changed. As of 2015, some overwintered. A red-tailed hawk lands at the top of the oak tree. Within minutes six crows come and chase it away.

    It is a hot, quiet afternoon. I am watching a downy attempting to make the entrance hole on a chickadee house larger, the hole is ringed with epoxy to prevent squirrels from making the hole larger. The epoxy is giving the woodpecker a challenge. When the downy hits the epoxy, the house resounds with a loud knock-knock-knock. The downy is squarely facing the house concentrating on pecking. Other small birds are at the feeders. Suddenly the downy goes flat against the birdhouse. The other birds are still feeding. Within five seconds all the small birds scatter. In a flash a sharpie comes swooping toward the birds at the feeder, they all scatter. It makes a sharp turn in front of me and lands on a dogwood branch. After a minute it flies on. The downy is still flat against the house. After another minute, the downy relaxes and flies away and the small birds return. The next day the downy is again pecking at the epoxy. How did the downy detect the sharpie?

Chickadee having breakfast

<u>Insects</u>: Bumblebees are on the few leftover loosestrife flowers. While trimming the arborvitae hedge, an exposed praying mantis quickly flies out and away. Crickets will be singing from dusk till dawn into October. While gardening, come across ladybugs and a wheel bug. The tool shed smells of stink bugs; they are looking for a place to hibernate. In 1998 after they were discovered in the Lehigh Valley they were everywhere. Now there are very few!

Stink bug

At the end of the week more no-see-ums attack; no escape, time to go inside.

<u>Lawn</u>: When the grass goes brown, it breaks off from lack of water when I step on it; I bring a lot of the broken grass into the house on the bottom of my shoes. If I am too late with the fungicide on the lawn, brown patches will appear. It takes hours to rake up the dead grass. I learned to put down deep-rooted fescue lawn seed in lieu of Kentucky blue. This is the time to spread lime; it takes 200 pounds to do the whole lawn.

Perennials: This week the perennials can look forlorn from the heat and dryness. Red hibiscus and phlox are out full. Klondike bush and cosmos have bright yellow flowers. The self-seeding spider plants are out. Black-eyed Susan that blooms all of the month used to wane during the first week of September; in 2018 they wane the third to fourth week. Yellow iris that had flowers in April-May are reblooming now.

Pond: Frogs are calling all night with a different call. Algae forms in the large pond; put a chemical solution in the pond that clears the water in a day.

Shrubs: Hemlock, japonica, and arborvitae have new growth which I trim the last week of August. The butterfly bushes are blooming. Holly and japonica have new growth. Woolly adelgid insect is on the hemlock hedge; spray with horticultural oil but it may be too late to be effective. The trimming of the holly, arborvitae, and rhododendron has also moved from the last week of August to the third or fourth week of September.

Rhododendron bushes have shriveled up young leaves from gall midge. Best preventive is to pick off the leaves and put them in the garbage.

Gall midge on rhododendron

Fall clematis shows a two-week change. It used to bloom the fourth week of August; in 2019 it blooms the first week of September. When out full, its strong fragrance encompasses the entire backyard. Surprisingly, with all the fragrance there are very few bees on the blossoms!

Fall clematis

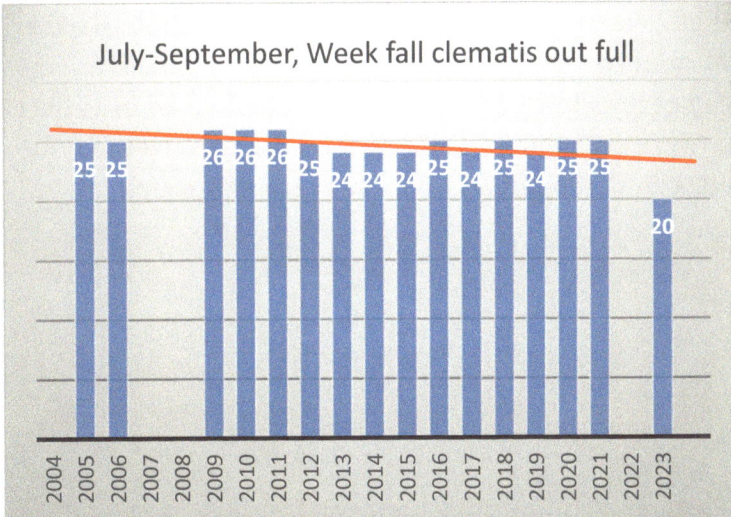

Week fall clematis is out, Bluebell hydrangea has new growth coming up along the first-tier border.

<u>Vegetables</u>: On the 21st, spade up red and white potatoes. Pick lots of basil, cherry tomatoes, peppers, pole beans, summer squash, and zucchini. Beets and pole beans planted eight days ago are up. If the earth has moisture, now is the week to put down fescue grass seed. Many summers the seeds for fall crops cannot be put down because the ground is cracked and hard as a rock.

Review of Data

Fauna

<u>Animals</u>: The quantity of bats has been decreasing due to the white-nose syndrome; 2005 saw four in the evening, 2010 there was population of eight. It was a very dry, hot month. In 2017 the quantity was one. In between there were years I observed none. In 2019 there were two bats. Those years with zero population all had extra monthly or year-to-date rain.

The quantity of rabbits goes from none to four. The years with no rabbits were again extra wet.

Squirrel population goes from three to zero. The zero population occurs when the monthly or year-to-date rain was extra wet.

<u>Insects</u>: Bumblebees are active across the whole month.

The evening rising of the lightning bugs from 2005 to 2007 was the fourth week of August. Then in 2009 there were only a few bugs. From 2010 to 2017 there was no rising, except 2016 had a few on the third week, in 2019 there were many. Amount of rain or high-low temperatures does not seem to be a factor.

Mosquitoes in 2005 through 2010 were out for a human attack the second week. By 2015 it was the first week. They are attacking earlier and earlier.

Spider webs and yellow jackets have not changed, they are both noticeable the third week.

There are 10 species of butterflies that I have seen in August. The red admiral, eastern tailed-blue, mourning cloak, and silver spotted skipper have passed through singularly on a regular basis. From 2005 to 2008 the monarch, black swallowtail, eastern tiger swallowtail, and Sulphur arrived the first week. After 2008, the first observation was the third week. From 2005 to 2009 they were watched for a period of 23 to 28 days. From 2010 the period has dropped to seven to 10 days. In each case the period goes to the end of the month. In 2012 there were no monarch, black swallowtail or eastern tiger swallowtail. No monarch was observed in 2013. The common factor is that July and August were dry months. In 2018 there were many monarchs all over the Lehigh Valley. The white and yellow Sulphur pass as erratic aerobatic couples.

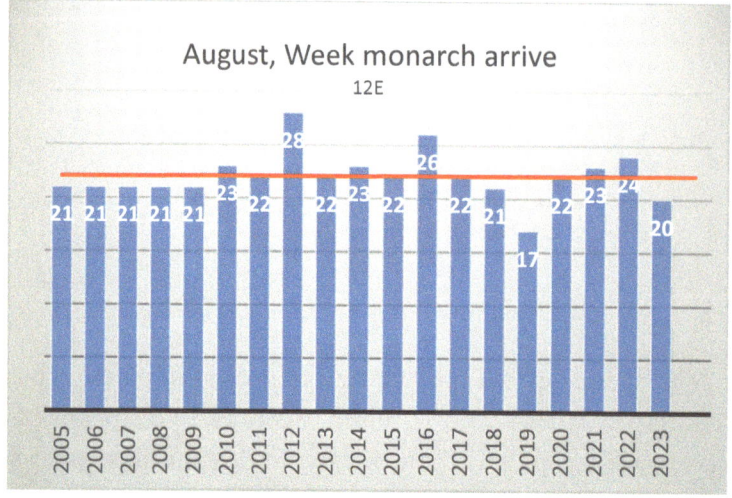

Weeks monarch arrive

Birds: I reviewed the number of species that made up the gang, the number of birds the yard was sustaining, and the number of birds in a species.

The number of species were 38 in 2005, then declined to 18 by 2012. The decline of migrating species consisted mostly of swallows, warblers, swifts, and bats. Then the number increased back to 28 by 2015, from the same species that had declined. Like the other months, the population of almost all the species dipped in 2010. The chart of blue jays is typical.

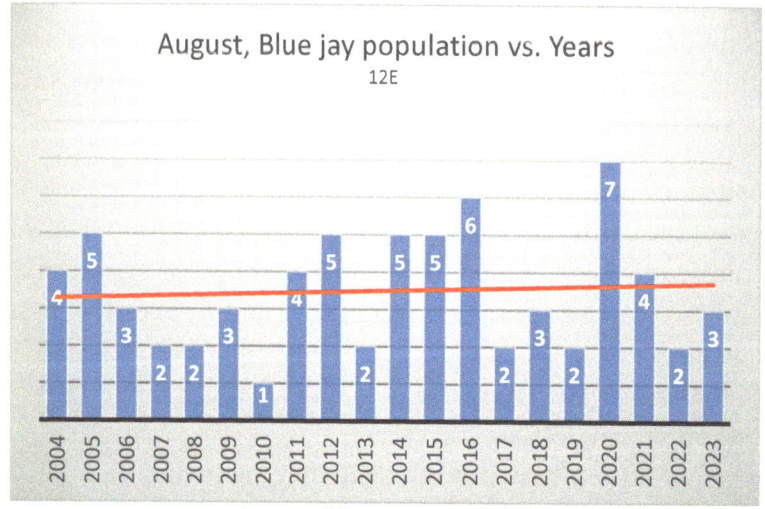

August, blue jay population

Table of the Gang

August, Table of Gang																	
Species/Quan.	04	05	06	07	08	09	10	11	12	13	14	15	16	17	18	19	20
Blue jay R, G	4	5	3	2	2	3	1	4	5	2	5	5	6	2	3	2	
Cardinal R, G	3	4	3	3	5	4	1	3	3	2	3	4	4	2	4	3	
Carolina wren R, G	3	2	2	1	1	1	1	1	1	1	3	2	2	1	1	1	
Catbird G	2	2	3	2	1	1	1	2	1	1	3	2	1	1	2	1	
Chickadee R, G	4	4	4	5	4	3	2	3	2	3	3	3	3	1	2	1	
Chipper sparrow G surrogate Cow	1 / 4	2	1	1	1 / 2	1 / 1	2	2	2	1	0	0	0	0	0	0	
Downy wood R, G	4	3	2	2	1	3	1	1	1	1	3	3	4	1	1	1	
Goldfinch R, G	2	4	4	2	0	2	0	2	0	2	2	2	2	1	2	1	
House finch R, G conjunctivitis date	7	4	4	5 / 23	2	2	6	2	6	6 / 2	6 / 19	6 / 19	2	6	8	8	
House sparrow R, G,	5	6	3	8	10	12	0	0	0	2	3	20	10	5	8	14	
House wren G leave	2 / 31st	2 / 31st	2 / 26th	1 / 30th	0	2	2	0	0	1 / 7	1 / 1	1	1 / 7	0	1	0	
Mourning dove R, G	7	5	6	7	7	8	1	4	4	5	3	6	7	6	4	8	
Red nuthatch	1	1	1	1	1	1	1	0	0	0	0	0	0	0	0	0	
Song sparrow R, G surrogate date	4	2	1	1	1	2 / 1	2	1	1	1	1	1	0	0	0	0	
Titmouse R, G	2	3	4	5	2	3	4	1	0	2	3	3	3	1	2	0	
Population	52	49	43	44	39	50	25	26	27	30	39	38	45	27	38	44	
Total gang	15	15	15	15	13	14	11	11	13	13	13	14	12	11	12	10	

The gang of summer resident species from 2004 through 2007 was 15, then there was a continuous decline to 10 by 2012, consisting of sparrows, finches, and wrens. After that, the number rose to 14, but declined again to 11 by 2017. The trendline is a decline.

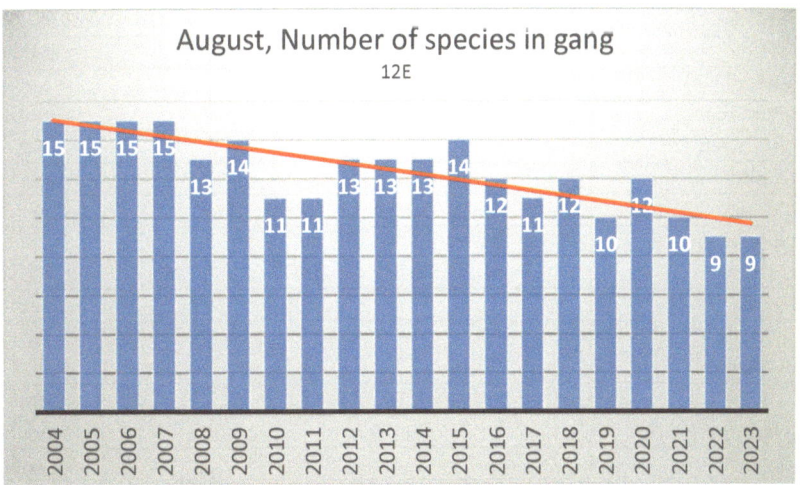

August, number of gang species

The cyclic population change of the house sparrow, house finch, and mourning dove has the biggest effect on the total gang population. All three species seem to cycle at the same time. The cause of the dove population is probably due to the return of the sharp-shinned hawks.

AUGUST

I feed the birds all year long. Some people say that there is enough natural food for the birds. All I know is they keep coming, especially for peanut butter. So, I counted the number of birds that visited feeders in the yard in August. In 2004, there were 51, this number declined dramatically to 25 in 2010. The number gradually climbed back to 45 by 2016, but declined again to 27 by 2017, and back up to 40 in 2019. Basically, the yard sustains 30 to 50 birds during the month of August.

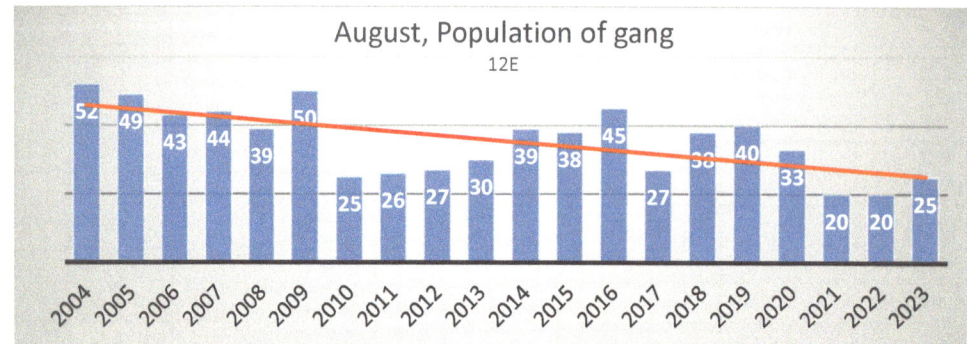

August, population of gang

The population of birds of a particular species over the years stays the same, two, three or four depending on the species. No particular species is prominent. The house sparrow may go from six to 12 but then crashes to none or two the following years.

The species that are no longer observed during August since 2004 are the barn swallow, chimney swift, great blue heron, ovenbird, red-breasted nuthatch, and yellow warbler. The reasons could be the change in my environment, township development, and climate change. When there is a very wet July and August, there may be a no-show for some gang species. The species bringing fledglings depends on the weather. The table gives a general indication on the arrival week.

These are the weeks in August when parents bring fledglings to the feeders, as well as have feather changes		
Species	How many young	Week of month
Blue jay	3	1st and 3rd
Cardinal teen turn red	1 to 5	1st to 4th 1st or 2nd
Carolina wren	1	1st
Catbird	1	2nd
Chickadee	1 or 2	1st to 3rd
Chipper sparrow surrogate	1	1st
Downy	1	1st or 3rd
Goldfinch change color	1	1st 4th
Hairy wood	1	4th
House finch	1 to 3	1st through 4th
House sparrow	1 to 3	2nd or 4th
Mourning dove	1 or 2	4th
Hummingbird	1	3rd
Robin	2	3rd
Song sparrow surrogate	1 or 2	4th
Titmouse	1	1st

Flora

<u>Perennials</u>: Over the period of 15 years some of the August perennials are blooming and waning later. August 2010 was very dry and hot, and many perennials did not bloom.

<u>Trees</u>: Leaves of the dogwood trees, which have shallow roots, would start to change to maroon during the third to fourth week of September. They changed the second to third week of August in 2013 through 2017. This may be due to the local very dry-high temperature conditions.

<u>Pond</u>: The data shows no trend of when the arrowhead white flowers bloom or when the white lily blooms. Both plants have bloomed from the middle of July to the fourth week of August. Frogs start to seriously croak all night from the first day to the 18th of August. The wide variation could be due to the water temperature. Unfortunately, I do not have that information.

I had cattail plants develop punks from 2000 to 2011, then suddenly no more punks. After 11 years growing in the sunken tote, the cattails produce no punks. I was told by an aqua plant producer that the soil they are in has run out of potassium. Adding supplements will not help; the plants should be replaced.

Overall

The weather of July and August drives the outcome of the flora and fauna. When it is abnormally wet, the plants rot from lack of sun, the insects starve or do not produce, so there is less food for the birds and fledglings. Sometimes it is the opposite, there is a drought and hot, the ground becomes like concrete and cracks. Again, plants die, no insects, no birds. This all became quite apparent from the plotted data of the flora and fauna, even when the frogs literally croak.

So, ends the 3H's of August.

Index of September

September Overall 225
Weather 225
First Week 226
Second Week 234
Third Week 238
Fourth Week 244
Review of Data 249
Fauna 249
Animals 249
Table of the Gang 249
Birds 250
Table of Migrating Species 252
Insects 253
Butterflies 253
Flora 254
Lawn 254
Overall 254
So, ends September 254

September

Week number for the year of 48 weeks					
Relative Month	Month/Week	First	Second	Third	Fourth
Previous	August	21	22	23	24
Present	**September**	**25**	**26**	**27**	**28**
Following	October	29	30	31	32

This is the month that is most prone to hurricanes. The first week can still be hot and humid with heavy thunderstorms. Large flocks of blackbirds and geese go south. Before 2006 many species of warblers would migrate through the last two weeks of the month. After 2006, there have been very few warblers. There are lots of other species migrating, some returning for the winter, others leaving to go south and others passing through. The quantity depends on the weather. If the temperatures stay warm in Canada, the geese will come later. September is when the migration of broad-winged, Cooper's, red-tailed, and sharp-shinned hawks starts, as well as the attacks of the small birds at the feeders. The catbirds will wait until the end of the month to go south. Crickets sing throughout the month. Dogwood berries are gobbled up by the migrants. Many types of butterflies glide through; spiders spin elegant webs. At the end of the month gnats and no-see-ums cannot leave you alone but you can look forward to picking apples and raspberries.

Weather

The average high temperature on the 1st is 78F and the low is 57F.

The highest temperature was on the 11th in 1996 and the 12th in 1983 at 96F.

The average high temperature on the 30th is 70F and the low is 49F.

The lowest temperature was on the 28th in 1947 at 30F.

Reviewing the high and low temperatures from 1927 to 2019, the decades with the highest percentage (23%) of high temperatures was the 1930-1940's and the 2000-2010's. The decades with highest percentage (27%) of lowest temperatures was also 1930-1940's.

The amount of daylight is 13 hours.

The full moon is called the Harvest Moon or Full Corn Moon.

The weather gets pleasant in the high 60s; we still can get hot days with thunderstorms and the 3Hs (hazy- hot-humid). Many years the drought condition of August goes into September. Normal rain for the month is 4.4 inches; one year the high was 12.7 inches and another year the low was 1.5 inches. Normal rain from the beginning of the year to the end of September is 34.8 inches. The extreme year-to-date high was 55 inches in 2011 and the extreme the low was 30 inches in 2016. 2024, September was the beginning of a record four drought.

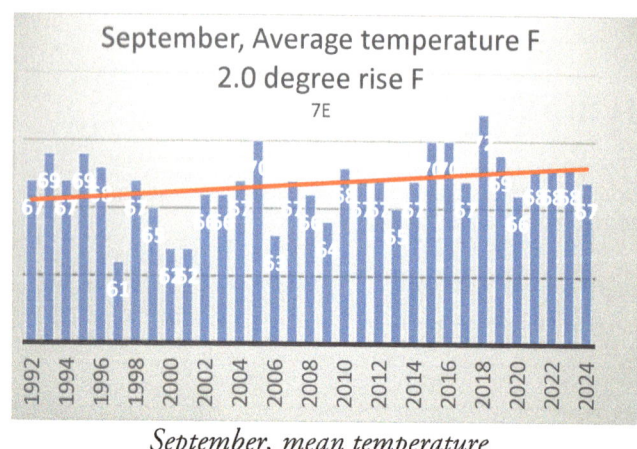
September, mean temperature

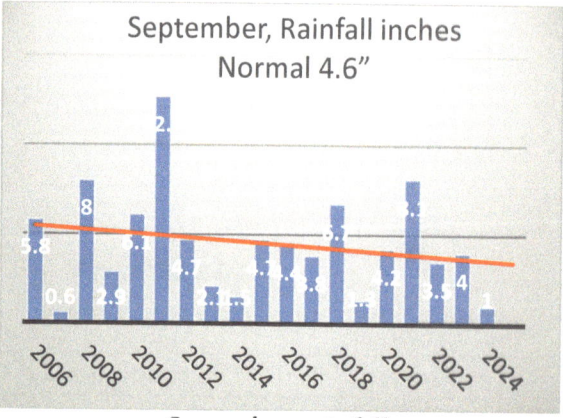
September, rainfall

First Week

<u>Weather</u>: The first week of 2005 it was 87F, migrating summer birds stayed on to the third week, crickets kept singing. This time of the year the temperature is like a yo-yo; it will continue this way until mid-November.

It is 7:00 a.m. when the sun first rose, its rays hit the dogwood west; it looked like a yellow-orange light bulb. As the sun got higher, it turned back to green with a maroon center. Just a nice sight to see.

7:00 a.m.

<u>Animals</u>: Meanwhile, a chipmunk is literally running under my feet to get spilled safflower seed. As it runs after another seed, its tail goes straight up as it calls out, "Hey, wait for me." They eat peanuts like we eat corn on the cob. They also like to eat grapes and apple slices. They climb up and over an 18-inch high chicken wire fence. How about a chipmunk jumping straight up nine inches to get onto a feeder!

Good apple

Birds: Brooding season is happening in September, jays and downy bring two; chickadees, house finch, titmice and robins bring one to two. House wrens will be leaving soon. Robins and song sparrows may become surrogate mothers to fledgling cowbirds. On warm evenings, bats are collecting insects.

First week September possible migrating species	
Species	Species
Yellow-throated warbler	Magnolia warbler
Red-bellied nuthatch	Blackbirds
Hummingbird	Cowbird
Chimney swift	Grackle
Hairy woodpecker	Ovenbird
Rose-breasted grosbeak	Total species 12
White-breasted nuthatch	Dragonflies

Goldfinch start to change color from bright gold to drab green. The start change color may be related to temperature. In 2005 the change started the third week of September; 2019 the change is the second week of October. The same upward trend as the mean temperature?

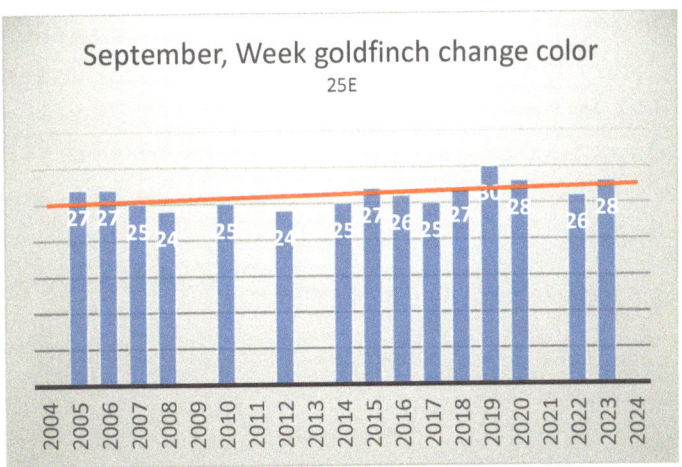
Weeks, goldfinch changes color

One year I observed a male goldfinch feeding two drab green youngsters. With hues of globe thistle and coneflower, the goldfinch's color of gold or green blend in making it difficult to observe them eating.

Cardinals are still molting, some have mange. They strip off brown cattail leaves for nesting. Flickers may return this week, but mostly in the fourth week.

Flicker looking me over

Sharp-shinned hawk returns

The sharpie looks down to see three young squirrels frolicking on the lawn, he dives, missing all three that scatter. Shortly afterward a broad-winged is soaring gently to the westerly breeze at 2,000 feet; a majestic sight. Suddenly it tucks its wings, plunges straight down; what a spectacle. I could not see if it was successful.

Flocks of 100 to 300 consisting of blackbirds, starlings, and cowbirds go over in the morning heading west. This will occur all during the month. Beginning in 2015 very few annual flocks were observed. Sometimes when a flock does go through, 30 blackbirds would stop to raid a house wren's birdhouse. None could fit in. I do not know what they had in mind; they were only passing through. In 2017, it was just twilight when a flock of fifty-plus blackbirds-cowbirds raided the main feeder.

Local geese mostly return during the first week, more will return all month. An hour after sunrise a flock of 15 to 20 will be go north for the day and return later in the day. Crows return for the winter. Like the geese, they fly in a loose flock to an unknown local destination just after sunrise. Crows form this loose flock for protection. They will repeat this act almost every day. They return in the scattered formation an hour before sunset.

Over the years I have seen a catbird, house wren, song sparrow, and Carolina wren with no tail feathers! They seem to fly okay.

Canada geese exercising

A returning sharpie lands in the dogwood; 15 minutes pass when suddenly it catapults straight up and seizes a dove that has just landed on a branch of the oak tree above it.

A hummer is still here but will migrate when the temperature gets colder. Leave the nectar feeder up until October which is when some youngster may be migrating through. The youngsters have built-in navigation, so they do not need their parent's assistance to make the trip to Central America. Watched a catbird chase a hummer from the nectar feeder three times. On a hot afternoon I put a piece of watermelon on the feeder. In a short time, a catbird was cobbling down the melon as fast as it could. Blue jays also like watermelon.

Catbird on hummer feeder

A lone starling makes a unique gargling-tingling sound, just like we do with salt water to relieve a sore throat.

Mostly chickadees eat seeds, but to my surprise I watched one get a little brown moth. Then another tried to pick up a moth that was lying flat on a window; no luck for the chickadee, the moth escaped. Two male chickadees are fighting over the same house. On a hot afternoon as I sit on the patio, a chickadee comes to the empty water dish three times that is three feet away. I take the hint and fill the dish. I overfilled the dish and the chickadee got wet feet when it landed so it flew off. After I lowered the level, it landed on the edge to get a drink. As soon as it leaves, a young Carolina wren is there for a drink.

On the 14th in 2015, I find five house finches on the roof with no head! I learn that a screech owl specializes in eating brains of small birds.

Decapitated finch

Screech owl

Blue jays are clever and choosy in what they want to eat. One was on the perch for the split peanut bowl when a nut spilled out; the jay immediately dropped off the perch and caught the nut in mid-air. Other times they will pick up and lay down two to four whole peanuts until they get the right one. They imitate a sharpie's call from a high branch to scare little birds from the feeder. They can gargle and screech and chirp; multi-lingual.

A downy is trying to enlarge the epoxy ring of the entrance hole of a house. Suddenly it goes flat against the house. Within seconds a sharpie flies through. How did the downy know of the presence of the sharpie when it was facing the entrance of the house? As soon as the sharpie left, the downy also flew off. The next day the downy was back at the entrance.

Jay getting a split peanut

Downy opening the entrance

When the cool, clear air of autumn fills the sky, turkey vultures do their very erratic acrobats; I wonder how they stay in the sky.

Fruits: Time to pick some apples. Some years they get a red blush during the first week.

Insects: I read that butterflies need 60 to 70 feet of straight space to fly! The backyard has that. Monarchs can only fly when their body temperature is 70F or above. They start arriving in August. September is now the prime month that they fly through.

Monarch on butterfly weed

Future monarch

Perennials: Pink phlox that I had deadheaded two weeks ago are now in full bloom. The reblooming will continue into October. The phlox can get powdery mildew on its leaves which dry and crumble when touched. Spray horticultural oil on the hemlock hedge for the woolly adelgid insect.

Perennials that are out are trumpet vine, black-eyed Susan, pink and white phlox, pink cushion, fall sedum is pink, blanket flower, yellow coreopsis, white and pink coneflower, spider plant, loosestrife, a few yellow irises, and Hosta.

Bumblebee on fall sedum

In spring the Jack-in-pulpit rises to a foot high to form the hood that has Jack in the pulpit. Then during the summer, the pulpit fades away and the top of the stalk has a cone cluster of red berries that develops during September.

Jack-in-pulpit beginning of May, turns to red berries in September

<u>Pond</u>: All the fish swim together in a school. They all swim around the pond then come to a rest all facing north. There are lots of tadpoles in the ponds, but not many adult frogs. A few frogs are still croaking. As the month cools down, they will come out to sun themselves, but not croak. The arrowheads have white flowers.

Arrowheads in flower

<u>Shrubs</u>: As the weather cools, I trim the roses so they will rebloom into October and maybe into November. The PJM and rhododendron are forming new growth and next year's buds. The butterfly bushes came out the first week in 2005, in 2016 it is the fourth week, in 2018 it blooms the second week.

Trees: Two years we had wet springs causing the canopy to flourish. Then at the end of September into October we had one or two heat waves of 90F plus. As a result, the canopy could not sustain all the growth; it rained brown leaves from the maples and poplar. Just as well, less resistance in a windstorm. The remaining leaves turned fall colors at the end of October into November.

The dogwoods are the first to show signs of fall. Late August to the beginning of the September, the dogwood berries start to change from green to orange. Even in this condition the cardinals will start to eat them. The leaves get a slight tinge of maroon. Then the leaves turn a rich maroon. This happened the first week at the beginning of the decade, by the end of the decade it occurs at the end of the fourth week. The berries did not get red until the third week and only advanced one week over time. Through the month when the berries turn red is when the big harvest takes place for many bird species and animals. A chipmunk is up in the branches eating the berries. In August 2015 due to the heat and drought there were hardly any berries and in 2016 there were none. The birds, chipmunks, and squirrels come for the harvest on the fourth week. Unfortunately, for those years the cupboard was empty for the migrators.

Red dogwood berries will be available until the end of the month into October

Birch trees are releasing feathery, light brown seedpods that look like a cross. Walnut trees are dropping their now-yellow leaves. This is the first tree of the season to drop leaves. During the first two weeks the squirrels will be running around with big black walnuts looking for a place to hide their treasures. The female holly tree flowers the first week; they become berries 10 days later.

Vegetables: When there is a warmer September, tomato plants grow so profusely that I must trim and thin the branches so air can pass through them to prevent blight. Still pick cucumbers. Malabar spinach has been struggling to grow in August, but it takes off in September. It cooks up like spinach but has a more mellow taste. The pole beans were doing fine until the chipmunks decided they wanted some too. It was so wet in 2018 that there were only a few tomatoes and no apples. Fall crops did not produce.

Malabar spinach

Second Week

Now that it is cooler at night roses are re-blossoming. Some of next year's blossoms on the PJM's are out. The grass is struggling to recover from the summer's heat. Because of the carpet of newly fallen pine needles, the hill has become slippery trying to walk up it. On an unusually sweltering afternoon, birds land in the coolness of the shadows that cross the lawn.

<u>Animals</u>: At dusk, I see more rabbits on the lawn. Later, as I cut the front lawn, the lawn moves in front of the mower like a wave. Out pops a little bunny and then another and another. I almost made minced rabbit. The female made a hole in the lawn overnight and dropped the youngsters.

Albert, the chipmunk, is still giving me a battle on getting safflower out of the feeder that has a horizontal dowel perch that can rotate. The feeder is mounted on an eight-inch-long vertical pipe. The pipe is put into a 12-by-12 wooden block so that the feeder can be positioned anywhere on the picnic table. Right now, it sits in the middle of the table. The weight of Albert should close the door of the feeder. Albert has no problem jumping onto the dowel from the right side of the feeder. He does this to distribute his weight. He straddles his feet so one is on the stationary part of the door and the other on the perch. So, I put a thin sheet of balsa wood on the right side on the feeder to block him from getting to the perch. I move the feeder on the table so Albert is in front of the horizontal dowel. Before this he would not jump straight up. I move the feeder to the edge of the plastic sheet so there was no room for him to jump up to the dowel. Now he makes it in one jump and swings around the dowel to get at the feed. I put a loose piece of nylon netting on the plastic sheet. He would not go over the net, instead he jumps up onto the back of the feeder and crawls along the side ledge that holds the moving dowel. He gets onto the dowel. I put grease on the side channel. He touches the grease and jumps down. He scampers into the dirt below and runs back and forth to remove the grease. Now he comes back and jumps up one foot onto the right side of the feeder in one leap. He also likes a little sweet, I see him one eating a half of a grape. We are having a drought, the pole beans growing on vertical netting are struggling to produce some beans. I go out the next morning to find not a bean on the vines. Albert was hungry and climbed up the netting to do a total strip job.

Trellis that Albert climbed

<u>Birds</u>: It is 6:30 a.m., light enough to see, I go out to replenish peanut butter in the drilled branch. There is skunk perfume in the air. I notice dark shadows in the lawn. As it became lighter, I discovered that a skunk had torn up the lawn looking for grubs.

A two-minute breakfast. It's 7:00 a.m., just light. A jay is on the perch getting split peanuts, a red-bellied woodpecker flies in to chase off the jay. A young cardinal lands on the other end of the perch waiting for the red-bellied to finish, it gets impatient and leaves. A male house finch lands where the cardinal was and waits, the young cardinal comes back and chases the red-bellied. As the cardinal leaves the house finch bends over to get a split peanut. As it does, a sharpie flies in and takes the house finch. All in two minutes.

A flock of 20 cowbirds attack the feeder. After 10 minutes of devouring seed, they fly on. A bully catbird chases a chickadee from the split peanut feeder. Soon the bully will migrate south, to the relief of the chickadee.

Heat index can still be over 100F. It is unusual, but I find crows walking in the lawn early in the morning. Later I see a flock of crows flying west. In the eastern sky, a flock of 30 local geese fly to get exercise. Two flocks of 100 starlings are going south. I hear the call and then see a returning flicker. This is the week that there may be four to six blue jays at the feeders, some are fledglings. Depending on the weather, the catbirds and hummers may leave this week or take an extension to leave at the end of the month. The male cardinal is still feeding two larger fledglings. Adult cardinals are eating the now orange berries of the dogwood trees. Cardinal and other species are molting. The crown of the head of the young male house finch turns red during the second and third week. Two male song sparrows go at each other for territorial rights. They go beak to beak, flying straight up from the ground for two feet, dropping down and then go up again, flapping wings and jabbing beaks.

A house finch and chickadee are both feeding from the same perch of the safflower feeder. They are four inches apart. The chickadee starts to inch up on the perch toward the finch, which is bigger. The finch turns and lowers its head at the chick; the chickadee quickly flies to the other side of the feeder. Friendly to a point, the pecking order still rules. Likewise, four house finches and a dove are feeding on the main feeder. A jay calls out that it is coming to land on the feeder. All the other birds scatter as the jay lands. At dusk, 20 mysterious mallard ducks circle the circle. Mysterious, because I do not know where they come from or settle. It is somewhere nearby. This is almost a daily routine from April through September.

Migrants this week can be ovenbird, black-throated warbler, yellow warbler, Connecticut warbler, and magnolia warbler.

Bats were observed feeding in the evening before 2006. Since 2009, bats throughout the east coast have gotten a fungus, white-nose syndrome, that has killed them in large quantities. Except for the fourth week in 2014, I have no longer observed bats in September.

At 8:00 p.m. it is too dark for a young sharpie to see properly. While trying to land, it bangs into the side of a dogwood. It shakes its head and flies away.

<u>Insects</u>: Yellow jackets like the sweetness of sugar and prevent the hummers from approaching the hummer feeder.

Yellow jacket nest

Since 2007, there is a noticeable reduction in bumblebees and honeybees. The population of bees has continued to decrease. Only a few honeybees are present in 2015. Some think the bee decline is due to an intestinal virus or to pesticide use. Gnats are plentiful throughout the month especially at the flower gardens. One myth I have heard is to wear a tall witch's hat because they will attack the top. The neighbors will really think I have gone over the edge.

I hear crickets. Some sound like a soft, slow police whistle, others sound like a loud, but not harsh, police whistle.

Lawn: I fertilize the lawn with corn gluten. With the ground dampness, mushrooms are coming up in the lawn.

Perennials: Red phlox are still full, bergamot is trying to re-flower, coneflowers are done, and globe thistle is coming back. Purple aster is just blooming. Yellow Klondike is out. French marigolds are doing poorly from a fungus, but the tall marigolds look good and so do the zinnia and dahlia. The goldenrod is turning gold. Rose campion has one or two flowers. Fall clematis is out in pure white; its strong fragrance is not noticeable until after the blossoms wane. Hosta leaves are riddled with holes compliments of the night activity of slugs. The autumn sedum has pink flowers which are loaded with bees. The Joe Pye weed is attracting pollinators.

Joe Pye weed

Annual butterfly weed

Butterfly weed flowers are out in bright yellow and orange behind the pond. When I bought these in late summer, I thought they were perennials. What a disappointment the next spring when not a one came forth. I guess they were an annual. I did not ask when I bought them.

Pond: Arrowhead leaves are turning yellow. Eight brown dragonflies are hovering around the top of a silver maple. They drop down to hover over the pickerel weed in the pond and then dart to the iris at the other end and hover again. When they are present, they come and go all month.

Arrowheads

Trees: When we have hot, dry summers, come September, maybe 10 to 20 percent of the leaves of oaks, maples, and poplar shrivel up and drop off. There is just not enough moisture for photosynthesis to continue. The remaining leaves retain their green, but it is a semi-drab green. Surprisingly, come late October-early November, when photosynthesis truly shuts down, the remaining leaves will turn to their respective bright fall colors before being released for their one and only flight to the good earth. Spray the evergreens with Isotox to get rid of the spider mites.

Third Week

Animals: Two young squirrels are on the chase; others are burying whole peanuts. A young kit (trainee) missed a jump high in the oak tree limbs and crashes to the ground. It gets up, shakes itself, and slowly walks away. The front lawn is dug up; a skunk was digging at night getting grubs. A raccoon took off the top of the metal garbage pail with the whole peanuts; the clang of the top woke me. I quickly go to the door. The coon looked at me defiantly before running off. Saw a chipmunk with a whole walnut, which is really an overload for its size.

Birds: Cardinals, goldfinch, robins, and flicker are eating red dogwood berries. When October temperatures stay above 80F the catbird and hummer will stay until the end of the month. Again, starlings unsuccessfully try to take over a house sparrow's house. The sparrows know how to hold onto their house from house wrens and starlings. Having fun or revenge, a house finch purposely dumps seed from a higher level of the feeder onto the head of a cardinal, who shakes it off as it lands on him. The 19th to the 24th a female red-breasted grosbeak is eating safflower. A migrating flock of robins is on the front lawn. A female house finch has conjunctivitis.

A pair of white-breasted nuthatch arrives at the feeder. In 2019 and 2020 they over-summered and overwintered.

Pair of white-breasted nuthatch

Red-breasted nuthatch

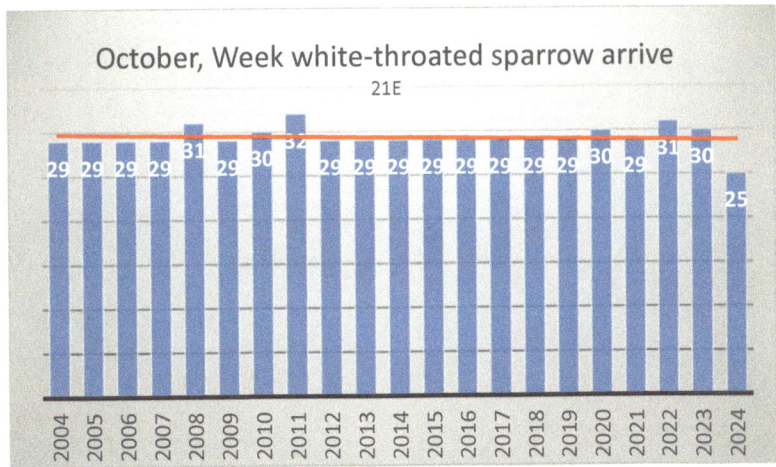
Weeks white-breasted nuthatch arrives

From 4:00 to 5:00 p.m. a small flock of tree swallows are overhead circling for insects. A goldfinch lands on a black-eyed Susan seed head. Red-bellied woodpecker is eating a split peanut. The house wren is still here. Until 2016, 2012 was the last time I observed a red-bellied nuthatch in the yard and that was in October. On the 18th in 2015, between 10:30 and 11:00 p.m., a pair of great horned owls is hooting in the spruce tree. What a surprise and enjoyable occurrence.

Goldfinch on black-eyed Susan

Red-bellied woodpecker at split peanuts

<u>Fruit</u>: Time to pick apples. September is when I get the best raspberries. I see a vole by the raspberries. They eat the roots of plants; it is time to set a trap.

<u>Insects</u>: A monarch passes through the yard. At 1:30 p.m., early in the day, two types of crickets are calling. It is 80F, white sulphur, monarch, and bumblebees are all active at the broccoli flowers. No yellow jackets. Spiders are mating; there are webs across the bushes in the morning. I catch a glimpse of the dew-laden spider-line as the sun rose to gleam on it. The longest line of a web was from the top of a 20-foot flagpole to a three-foot bush that was 10 feet from the pole. I wonder how they obtain the long distances of their webs.

Garden spider's web

Host garden spider

Web in pine shrub, the black hole

<u>Lawn</u>: Grass seed I planted a week ago is up. Lawn moths rise as I cut the grass. Fertilize the lawn with pre-emergent herbicide corn gluten.

<u>Perennials</u>: Goldenrod is in full gold and covered with bumblebees whose pockets are full of nectar. Goldenrod, orange trumpet vine, and spider plants bloomed the last week of August at the beginning of the decade. Now they bloom the third to fourth week of September. Fall sedum has pink blossoms, it too has bumblebees and honeybees all over them. Cut back the iris.

TRENDS DUE TO CLIMATE CHANGE

Goldenrod

Wax begonias are full red, purple aster is out full, globe thistle has new blue flowers, red phlox is still out. Red hibiscus, and pink and white phlox that were deadheaded two weeks ago are reblooming. A few black-eyed Susan are still out but starting to wane. Grape hyacinth lush greens are up 6- to 8-inches. Dahlia are still coming out strong. Except for some deletions and additions, the cycle of the perennials is still the same. Trumpets appeared for two days.

Early summer I planted morning glory seeds in a pot next to the silver maple. The vines grew up the tree and a few yellow-trimmed purple glories opened. Then one morning the third week of September as it was just light, I saw a larger than a saucer white trumpet disc on the maple tree. It was a pure white glory. It closed shortly after sunrise. Now I know why they are called "morning" glory.

White morning glory

The blue ageratum and purple aster come to full bloom and the fall sedum turns maroon during the last week. The black-eyed Susan waned in the first week in 2005 and now wanes into October.

SEPTEMBER

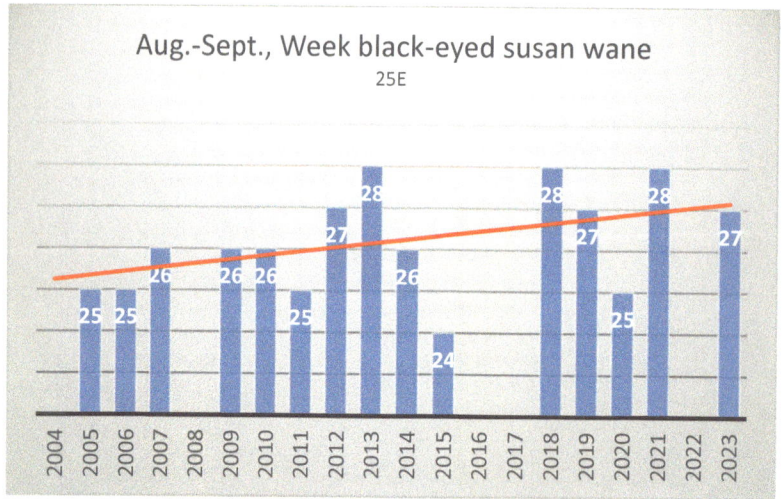

Weeks black-eyed Susan wanes

Fall sedum in pink phase

Pond: Most plants in the pond have yellow-brown leaves. If the temperature goes to 70F the frogs will be out, but not croaking.

Frog sunning on a September day

Shrubs: Hydrangea looks good. Verbena on the hill has blue-pink berries. Winterberry has red berries, ready for a bird attack. In 2016 with the heat and drought there no berries. Trim the holly bushes.

Winterberry

Trees: It took me a while to realize that there is an annual dropping of the needles from the pine trees on the hill. The pine trees dropped their needles the last week of September in the beginning of the decade, in 2017 they fell in the last week of October, a four-week difference. In 2019 it's back to third week of September.

Needles in the pine are ready to fall

The birch tree is dropping its little brown crosses.

Vegetables: When the lower leaves of tomatoes have black circular spots on them it is from late blight; the plant will die in two days. There are only a few ripe cherry and grape tomatoes left on the vines. The three large tomato plants have lots of green fruit. I will wrap them in newspaper where they will ripen in two weeks. Broccoli planted in the spring only has a few small heads. Depending on the weather the pole beans are either coming on strong or struggling. Beets are small. Red cabbage is ready to pick. Plant young red cabbage plants, they will overwinter to be picked late next spring. Pick the last two cucumbers. The cauliflower plants are forming heads. Yam vines are growing great and have nice white-violet flowers. Brussels sprouts are developing. With the trend of warmer falls, crops are still being picked later in September and maybe into October. Four-legged critters like to eat the tomatoes. From 2007 to 2012 a groundhog would attack the fall vegetable garden. Not complaining, I have not seen one since.

Fourth Week

<u>Animals</u>: Two young squirrel kits and two chipmunks separately are on the chase. After the chase, three squirrels are rolling on the grass together, a fourth comes and they all roll on the grass.

On the chase

Three chipmunks are mindlessly running around playing for quite a while, when bang, a sharpie takes the middle one. It squeals as the hawk flies off, the other two quickly disappear.

<u>Birds</u>: At 7:30 a.m. a "V" of 50 Canada geese go north with six robins right next to them! On the 29th of 2004 a flock of 50 blue jay are going west. I've never seen such a large flock of jays since. A male house finch has conjunctivitis. Again, I observe the male cardinal feeding the fledglings. Individual robins are still gathering to form flocks of 20 to 30 for migration. The gathering will continue to the end of the month. A brown thrasher is here for a day. A dove that is trying to escape a sharpie slams into a window and dies. A sharpie is sitting on the split rail fence when a squirrel tries to get on the fence; the sharpie opens its wings and tail and looks at the approaching squirrel. The squirrel jumps to a nearby tree and the sharpie settles down again. Not only sharpies, red-tailed hawks also go after doves. The red-tailed was perched on the very top of a spruce, surveying for a dinner. It suddenly lifts-off and dives. Next morning, I found a dove alive with a big headache under a window; after an hour it flies off.

Blue jays have a gullet. They can put a whole peanut in the gullet and put another in its beak before flying off. I counted a jay deposit 36 sunflower seeds in its gullet. It is a wonder it could still fly with its head up. A pair of jays are fighting each other for a maiden. Jays are still chasing doves from the feeder.

White-breasted nuthatch tries to land on a blade of the cattail, which promptly bends over. Unbelieving, the nuthatch gives out a kaw-kaw as the blade bends.

Fox sparrow arrives to do a bed and breakfast before proceeding south. It may stay three to five days. Depending on the weather it may not arrive until November.

Fox sparrow

Species of birds that are pairing up for the winter are downy, white-breasted nuthatch, house finch, house sparrow, mourning dove, blue jay, titmouse, chickadee, and sometimes Carolina wren. A woodpecker is knocking 20 times a second giving mating calls from trees, telephone poles, and house siding or anything else that can transmit its call.

There were flocks of 200 grackles, red-winged blackbirds, and starlings that roosted on tops of oak and maples in 2006. What a racket they would make. In 2015 there are no more flocks! In 2017 I see one lonely crow.

Midmorning flocks of Canada geese are up high going south. At 7:30 p.m. a flock of Canada geese fly in front of a full moon; a real fall sight.

<u>Insects</u>: Three types of crickets warm up their repertoire, the conductor lifts the baton and they play most of the night. Another evening I can hear the crickets going he-he-he-he or chip pause chip pause chip or as a continuous din in the background. It is still dark when I go for the morning newspaper and the crickets are still playing. They sing throughout the month. Grasshoppers are also chirping.

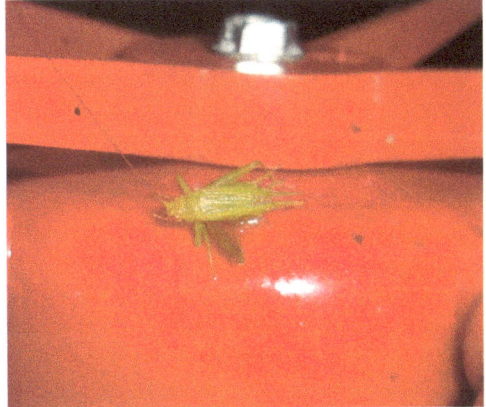

Green cricket, quarter of an inch-long

Brown cricket

I find a walking stick on the patio. I looked twice to realize what I was seeing. This is the only time I ever saw one. They eat oak leaves.

Walking stick on a yardstick

With a warm fourth week, white and yellow sulphurs are fliting around; little lawn moths come up from the lawn as I walk on it. Gnats are still attacking.

When the temperature is above 70F, yellow jackets are active at the hummer feeder. They also like grape jelly and orange sections. If you have an open can of soda or beer sitting outside, be careful when you take a swig, a yellow jacket may be inside. You can get stung in the mouth. That is a big OUCH! They are active the whole month.

Large dark bands on a woolly bear caterpillar indicates a cold winter is coming. In 2018 the woolly's had wide bands. Yes, it was cold. The boxelder bugs sun themselves on the warm front bricks of the house. An eastern tiger swallowtail flits through the backyard. If you see a wheel bug do not kill it, it kills stink bugs. By 2018 there were only a few stink bugs!

Perennials: Perennials that are still out are loosestrife, globe thistle, purple aster, goldenrod, fall sedum which is now dark maroon, pink cushion, and Montauk daisy, all with lots of different types of bees. Annuals of snapdragons, herbs, and marigolds may still be flourishing.

Purple aster and grape hyacinth foliage

Purple loosestrife

TRENDS DUE TO CLIMATE CHANGE

Fall sedum, maroon stage

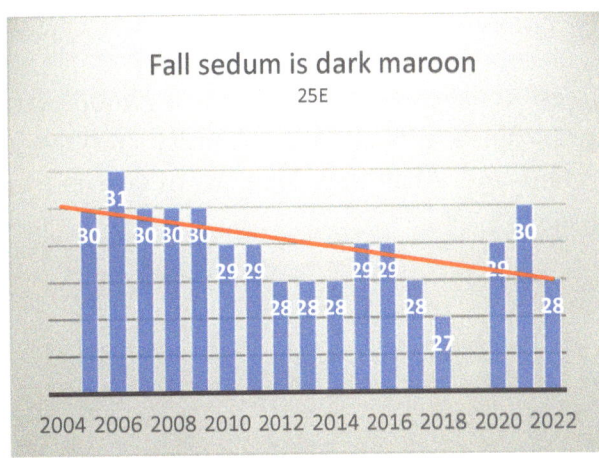
2019 very warm October

Marigolds will rejuvenate with cool weather; the foliage will thicken into darker colors of green. Time to cut back the loosestrife. Seedpods develop on the milkweed. It is the only host for monarchs.

Milkweed with pod

When there is a warm September, the tall goldenrod will still be blooming. The hosta has a second set of violet flowers. Zinnia flowers are bright. With a cool September it is time to cut back the perennials.

Pond: Dragonflies are still hovering over the pond. A monarch had landed on a butterfly weed flower that was 18 inches above the ground alongside of the pond. Suddenly there was a flash as a frog jumped straight up trying to get the monarch, but it was just a little too high. It nonchalantly flew away.

I used to have jewelweed next to the pond. It had an orange flower that would be out now. The juice in the stalk is good to put on a poison ivy rash. It is ironic that in the wild the two plants may be growing next to each other.

The few pickerel weed in the pond still have white flowered spikes. Cut back iris and waterlilies in the pond. With a high temperature of 89F, frogs and tadpoles are still very active.

Shrubs: The holly has white flowers and red berries. Trim the hemlock hedge.

Trees: At the end of this week the dogwood berries will be ripe; the species of raiders of the berries are cardinals, robins, flickers, titmice, starlings, chipmunks, and squirrels.

One-and-a-half-inch-long brown samara from the sugar maple are all over the lawn.

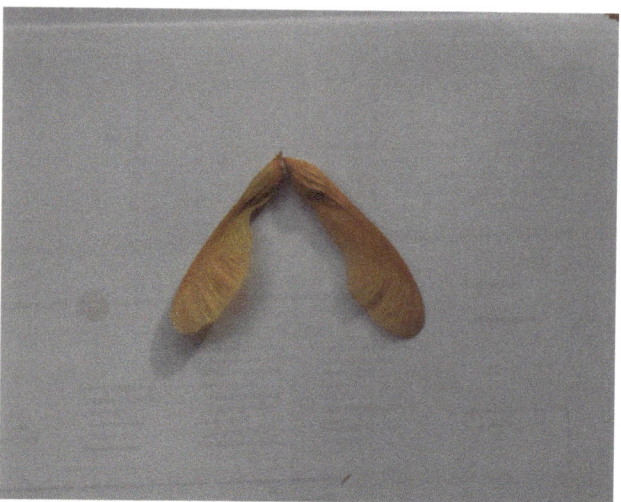

Samara from a maple

As a kid we called the samaras from a maple tree pollywogs. We would split the pollywog in two then pry open the end and put it on our nose. The dictionary says a pollywog is the same as a tadpole. I don't remember seeing any tadpoles in trees, but then again there are tree frogs; we did not have tree frogs where I grew up. Years later I learned that the so-called pollywogs are called samaras. Mt. Cuba Center in the state of Delaware has a four-foot bronze statue of a samara.

My neighbor's maple is always one of the first to start to change color. The change always starts at the top and works its way down. Silver maple leaves are starting to change color. After the walnut tree, this is the first true sign of the fall leaves. The stand of birch trees has yellow leaves.

Vegetables: Take down the pole bean trellis. Pick green and red peppers and Malabar spinach.

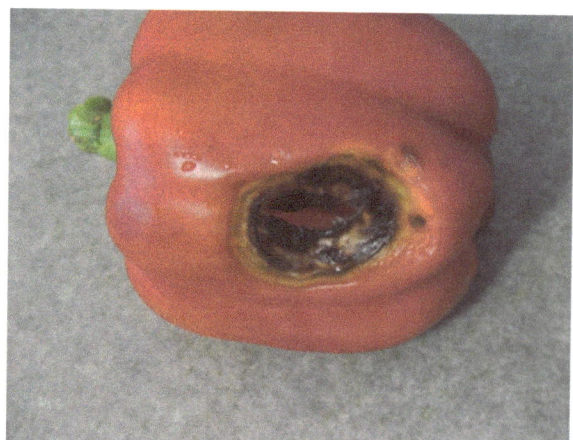

Red pepper   *Red pepper with fungus*

I grow potatoes in pots. Each pot produces three to five pounds of small to medium potatoes. Some years harvested them in September, other years I may wait until October.

Potatoes from the pot

Review of Data

Fauna

<u>Animals</u>: Through the decade, I have observed 11 species of animals in September, including the bat. Most were recorded in 2005-2007. The chipmunks and squirrels were consistent. Bats, groundhog, mouse, raccoon, skunk, and vole were no-show until 2014 when I spotted one bat, a skunk, and a vole. 2024, raccoon and skunk still show in September. However, we have gained red foxes and still have them in 2024.

Table of the Gang

September, Table of Gang																
Species/Quan.	04	05	06	07	08	09	10	11	12	13	14	15	16	17	18	19
Blue jay R, G	50*/5	9	2	5	1	2	3	2	1	4	1	5	6	2	4	2
Cardinal R, G	4	3	4	3	9	3	3	2	7	3	9	6	5	4	6	2
Carolina wren R, G	-	3	1	2	1	1	2	1	3	3	2	1	1	1	1	2
Catbird G	-	2	2	1	1	2	1	1	2	2	1	1	1	1	1	2
Chickadee R, G	5	5	4	4	4	1	3	2	1	4	3	3	6	3	1	2
Chipper sparrow G	-	1	1	1	-	1	-	-	1	-	-	-	-	-	-	-
Downy wood R, G	4	2	2	2	3	3	1	-	1	-	3	1	2	2	1	1
Flicker	1	1	1	1	2	1	-	-	1	-	1	1	-	-	1	-
Goldfinch R, G	-	4	2	4	3	5	3	-	2	1	4	1	2	1	4	2
House finch R, G	10	8	12	4	4	5	3	1	6	5	6	8	6	9	4	4
House sparrow R, G	4	5	4	12	4	14	2	-	4	2	6	4	10	2	10	5
Hummer G	1	1	1	1	1	1	1	2	1	1	2	1	1	2	1	-
Mourning dove R, G	6	5	8	3	9	5	1	5	5	7	7	5	10	1	8	4
Red-bellied wood R	-	-	1	1	-	1	-	-	1	1	1	1	1	1	1	-
Red nuthatch	-	1	1	1	1	1	-	-	2	1	1	-	1	1	1	-
Song sparrow R, G	-	5	2	1	2	2	1	-	1	-	-	-	-	2	-	2
Titmouse R, G	-	3	4	2	3	2	2	1	1	3	2	1	2	-	2	2
White nuthatch R, G	3	4	2	2	2	2	2	1	1	2	2	1	2	-		
Total species 18	10	17	18	18	16	18	14	10	18	14	16	15	16	15	16	12
Population	38	62	54	50	50	52	26	18	41	39	51	40	56	30	37	29

*large flock of blue jays

Birds: The number species in the gang went from 17 for 2005 to 10 in 2011, up to 18 in 2012, back to 15 in 2017. Not only the number of species changed, but also the population. In 2005 it was 62, dropped to 18 in 2011, back to 56 by 2016, and dropped to 29 in 2019. The trend is a constant 16.

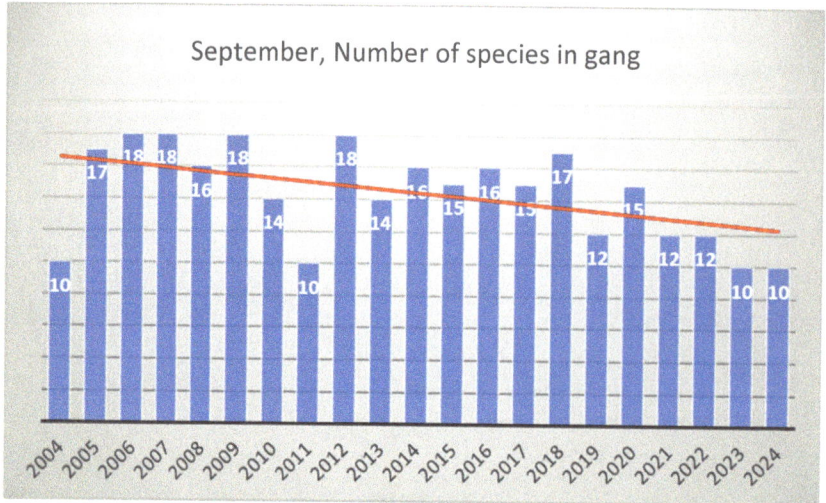

September, number of species in gang

The quantity of gang of some species stays the same from year to year, but others like the house finch, house sparrow, and mourning dove go from 12 to two and back to 10. Some years the quantity is down due to heat, not enough insects or it was very wet that plants rotted; 2016 was hot and dry yet the quantity of many species went up!

The gang population for September has dropped significantly from 2005 to 2019. The dip is sharp for 2010-2011.

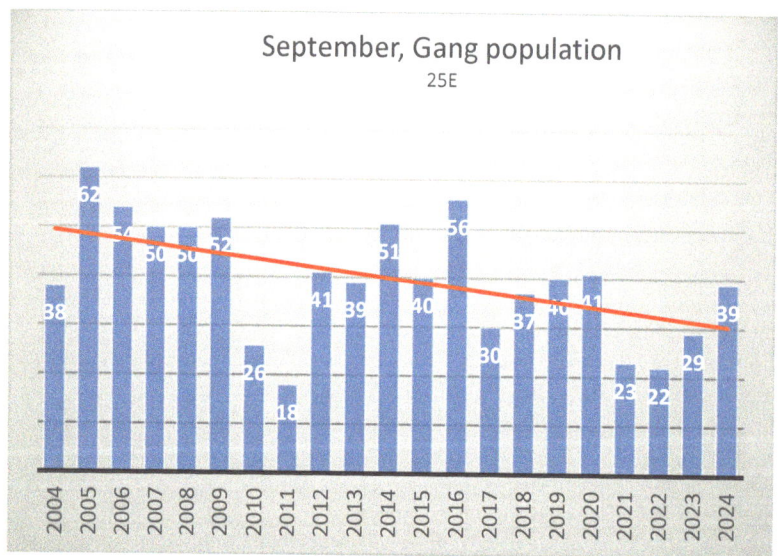

September, gang population

Overall Quantity of All Species: The number of accumulated species in September that at one time or another have passed through, flew over the circle or are residents for the 15-year period is 61. The number of species in September during a year varies from 11 to 37. Like other months, many of the species have only been spotted once or twice. The general trend has been decreasing. From 2005 to 2011 the count went from 37 to a low of 12, rising to 30 in 2012, and in 2016 dropped again to 19. Main losses have been ducks, warblers, vireos, swallows, ovenbirds and flocks of Canada geese, crows, blackbirds, and starlings.

Migration: How many species migrated through for each week of the month? The first week were 11, second 17, third 20, and the fourth 39. That is quite impressive, until I looked closer by the year they migrated through. Many only passed through once or twice in the 15-year span, not to be seen again. Looking closer, I find that on the first week, one specie may pass, second two, third two, and by the fourth week there are four species. Far cry from 39. The main thing is the data shows how the number of migrating species picks up from week one to week four. Between 2008 to 2015, 19 species are no longer seen.

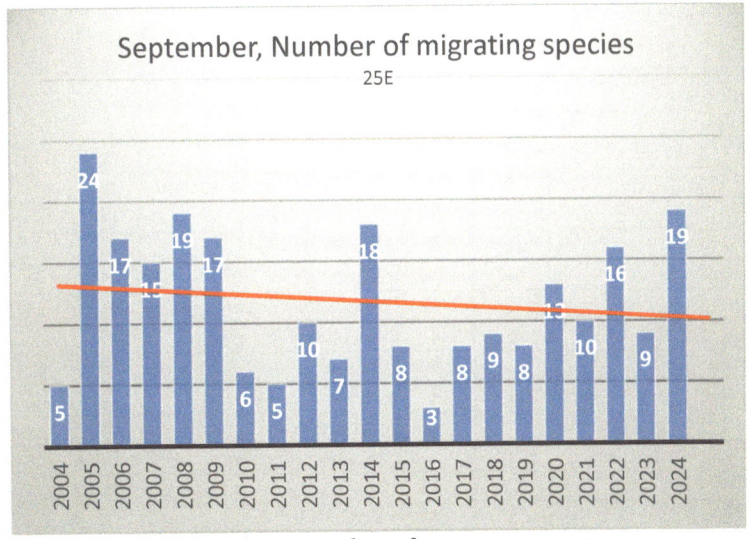

September, number of migrating species

The migration is not only passing through from the north, but includes those that leave my area, like the catbird, hummingbird, and some goldfinch. The 2010 dip shows this, and another dip in 2016. The fourth week is the one to be diligent in observation. On the 11th in 2017 a new species was the nighthawk from 5:00 to 6:00 p.m.

Ten of the species in the gang are still having broods in September. The blue jay, Carolina wren, chickadee, downy, goldfinch, house sparrow, house finch, and titmouse bring one fledgling in the first week, song sparrows bring two fledglings the second week, the Carolina wren and robins bring one on the fourth week. But the cardinals are the most productive. They bring fledglings every week of the month; mostly just one, sometimes two. However, one week they brought four fledglings!

SEPTEMBER

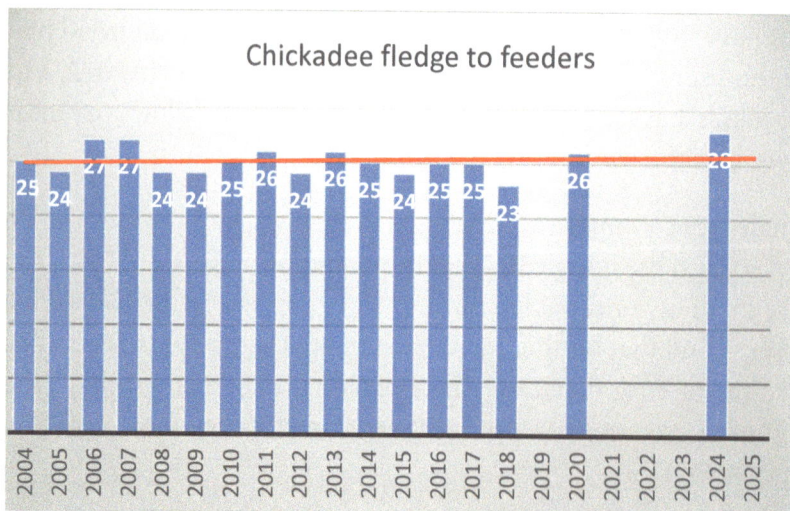

Trend shows chickadee fledgling earlier

September, Migrating Species from 2004-2018 Yr. = last year observed in the 2000's									
Species	Yr.	Week of September observed			Species	Yr.	Week of September observed		
Bat	14		2	3	4	Yellow-bellied sapsucker	12				4
Barn swallow	08		2		4	Eastern phoebe	17	1			4
Blackbird flocks	17	1			4	Starling flocks	12				
Broad-winged hawk	18				4	Vireo	14			3	4
Brown thrush	14				4	Magnolia warbler	18	1	2	3	4
Cooper's hawk	14				4	Wood thrush	15				4
Hairy woodpecker	15				4	Flicker	18	1		3	4
Canada geese 20-30	13			3	4	Mallard ducks	09		2	3	4
Grackle flocks 200	12	1			4	Catbird leave	18				4
Nighthawk	17		2			Hummingbird leave	18	1	2	3	4
Ovenbird	08	1			4	Snow geese flock	08			3	4
Cowbird flock	18	1		3	4	Red-headed wood	12				4
Sharp-shinned hawk	18	1			4	Carolina wren	18				4
Winter wren	05				4	Young robin	14				4
House wren if it has been warm	09			3	4	Yellow warbler	14		2	3	4
Pewee	14				4	Mockingbird	14				4
Screech owl	07				4	Yellow-throated warbler	17	1			
Fox sparrow	12				4	Red-bellied woodpecker	18	1			4
Chipper sparrow	09		2		4	Rose-breasted grosbeak	17	1	2		4
Chimney swift	18	1				Red-breasted nuthatch	18		2	3	4
Black-throated warbler	14		2		4	Tree swallow	18	1		3	
Connecticut warbler	14		2			Purple finch	18	1			
Red-winged blackbird	14			3		Savannah	08			3	
						Total species		15	12	15	37

Blue jay and cardinal molt the first week. The process may continue through the fourth week.

The black-throated warbler's last arrival was the second week of 2017, the Connecticut warbler arrived the second week and its last arrival was in 2014. The magnolia warbler was the second or third week with its last arrival the fourth week of 2017. The yellow was third week in 2008, earlier to the second week in 2014, 2015, 2017, and 2018, and the yellow-throated warbler was the first week in 2017.

The winter wren arrived the first or second week.

Goldfinch change color any week during the month, but the third and fourth are more prevalent. Male goldfinch seemed to be changing color earlier.

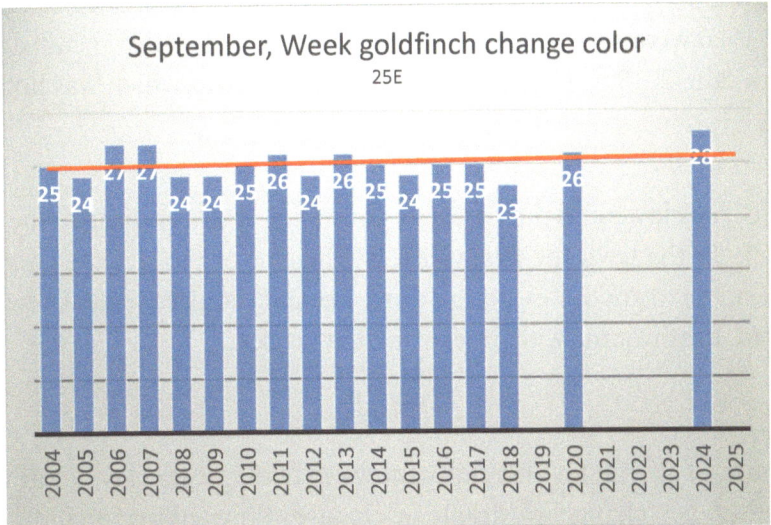

Young male house finch gets a red breast the second week.

Insects: Unless I am specifically looking for or by happenstance or obvious, it is hard to be aware when many insects are present. Over the decade 16 species have been observed during September. I'm not complaining but no yellow jackets have been observed in September since 2015! In 2019 they were back. Honeybees have been a few to none. Boxelder bugs came out if the days were warm. Gnat, katydid, and locust make their presence known when there is a hot day. Stink bugs seemed to be everywhere until 2009, then they dwindled down; in 2016 I only spot a few! The only reason I observed a walking stick is that I was sitting on a bench and saw the stick sitting on a stick. Crickets sing through the month into October. Spiders are making their nests at the beginning of the month. After 2011 there has been a reduction of all insects.

In 2014 we are made aware of the spotted lanternfly. They kill fruit and lumber trees. They tried to quarantine them to the known locations. That did not last, and they have spread to many other counties in Pennsylvania and in 2020 they are still spreading.

Butterflies: From 2005 to 2010 eight species of butterflies flit through the backyard including the white sulphur. Depending on the reference, some call the white sulphur a cabbage moth. From my chewed cabbage-type plants, I lean to calling it a moth. They are still present at the end of the month. Then beginning in 2011, the number of species was down to three. The admiral, blue streak, and silver-spotted skippers were gone. Black swallowtails are seen the first week.

Pond: Life observed around the pond are the cattail, dragonflies, frogs, tadpoles and arrowhead plants. Used to watch the dragonflies hover over the pond, that stopped in 2010. One showed in 2014, then one or two until 2018 when there were many. Some years, the same thing happened with frogs and tadpoles?

To summarize the fauna, there has been a decrease in all the fauna from 2011 to 2014, then a slight increase. The weather (average temperature and year-to-date rainfall) from 2005 through 2009 was close to normal. Then in 2010 we had a bad drought year followed by an excessive wet year; both situations fouled up the wildlife. In 2012 and 2013 the wildlife started to recover, which showed in 2014, but 2015 and 2016 were again drought years. At the end of 2016 we ended with a deficient of minus nine inches of rain, 2018 was hot and wet, followed by 2019 which was hot and wet.

Flora

Lawn: Prominence of red clover and purslane has gone from the beginning to the end of the month. Brown spots show up in the lawn the first week. By the end of month, the cooler weather makes the lawn look presentable to good. Reseed the third week; it requires seven to ten days for the seed to come up. If humid, mushrooms will appear the first week.

Overall

The overall picture of the fauna and flora shows that many species are affected in the same manner by the weather. Is it change in climate or change in my environment?

The dramatic year was 2011, being the wettest on record. The year-to-date rain for September was 55 inches compared to the norm of 35 inches. People continuously had umbrellas up and raincoats on, not concerned on what was happening with nature. A lot did not happen. There were fewer species of birds, it had the lowest population of birds, perennials did not bloom, there no frogs or dragonflies, honeybees, bumblebees, squirrels, and raccoons. No berries formed in the dogwood trees, and there were few tomatoes.

Miraculously, in 2012, silently, unnoticed by most, the flora and fauna started to recover, some back to where they were but many are limping along. We had a very hot, dry year in 2016, which a lot of species needed. Things grew and were lush, but 2017 was another disastrous year and again the flora and fauna did not do well or even knew how to react. People asked why don't the leaves fall or change color? Where arc the birds? The population of many bird species has and still is decreasing.

Then 2018 and 2019 were almost a repeat of 2017 with 48 inches of year to date rainfall and the same complaints. 2024, Second month of drought. Grass is like brown hay.

So, ends September.

Index of October

October Overall	256
Weather	257
First Week	258
Second Week	264
Third Week	269
Fourth Week	271
Falling Leaves and Peak Color	275
Climate Change on the Leaves of Fall	281
Review of Data	282
Weather	282
Fauna	283
Animals	283
Squirrel War	283
Birds	285
Table of the Gang	285
Table of Migrating Species	288
Quantity of Species	289
Fledglings	290
Insects	290
Pond	290
Flora	291
Perennial	291
Pond	291
Trees	291
So, ends October	292

October

Week number for the year of 48 weeks					
Relative Month	Month/Week	First	Second	Third	Fourth
Previous	September	25	26	27	28
Present	**October**	**29**	**30**	**31**	**32**
Following	November	33	34	35	36

Weather

The average high temperature on the 1st is 69F and the low is 47F.

The highest temperature was on the 2nd in 1927 and the 6th in 1947 at 92F.

The average high temperature on the 31st is 59F and the low is 37F.

The lowest temperature was on the 22nd and 23rd in 1940, the 28th in 1936, the 29th in 1940, and the 31st in 1980 at 21F.

Reviewing the high and low temperatures from 1927 to 2018, the decades with the highest percentage (39%) of high temperatures was 1930-1940's; the 2000-2010's had 23 percent. The decades with highest percentage (26%) of lowest temperatures was also 1930-1940's; 2000-2010's had 16 percent.

The amount of daylight is 12 hours.

The full moon is called Hunter's Moon or Harvest Moon.

Overall Summary of October

October is a transition month between summer and real fall. There is not much going on with the garden flora except the picking of the pumpkins and apples and planting mums. Farmers harvest by the light of the moon, hence a "Harvest Moon." The dogwood leaves turn maroon. The main change to autumn leaves of all the trees starts to occur at the end of the month.

A lot is going on in the bird world on migration going south. Winter residents of junco, white-throated sparrows, white-breasted nuthatch, and sharp-shinned hawks arrive. Goldfinch will turn drab green. Larger hawks, flocks of geese, and warblers pass through. The barred owl will be hooting for a mate. On a warm evening crickets can have a symphony.

The first and second weeks can still get hazy, hot, and humid, like 86F with thunderstorms. October is the month that there is a lot of migrating birds going south and returning of birds to take up winter residence. There are now heavy layers of pine needles on the hill. The flocks of migration can be in the hundreds. Perennials and annual flowers are still strong. The weather from April to October will determine when the apples are ready to pick in October; sometimes it is the first week others it is the fourth, but frequently the second week. At the end of the month the leaves of the trees and shrubs start to change color. If it is warm, butterflies still flit around while the flowers and frogs bask in the sun. It can be humid.

Weather

October brings the first frost. Some years it may be the first week, other years the fourth, usually it is the third. Indian summer is after the first frost. The weather of October determines when the apples turn red. A nor'easter may come up with 50 to 60 mph winds. Not to be forgotten, a strong hurricane can still occur.

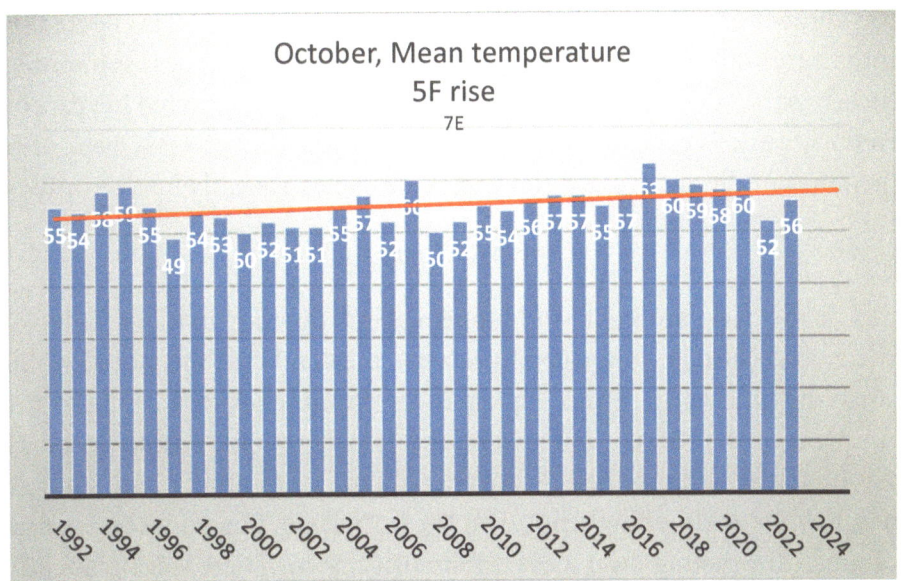

October, temperature

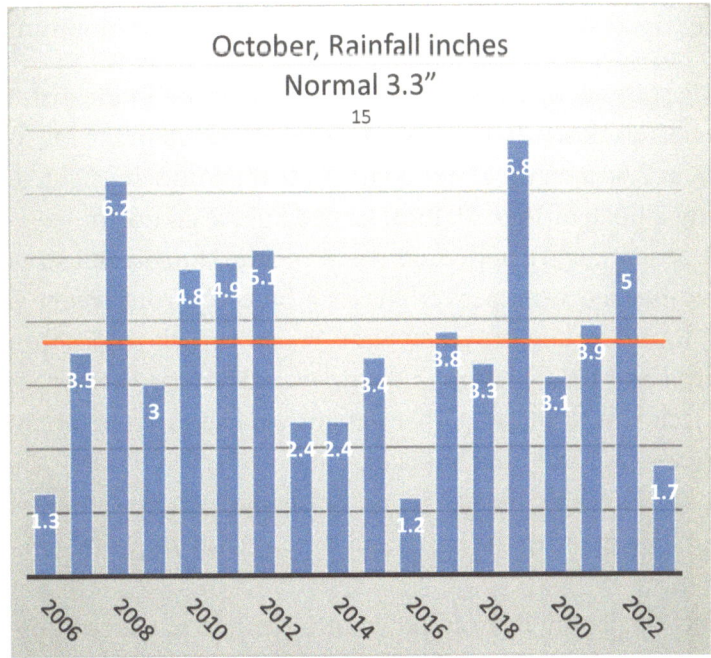

October, rainfall

Effect of Weather on Fall Migration in October

Reviewing the annual arrival of winter residents it becomes evident that the in-route weather affects the arrival date. Some years they all arrived two to three weeks later than normal, others one or two weeks early. With a wet year, especially in August-September going into October, the migrating winter residents arrive two to four weeks late (2005, 2011, and 2018). Likewise, when September-October is hot and dry they arrive three weeks earlier.

When it is very wet the number of gang species drops as well as the overall population probably due to the lack of seeds or rotting vegetation. The reverse is true when September and October are dry and hot. The number of passing through migratory species observed is affected by the extreme wet or dry months but is not consistent. In years with locally wet Octobers sometimes the number increased and other years it decreased. The same non-consistency occurred in years of hot and dry. Similarly, when the ground is exceptionally wet, the plant growth cycle is a week or two later than normal and when the conditions are dry and warm the cycles are early by one or two weeks.

First Week

Animals: Turned on the backyard lights and found a skunk digging up the lawn looking for grubs. Not to be outdone, a raccoon has managed to remove the can lid of the stored whole peanuts. In the morning shells are all over the patio. Squirrels are running around trying to find soft soil to bury the walnuts. Of course, there is a soft spot of ground where I just buried my new crocus bulbs. Sometimes they husk the walnuts before burying them. Chipmunks and squirrels are eating dogwood berries. Rabbits are hopping around on the front lawn. I had not seen a possum for years, then in 2014 I found three young dead joeys on the front lawn. There was no apparent outwardly cause of death!

Annuals: Impatiens, marigolds, zinnias, and red dahlias may still be blooming.

Birds: At 6:00 a.m., 41F, stars are out. The moon is at a lower crest in the east. Or, at 7:00 a.m. there can be a morning fog. I listen closely to hear the local geese take off in the fog for their exercise. That is some navigation! Or, at 7:00 a.m. the sun is coming up; it is still semi-light. The gang is here.

Early in the morning a flock of 10 to 20 local Canada geese go north.

Just after sunrise I observe what newcomers will be here for breakfast. Ah, there is a yellow-throated warbler; a red-headed woodpecker chases a blue jay from feeder (last one observed in October was 2008). A loose flock of crows go west. Wood thrush passes through, it is scratching under the rhododendron and then goes up in the dogwoods eating the red berries. It also likes the red berry cone of the Jack-in-the-pulpit. Other passing species are rose-breasted grosbeak couple, chipper sparrow, hairy woodpecker, and a flicker.

A flock of starlings was attacking the berries on a dogwood when a broad-winged hawk flew out of a nearby maple. The starlings spot the hawk and fly from the dogwood to chase the hawk. What a reversal!

A house finch drinks by tipping its head down into the pan of water, brings it up horizontal to swallow; it does this three times. A dove gets water from the pond. It dips it head down and keeps its head down until it is done. A drab colored goldfinch dips down, brings its head horizontal, swallows; a downy does the same thing, but four times. A titmouse dips down to take a drink, then raises its head up at a 45-degree angle before swallowing; it does this three times. Blue jay dips down and puts its head up to swallow.

Goldfinch in drab color

House finch chases a chick off the safflower feeder. Blue jays take eggshells and whole peanuts. Four crows are cawing from the top of the spruces. Hear a noise; all the birds are gone except a downy that it is frozen to the peanut flake tube. I don't see it but suspect that a sharpie has made a hit. Half hour later the gang is back.

A flock of 50 mixed blackbirds go west, a flock of 20 cowbirds land and hog the feeder. Mid-morning to noon flocks of 200 to 1,000 starlings go west. All of them turn together and they do it again. How do they all know to turn at the same time? There is no lag time when they all turn.

Two young cardinals go at each other hissing while trying to get territorial rights.

Looking up real high, large flocks of snow geese are going south.

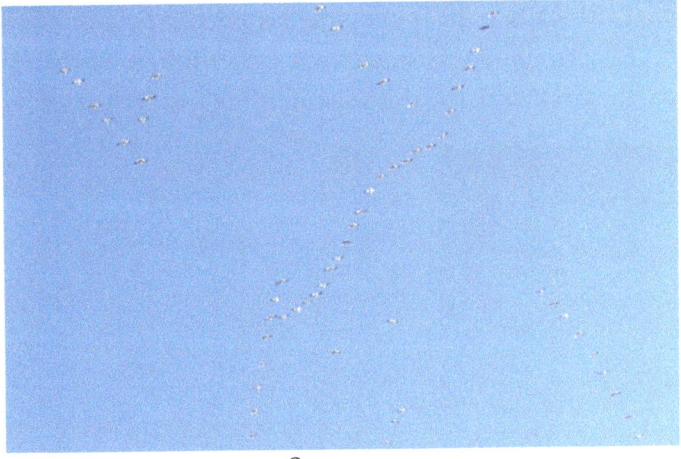
Snow geese

There is a song sparrow and a Carolina wren in the afternoon. Catbirds have left to go south. A red-breasted nuthatch chases another from the peanut flake tube. A red-bellied woodpecker is taking insects from behind the old shoe birdhouse nailed on the maple tree. It is going kay-kay-

kay while it is having this feast. Those shoes were nailed to the tree six years ago but no takers to make a nest. My shoes must smell too bad. Almost dark, a pair of cardinals are at the feeder. A titmouse and chickadee are getting a drink, when suddenly a sharpie swoops down, misses the birds, hits the window and then is gone; so is everyone else. Not very frequent, almost rare, a flock of cedar waxwing momentarily stop at the winterberry for a feast. At 6:00 p.m. a young Cooper's hawk lands on the feeder.

Cooper's hawk

Adult hummingbirds migrate as early as August but that does not mean it is time to take down the nectar feeders. Young hummingbirds may not migrate until the middle of October. They have a built-in navigation system and do not need their parent's guidance but need the feeder nectar for energy.

A female house sparrow waits its turn on the perch for split peanuts while a female cardinal finishes, then the female card waits until the female house sparrow finishes. House sparrows are eating peanut butter! An eastern phoebe is scratching under the feeder. Some years the first white-throated sparrow arrives the first week for the season, other years they arrive the second week.

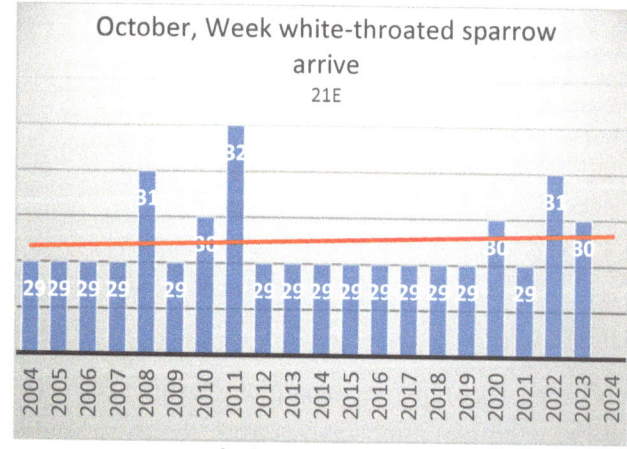
White-throated sparrow in winter drab

A Cooper's hawk perches in a dogwood with a squirrel sitting right next to it! A red-tailed hawk lands on the very pinnacle of the tallest spruce. It is nice and warm and sunny, the turkey vultures, broad-winged, and red-tailed will be soaring.

As evening approaches a big owl flies over a terrified rabbit that is grazing in the middle of the front lawn. The owl did not spot it.

Forty-one bird species were sighted the first week.

The normal life span of a bird is only three to five years. People ask why don't they find dead birds? I have no good explanation. Unfortunately, house finches get conjunctivitis which blinds and deafens the birds which leads to its demise. I have found dead finches next to my birdbath that have died from this disease or maybe from starvation.

<u>Insects</u>: An assassin bug is looking to prey on a stink bug. During a warm afternoon, boxelder and stink bugs come out to sun. On a warm evening, the crickets can still be heard. With a warm October, it gets difficult to do gardening because the gnats are out in full force.

Boxelder bug

Bumblebee at hummer feeder

White and yellow sulphurs flit among the perennials. A thousand small ants have emerged from their nest along the driveway. An adult praying mantis is solitude on a perennial. Bumblebees and yellow jackets will also be taking advantage of the warm weather to gather nectar.

Fruit: The apples that I removed from the bag a week ago are turning from light green to red. The Empire apple first produced in 1995 in September. In 2007 on the 12th I picked 30 pounds, in 2015 40 pounds. In a hot, drought year like 2016, I hardly picked any. In 2018 the 26-year-old tree died.

I took time to bake an apple pie. I used eight three-inch diameter Winesap apples; it took one-and-a-half hours to make it, but it was worth every bite. Still picking some raspberries.

Empire apple

Protective bag

TRENDS DUE TO CLIMATE CHANGE

Winesap apple

<u>Lawn</u>: Grass seed is up.

<u>Perennials</u>: The purple aster is out full. Perennials that may still be blooming are red hibiscus, white phlox, purple aster, black-eyed Susan, Stella de Oro, butterfly weed, red-flowering fern, pink cushion flower, pink phlox, goldenrod, and a few roses. Fall sedum is dark maroon.

<u>Pond</u>: Cattail stalks have their punks. A brown dragonfly hovers over the pond. Several silent frogs are basking along the edge of the pond, many tadpoles are swimming in the pond.

<u>Shrubs</u>: I planted some winterberry bushes but over three years they produced no berries. Then I was advised that I needed a male bush. Okay! Next year all the females had bright red berries on the first week. Amazing! What will they think of next? Now the berries are red the second week. Like the apple trees, there were few to no berries on the bushes in 2016. The burning bushes have turned scarlet red. At 65F, the boxwood gives off a strong cat scat fragrance.

Fall roses

<u>Trees</u>: The hawthorn tree develops red berries the second week.

Lots of robins and starlings are migrating; they land in the dogwoods to eat the berries. Who eats the berries by the week? First week is robins, chipmunks, squirrels, and starlings. The second week is robins. The third week is robins and squirrels. The fourth week is chipmunks and squirrels. Not only are squirrels eating the berries, now crows have moved in to feast!

<u>Vegetables</u>: There may still be lots of green tomatoes and peppers. Wrap the green tomatoes separately with newspaper then put them in a brown bag. They will be red ripe in two weeks. If you forget to inspect them after that they will rot and stink!

Pull beets and pick green beans. We had such a mild fall and winter in 2016 that beets were not pulled until February 10, 2017!

<u>Weather</u>: Nor'easter moves in at night with 50 to 60 mph winds.

Second Week

<u>Animals</u>: Squirrels are burying walnuts. Squirrels eat the seeds in the silver maple's pollywogs. At 9:00 p.m. a skunk may be under the feeder, meanwhile a rabbit munches on the recovering lawn.

Rabbit enjoying the lawn

<u>Annuals</u>: Still blooming are marigold, impatiens, and petunia. Put elephant ear plant and amaryllis in garage.

<u>Birds</u>: On the 8th in 2004, there was no rain for nine days and the ground was hard. A year later to the day there was a downpour of many inches of hard rain! When it stopped and the sun shone, the broad-winged hawks were soaring.

There are five to nine species of birds at the feeders every day; most species are in couples. Know a sharpie is around because I find a pile of feathers on the lawn. A pewee and ovenbird are here. A white-faced house finch has been around; the unusual markings makes it easy to track. A Cooper's hawk tries to snatch a house finch, misses. There are three jays at the feeder, two are going at each other.

An example of climate or environmental change. In 2009, noon, 300 blackbirds were on the ground and in the trees. Every time I clapped my hands 50 more flew from somewhere. There were flocks of 100 starlings, 20 to 50 crows; by 2010 no more large flocks were observed! None.

Flock of Canada geese go west. A flock of 100 snow geese go south. A flock of 10 robins has gathered on the front lawn; two young ones are having a fight. Jays are screaming at two sharpies sitting in a dogwood. The juncos arrive.

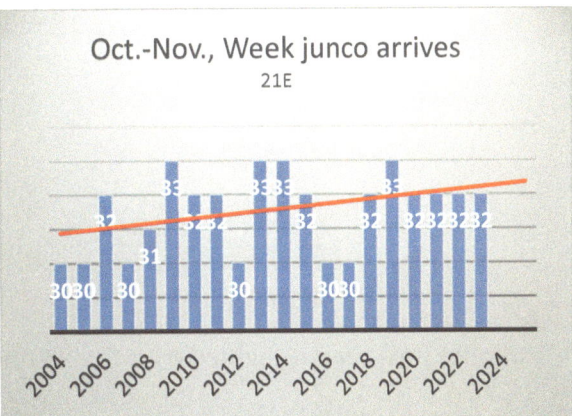

Junco

Eight to 10 house sparrows are active at the houses. Three house sparrows and a Carolina wren are fighting to get into the same house, again the sparrows win.

House sparrow guarding her house *Blue jays*

A mockingbird is singing its repertoire. Red and white nuthatch and a red-bellied woodpecker are at the split peanuts for breakfast. Three jays are looking for whole peanuts.

The pecking order is in place. A bull dove stands its ground against a jay. Song sparrow prevents a white-throated sparrow from landing on the feeder. Titmouse chases a chickadee from split peanut globe.

Some warm years the catbird, house wren, and goldfinch may still be here. With the climate warming, some goldfinch and robins have been overwintering. Magnolia, pine, and a yellow warbler pass through the third to fourth weeks. Red-bellied woodpecker and flicker go for suet. The flicker may stay a week or so before continuing south.

Flicker

Ten more bird species sighted over the second week.

Fruit: There may still be some raspberries to pick.

Insects: Gnats are bad. Bumblebees are at the perennials that may have a few flowers. Yellow jackets go to the open bruises of fallen apples. In the evening, 6:30 p.m., I find a big, black house cricket on the patio. At 8:00 p.m. the little green crickets are calling.

Lawn: If the lawn had turned brown in September, now is the time to rake up the dead grass. If the grass is in good condition, put fertilizer on the lawn. Reseed bare spots but do not fertilize newly seeded areas. Put lime on the lawn.

The soil gets damp from the night moisture; consequently, mushrooms that developed in the lawn in June are again forming in in the same spot. Once the spores are there, they live there for a long time. They also form on decaying tree stumps.

Fungi chocolate lenzites

Perennials: Still in bloom are Montauk daisy and dark maroon fall sedum. Plant windflower bulbs. Tall ferns turn brown.

Pond: When there is a warm 70 to 78F day, a green dragonfly may hover over the pond. Frogs sit on the rocks edging the pond. Many tadpoles can be seen skimming across the bottom. Put liquid-granular winterizer in the water.

Protection from the great blue

Trees: Take wheelbarrows of compost (black gold to a gardener) from the bin on the hill. As I dig out the compost that has been generating for two years, feeder roots from the adjacent spruce tree were uncovered. Even though I had put down a tarp to prevent roots from entering the pile, they managed to go around and over the tarp to get the nutrients of the pile. This is a good demonstration of what will happen when mulch is mounded against tree trunks (volcano mulching) as done by many landscapers. The mulch should be kept six inches from the trunk. Mulch against the trunk is also inviting small critters to gnaw the bark in the winter resulting in girdling the trunk.

Compost pile next to spruce trees

Exposed feeder root

Weather dependent, some years the birch tree leaves are just turning yellow; other years it may have already dropped all its leaves. Top half of neighbor's maple is red, lower is still green, no drop yet. Silver maple by steps is 20 percent red. Leaves of back dogwoods are dark maroon.

<u>Vegetables</u>: With a warm fall, cherry tomatoes are still producing and ripening to a bright red. Pick giant Swiss chard, Yolo Wonder bell pepper is very sweet. Depending on the weather, may want to wait until October to dig up carrots, white potatoes, and yams. Asparagus ferns turn yellow and have red berries. Brussels sprouts are forming. Pick the last of the Malabar spinach.

Yam

Third Week

In the third week it can still get above 55F which is warm enough for the honeybees to come out and get a few remains of nectar. Then again it can get cold enough that when I open the door in the morning there is frost on the roof and lawn. The first frost can occur during the end of the third through the fourth week. Some years we wake up to see frost the first week.

Possible frost prediction from Penn State is October 18th 25 percent, October 26th 50 percent, November 3rd 75 percent, and November 10th 90 percent.

Also, might open the front door to the smell of a night visit from a skunk. Crickets are singing, frogs are out but not croaking. In the morning there can be six to 11 species at the feeders.

<u>Animals</u>: Find a headless adult rabbit on the hill, probably hawk kill. A vole is running around the rock wall of the pond. The long inhabiting (20 years) vole descendant scurries across the step. A skunk was under the feeder at 10:30 p.m. digging up the lawn for grubs.

<u>Annuals</u>: This is the week to put basil, rosemary, and geranium plants into the garage.

<u>Birds</u>: Flocks of robins are flying south. Jays are taking cracked eggshells and whole peanuts. Due to the 3H's (hazy-hot-humid), bird seed in the feeders can go moldy. Cardinals bring their fledglings to the feeders. Downy are paired up. Turkey vultures are soaring. Six turkey vultures land in at the top of an evergreen. Unusual, 10 crows chase a low flying turkey vulture. Towhee, mockingbird, and yellow warbler are here as a bed and breakfast. A sharpie gets a dove. Goldfinches land on the globe thistle. At 5:00 p.m. visiting starlings gargle, many flocks of starling's fly over. A pair of purple finches is at the feeder. Downy has opened the hole in the cuckoo on the front porch.

Downy's work

Male purple finch

Both a sharpie and Cooper's hawk survey the feeders. House finch have conjunctivitis. A white-crowned sparrow comes through for a short visit, 30 local geese fly north, a flicker is calling, a Carolina wren sings all night! A red-tailed hawk soars at 1,000 feet, three crows go west.

A titmouse lands next to a white-throated sparrow on the feeder. The sparrow did not like the proximity. It raises a wing and goes toward the titmouse which promptly takes off.

Three more new bird species in the third week.

Fruit: Yellow jackets are still eating fallen apples. I almost stepped on them while cutting the front lawn.

Lawn: Grass is greening up; still not too late to put down grass seed. Mushrooms may still develop. Fertilize and put down lime.

Perennials: There may still be blooming of pink phlox, globe thistle, hardy aster, blanket flower, Montauk daisy, wax begonia, dark maroon fall sedum, yellow iris, and a few Stella de Oro. After a heavy frost is when I cut back the perennial flowers and hosta leaves. Plant crocus bulbs; place chicken wire or a small oven rack over them so the squirrels will not dig them up. Put a light coating of bone meal on the iris and clematis to strengthen the stems.

Honeybee on globe thistle

Pond: Put the net over the pond; learned not to wear anything with buttons, inevitably the netting will find them. Take the pumps out of the pond.

Trees: This week is when the leaves may start to change. The color and drop time all depend on the weather, not just in October but throughout the year.

Fourth Week

This week is when many trees have peak leaf color and others drop all their leaves.

Animals: Squirrel population has increased from three to five. If it has cooled down, I will see no chipmunks nor hear crickets.

Annuals: Pull out the impatiens and marigolds. Dahlias are still blooming.

Birds: A surprise. At 7:00 a.m. I hear a great horned owl in the poplar tree. A house sparrow takes a bath by dipping its feathers into the water five times to get them wet. The great blue heron comes in to dine on a big, fat, old white fish I had for 14 years.

No, it was not pregnant, it was just fat

A magnolia warbler is migrating. A junco lands on the lawn, fluffs its wings wide and does semi-rolls in the grass similar to taking a sand bath. A Carolina wren sings to charm a mate. Song sparrow chases a white-throated sparrow from the feeder. A cardinal helicopters to jab at the peanut butter log. In 2005, 2006, and 2007 a ruby-crowned kinglet arrived on the 25th. I have not observed any since.

This week cardinals start to bring their fledglings to the feeder. This goes on through December. There are four male bright red cardinals plus four teenagers.

Young cardinal

On the 26th for five years a pair of red-breasted nuthatch arrived for the winter; that ended in 2008. A pair did not return until 2016-2017 when they overwintered. They overwintered again 2018-2019; none 2019-2020. This may be due to climate change. A pair of white-breasted nuthatch also overwintered in 2018-2019 and 2019-2020.

A Cooper lands in a dogwood; a squirrel in a bush below wags its tail and goes up the tree to chase the hawk who raises its wing signaling the squirrel to back-off, which it did.

Unusual, a house finch eating an apple core. A song and house sparrow are having an argument, the song wins. A lone pigeon flies over! Northern waterthrush arrives and usually stays for a week. A female house finch has conjunctivitis.

A mystery. Sitting at the kitchen table at 5:00 p.m. Suddenly I hear a loud crash and the sound of a lot of falling glass behind me. My first thought was what let loose, what didn't I put up securely? I got up and looked through all the rooms downstairs, then upstairs, the bathrooms, the closets, the cellar; nothing. I did this three times. Could the noise be my imagination? It is still light, so I go out the front door to look at all the windows, still nothing. I decide to go in the backdoor. As I start to open the storm door, I hear a noise between the storm and backdoor. Then something catches my eye and I look at the upper pane of the storm door. It is shattered. Long sharp pieces are pointed toward the center like a glass spider web. I slowly open the door and a mourning dove flies out. It hit the windowpane with such force that it smashed the pane. A hawk was chasing it. The remarkable thing is that there was no blood anywhere, but there was lots of glass.

Dove got a big headache but survived

Six crows fly toward a spruce tree to chase (called mobbing) a broad-winged hawk from the area.

In Bethlehem, Pennsylvania, there is a huge rookery of over a couple of thousand crows. In the morning they go to fields north of the Lehigh Valley Airport to get seed. These flocks can cause a big concern for planes landing or taking off. Then an hour before sunset they fly back to the outskirts of downtown proper where they defecate on sidewalks, cars, and whoever passes beneath, all the while cawing to each other. It takes an hour before they all get back and another couple of hours to settle. They have been roosting in Bethlehem for over 40 years. The town has tried in vain to get rid of them. When inflight, the crows steadily flap their wings at a medium pace. Now and then they will glide, but not often. Their path always seems straight to go from A to B. One or two of them caw in flight like giving an order.

A Carolina wren chases a goldfinch from the thistle tube. A white-breasted nuthatch, three chickadees, and a squirrel are at the feeder at the same time.

From the 28th through the 31st, 11 species including a red-bellied woodpecker are at the feeders. A titmouse chase another off the feeder. Flocks of 30 to 200 starlings go over, some stop to eat dogwood berries. Four wood thrushes are under the holly bush. A Savannah sparrow arrives.

Sometimes they go north, south, and others west! A murder of 100 crows goes west. A flock of 20 to 100 robins and 30 starlings go south. Observed a flock of Canada geese a couple hundred feet above a flock of crows going 90 degrees from the crow's direction. Thirty to 50 local geese go north for the day.

Two to eight juncos are going for millet. Jays likes Raisin Bran cereal. Robins and jays are at the winterberry bush on the hill. In November the winterberries are bright red, robin flocks are coming for a feast. Other years the feast takes place in October.

Robin going for winterberries

A red-tailed hawk and a pair of turkey vultures soar. A flycatcher is here for two days.

Three more new bird species in the fourth week which gives a total of 57 species sighted in October from 2004 to 2019.

Insects: Leave the hummer feeder up in October because some young hummers may still be migrating. However, the feeder will attract yellow jackets. Counted 13 yellow jackets raiding the hummer feeder.

A warm 70F October day brings out the insects. The boxelder bugs like to sun on the siding. Hundreds of ants are on apple cores, also brings out flying ants. Bumblebees are at the globe thistle. At 10:00 p.m. crickets are chirping.

Perennials: Sometimes the Montauk daisy is still blooming, purple asters are done. Put compost on the perennials. The butterfly weed has green pods.

TRENDS DUE TO CLIMATE CHANGE

Dig a long trench for tulip bulbs. Put bone meal down in the trench. Plant a hundred red and white tulip bulbs. Lightly fill the trench and put down more bone meal before completely filling the trench. While I was bending over planting the bulbs, a chickadee landed on my hat.

Falling Leaves and Peak Color

The following is a representation of what may occur the fourth week with falling leaves. The time of change depends on the weather starting in spring. First to change color is maple A, it may be 95 percent red this week. It is also the first to drop all its leaves. Then maple B turns red at the top and works its color down with a 40 to 70 percent drop this week. The silver maple by the back step goes to full orange or orange-red. The sugar maple may still be green or going to semi-yellow. Dogwoods which are dark maroon may be 40 to 98 percent down. The small birch tree is 50 percent yellow leaves or 40 percent drop, the tall birch is 90 percent drop. The red oak has a few yellow leaves. Poplar is turning yellow. Another silver maple is turning orange or red. The pin oak can have 20 percent yellow or red leaves. The hawthorn has 90 percent yellow leaves.

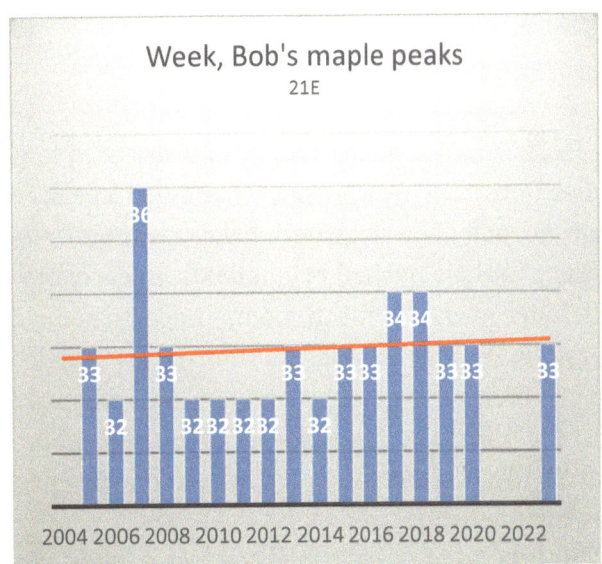

Bob's Silver maple

A review of the data from 1985 to 2019 shows that the week that peak could occur has moved to a later date by two weeks.

Month-Day Peak Color Occurred					
Year	Maple	Poplar	Pin Oak	Dogwood	Serviceberry
1984		Oct 30			
1995	Oct 23	Oct 26	Oct 22	Oct 12	Oct 20
2010	Oct 28	Oct 28	Nov 2	Oct 24	Oct 22
2016	Nov 3	Nov 8	Nov 8	Oct 20	Nov 8

The peak color time will occur later when there is an abnormally high temperature-dry spell in September.

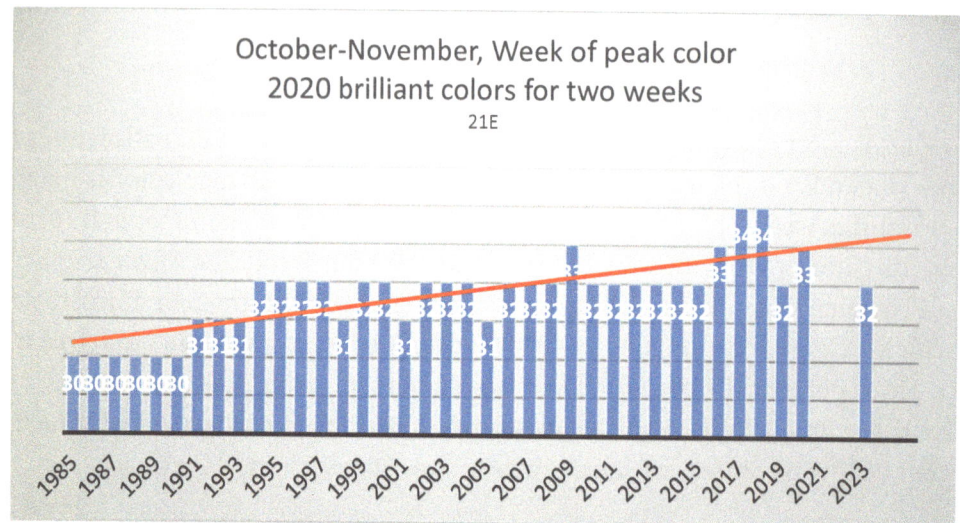

*Week when overall peak color forms.
The trend shows climate change.*

In 2007 and 2017 September was dry with temperatures during the day and night of 10 degrees above normal. The process of changing to peak color for most trees occurred seven to 10 days later than normal. The dogwood, oaks, and maples occurred 20 days later! The maple leaves did not change to their normal bright color. In early October 20 to 25 percent of the leaves withered and fell. Later in the month the remainder turned a drab brown and finally dropped. With the warm days photosynthesis was still in progress. Finally, photosynthesis had to shut down and the oaks turned their reds, oranges, and tans and the leaves were released. I was raking leaves in December, past the scheduled last week for township leaf pickup. As I put the last pile of leaves on the curb it started to snow. Fortunately, the next day it warmed up enough to melt the snow. Whew! The next day the township did pick up the leaves.

So how long does it take for leaves to fall from peak to all dropped? It is not consistent. It depends on the weather conditions and moisture in the ground.

Weeks for Leaves to Fall after Peak Color					
Type tree	Back Maple	Pin Oak	Poplar	Serviceberry	Dogwood
2005	2	3	3	1	2
2009	1.5	-	2	1.5	1.5
2015	1	3	1.5	1	1.5

2005 to 2015, it is taking leaves to fall a half a week to a week less to fall

Like the azalea in May, do the various tree species have a sequence of change to peak color from year to year? Yes, they do. The sequence remains the same but the date changes with the weather conditions from spring to now.

TRENDS DUE TO CLIMATE CHANGE

Sequence of Peak Color	
Tree	2010 Date of Peak
White Oak	Nov 18
Pin Oak	Nov 2
Crimson Maple	Oct 30
Poplar	Oct 30
Serviceberry	Oct 28
Dogwood	Oct 26
Maple A	Oct 20
Maple B	Oct 23

Start raking leaves on the 23rd. In 2010 the maple samaras that were dropping in September are now dropping in the first week of November.

The following are examples of how the leaves fall from the various trees and the change in color dates from year to year.

Maple B has a 40 to 70 percent drop by the end of the third week. It always starts to turn red at the top and works its way down. For five years, it had 90 percent drop between the 21st and 24th and 100 percent by the 31st. An exception was in 2010 when it only had a 25 percent drop on the 24th but caught up to be 100 percent by the 31st.

October 20, 2010 – Maple

The sugar maple leaves are dropping fast on the 22nd, on the 23rd 90 percent drop, and 100 percent drop by 31st. An exception was in 2010, it had no drop on the 24th, 50 percent on 27th, and 100 percent by the 31st. It was warm and dry in October 2015 and there was only a 60 percent drop on the 31st.

Sugar maple

Silver maple by the steps will be all orange on the 24th and then change to yellow-orange on the 27th; most years there is a 100 percent drop in the first week of November.

Silver maple by the shed turns yellow-brown and then to a bright yellow. Most years it is only a 70 to 90 percent drop by the 31st. On warm years, it may only be a 40 percent drop on the 31st. In 2010 it was only a 30 percent drop.

October 1996

November 11, 2016 – 2016

Silver maple

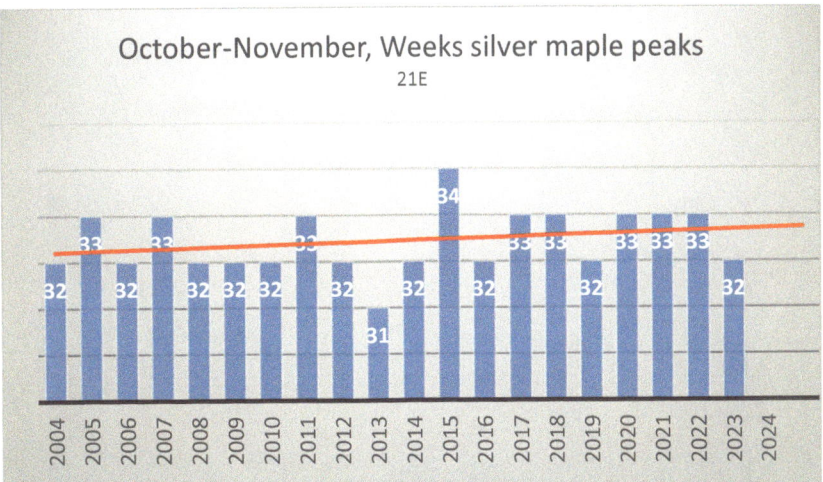

Silver maple – 2019 peaked and dropped leaves in a week

The pin oak may be brown or red the last week of October. On the 27th in 2010 it was a green-brown. In 2012 by the 29th it had a 50 percent drop. Most years it is the one that has the last leaves.

October 1982 – pin oak, planted 1972

1995 – pin oak

2000 – pin oak

October 25, 2015 – pin oak

OCTOBER

October 30, 2015 – pin oak

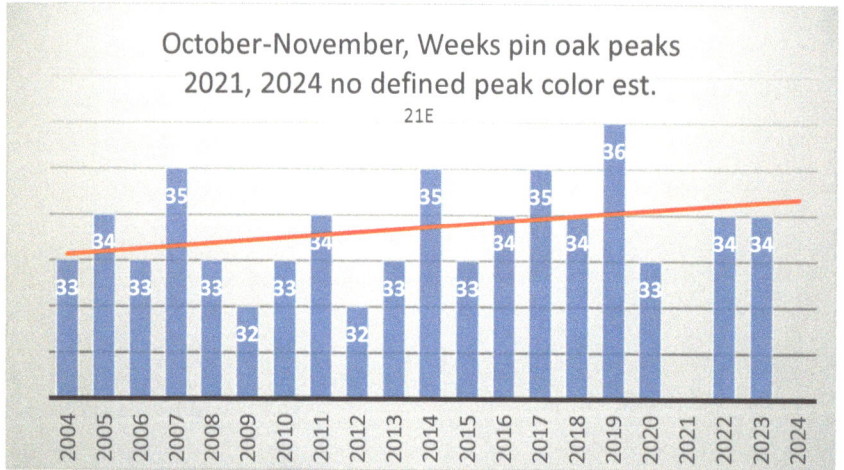

Weeks pin oak has peak color

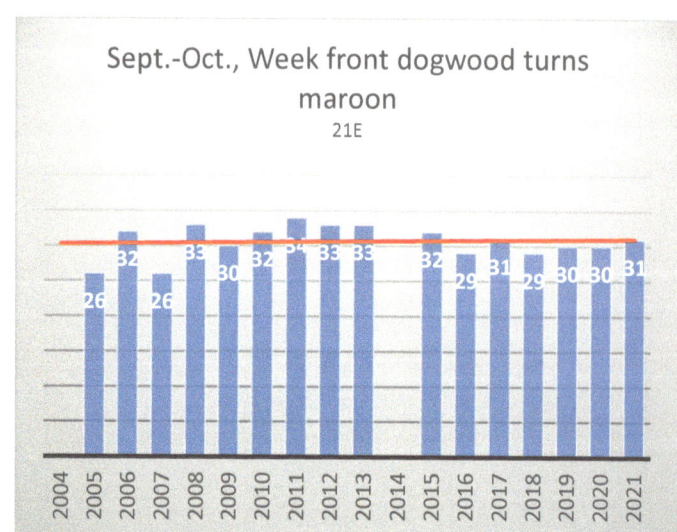

Dogwood, turns maroon

Climate Change on the Leaves of Fall

The east dogwood is dark maroon. Most years by the 28th there is a 100 percent drop. In 2015 only a 40 percent drop. The west dogwood is dark maroon. Most years there is a 100 percent drop by the 28th, in 2015 only a 60 percent drop on the 31st. The middle dogwood middle is dark maroon. Most years there is a 100 percent drop by the 28th, in 2015 only a 60 percent drop on the 31st. The front dog is dark maroon. Most years there is a 100 percent drop by the 31st, in 2015 only a 60 percent drop on the 31st.

The small birch usually has a 100 percent drop by the 28th. In 2010 it got off to a late start with no drop on the 24th, but had 100 percent drop by the 28th.

Pine and spruce needles and cones are falling on the hill on the 20th through the 23rd.

Serviceberry most years by the 31st it is orange-red with a 40 percent drop. On the 27th in 2010 there was no drop.

The poplar most years it is yellow by the 31st with a 50 percent drop. The 100 percent drop went into November. In 2010 the 31st had a five percent drop.

Poplar at peak yellow

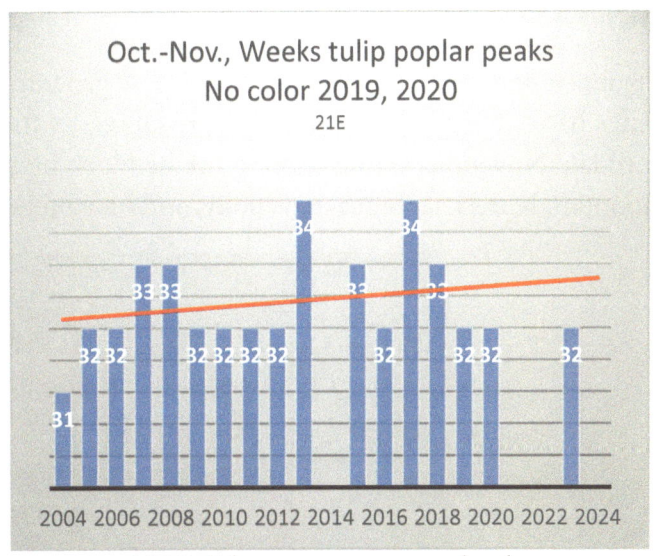

2019 did not turn a true peak color, leaves were a rusty brown

Maple B most years by 31st is orange or red with a 100 percent drop. In 2010 on the 31st there is a five percent drop and in 2015 on the 31st there is no drop.

Maple C in 2010 was in full color on November 1st.

Maple C in November

Maple D in November

Trees that have 100 percent drop in November are crimson maple, red oak, pin oak, hawthorn, maple C, and maple D.

<u>Shrubs</u>: Japonica out front has a lot of new growth and clusters. Holly tree within the hemlock hedge has red berries. Burning bush are all red by the 31st. The bush in the back may not change until late November. Holly among the hemlock hedge has berries. Put aluminum sulfate on the mountain laurels, japonica, new holly, and blue boxwood.

Burning bush

Review of Data

<u>Weather</u>: On the 29th-30th in 2011 we received a record six inches of heavy snow. The leaves were still on the trees. The snow weight broke off six- to eight-inch diameter branches of the poplar.

Spruce, dogwood, and maple trees also lost branches. The next day it warms up. As the snow melts, leaves start to fall.

A year later to the day, the 29th, Hurricane Sandy came with 70 mph northwest winds, the barometer drops to 28.0 inches. It caused massive destruction along the coast of New Jersey and New York. We had no power for a week. Lost two big spruce trees on the hill. One leaned on the telephone pole transformer causing a fire. Two days later the bird feeders were very active.

With the yo-yo of temperatures, we can get heavy fog at night which usually forms between 10:30 to 11:30 p.m. To the other extreme, we can wake up to a heavy frost.

In 2015 on the 29th we had a short, cold spell, the temperature went down to 25F at night. Then a warm spell occurred, and the temperature went back up to 70F resulting in flowers opening on the rhododendron, azalea, and PJM. The rhodie did not have flowers the next June. Then a 25-mph wind came blowing the leaves off the trees.

Fauna

Animals: A squirrel eats Mollis azalea buds! A pup squirrel jumps toward the safflower feeder, falls off; tries to go out on the wire to the peanut butter log but again falls off. Practice will make me perfect. A squirrel has mange in its right eye. It came around for a few days and then I do not see it any longer. Three chipmunks are running around with their jowls full of safflowers seeds. A rabbit nips a bunch of new branches off the winterberry, probably sharpening its buck teeth. In 2013 I spot a red fox under the feeder. At 6:00 a.m., 44F, with some light clouds and a sliver of a moon, I spot a skunk calmly walking down the circle. It must have had a late night. At 9:00 p.m. another, or the same skunk, is under the feeder. While warm, one to three chipmunks will still be collecting their winter larder. Rabbits are out in the evenings in the third week. Raccoons attack my seed cans at 3:00 to 4:00 a.m. in the fourth week. As it cools, I may spot a red fox late in the evening. Squirrels pair up during the third and fourth week. Like the raccoons, skunks come out late at night; I find droppings under the feeder or a dug-up lawn in the morning.

Squirrel Wars: An older squirrel who is a good acrobat goes to the top of the two-by-four post and has been jumping upward across five feet to land on the feeder cable. It twirls around and gets on top of the cable, a feat, and goes out to the globe peanut flake feeder. I put grease on the cable, but it has no effect. I put vertical nails in top of the two-by-four. The squirrel still finds enough room on the top to jump to the cable. Put a five-inch-wide plastic funnel slip over the cable, it lies to one side. It must be supported. Get a piece of stiff, bendable wire, bend a loop at the middle to go around the cable and stick its ends through holes at the edge of the funnel. This supports the funnel around the cable. The funnel is pointing so the narrow end is where the squirrel comes from. Now I wait. At 5:00 p.m. my friend comes and jumps onto the cable and goes out six feet to the funnel. It realized it cannot get past the funnel. Get this, it turns around on the cable and starts to go back the other direction. It goes about a foot and slips off the cable. It has not come back since. I put more vertical nails in the post; it should not have enough room to launch a jump. Wrong! The squirrel went up the pole and stopped below the nails. It is still gripping onto the side of the pole as it looks the situation over. It launches from the side and lands on the cable beyond the funnel. I give this squirrel a lot of credit. It proceeds to munch on some peanut flakes. I hang a strip of wire fencing from a bracket on top of the pole. It should act as a barrier when the squirrel launches. Wrong again, somehow it launched and got on the cable. Enough is enough. I take down the whole pole. Six days have passed and no squirrel on the wire. I am waiting again.

Second battle of the peanut flake tube by the kitchen window. The tube now hangs down from the gutter by a sixteenth diameter wire. Found that a squirrel can slide down an eighth-inch wire. The tube is only two feet from the rhodie bush and two feet above the raised feeder. A pup squirrel found it could jump from either direction to land on the wire and tubes. The tube would go around in fast circles and swing back and forth when the squirrel landed on it. After it slowed down the squirrel would feast. Put a four-inch-tall wire basket on the bottom. The vertical wires of the basket are wide enough for the chickadees and titmice to get through. Then I wrapped a five-inch-wide piece of roof flashing around the upper part of the tube, so my friend had nothing to grab onto. I wait; the squirrel comes and looks the situation over. It tries to slide down the wire from the top, but the wire is too thin, it slips off and lands on the ground. Next it jumps onto the edge of the basket, puts its head down between the edge of the basket and tube and feasts. Made a cage out of chicken fencing to cover the top. The corners are open enough for the birds to enter. Gotcha squirrel! No, an hour later the squirrel is feasting; it jumps and squeezes between the basket's uprights. It looks like a little ball of fur in that tight space. Reduced the size of the openings in half. Waited to see how this affected the squirrel and the birds from getting to the peanut flakes. The squirrel squeezed between the basket and tube on the other side. However, its tail was sticking out. I slowly opened the window and pulled on its tail. It squealed and jumped out. I have not seen it since. I tightened up where the squirrel got in. Other squirrels have tried to get the peanut flakes but have not been successful.

Third battle. A friend wanted some bird feeders. There was only one place to put them so the squirrels could not get them and that was on an old clothesline. The line is 25 feet long and 3/16-inch in diameter. It is attached between a tall arborvitae and a seven-foot clothesline pole. I hung a peanut flake tube and a platform feeder from the middle of the line, filled them, and waited. The birds came and my friend is delighted. About an hour later a squirrel is sitting on the platform enjoying the seeds. Just like at my house, this is quite a feat. I observe that it twirled its tail back and forth like a trapeze artist as it traversed along the top of the wire. I got a liter soda bottle and drilled a hole in the bottom and the cap. I then strung the wire so the bottle is horizontal and spins on the wire. I read about this in a bird book. The first time the squirrel came it tried to mount the bottle. It spun around and fell off. So that was that. Wrong. After another 15 minutes the squirrel was back on the platform, I still do not know how it got there. I had a plastic pot that was 10 inches in diameter. I cut the height down to three inches, and I screwed the bottom of the pot to the bottom of the liter bottle; 15 minutes later the squirrel was back on the platform. I added another liter bottle after the baffle. Within a half hour the squirrel is back on the platform but this time I saw how it did it. This is quite a feat; it deserves the seed. It went to the top of the post, then hunched up and launched over the liter bottle 10-inch baffle and next liter bottle and landed on the wire and balanced itself and skipped down to the platform feeder. How about that! I tried one more thing. I put Vicks VapoRub on the pole; it did not like that. It just went down the pole and went way. I won at that location, but the squirrel was quite an adversary.

TRENDS DUE TO CLIMATE CHANGE

<u>Birds</u>: The number of species in the gang is declining, with 2011 the lowest, which was a very wet year.

Table of the Gang

Species/Quan.	04	05	06	07	08	09	10	11	12	13	14	15	16	17	18	19
Blue jay R, G	2/28	3/29	3/29	2/23	2/29	2/29	3/7	2/22	5/11	2/1	4/4	2/1	2/2	2/10	2	4/16
Cardinal R, G fledge	5 1/12	2 2/11	4 1/9	2 1/16	6/26	2 1/31	2 2/16	2	7/30	2/2	8/14 2/2	2 1/1	2 1/1	1/2 1/1	3	3/23
Carol wren, G arrive	1/1	2/10	1/4	1/8	1/7	1/16	2/16	1/22	2/3	1/2	1/3	1/6	2/4	2/10	1	1/2
Chickadee R, G	2/1	4/29	3/11	3/8	3/29	2/16	3/4	3/2	2/1	2/1	3/1	2/1	3/26	3/29	2	3/3
Chipper sparrow G leave		1/7	16		10		18	22	1							
Downy wood R, G	1/1	1/8	1/9	2/1	2/1	1/14	1/15	1/3	2/1	1/2	1/2	5/3	1/2		2	1/8
Flicker G arrive		20	3	28	25	13				5	3/11	2/9	1/6			
Goldfinch still here change color	1/1	2/29 1	3/27 7	3/26 5	3/25 5	1/29	1/4 7	1/26	2/31	2/27	3/3	1/31	1/4	2/29	4/13	1/1 1/1
House finch R, G conjunctivitis	5/8	8/8	2/3 13	1/4	2/1	2/15	4/4 18	-	6/1 31	4/10	4/1	4/4	4/5	4/1	4	2/23
House sparrow R, G	4/4	10/13	6/11	8/10	8/12	6/15	1/2		1/5	1/10	6/2	8/31	4/5	4/6	8	8/2
Junco arrive	14	2/26	25	13	21	Nov 2	30	26	13	Nov 5	Nov 1	4/27	1/2	Nov4	1/21	Nov 2
Mourning dove R, G largest #	6/28	4/1	8/9	2/24	3/9	2/11	4/7	4/3	6/13 8/24	6/28	3/2	12/4	3/9	11/13	10	6/1
Red-breasted arrive	2/26	1/1	2/6	2/2	2/7	1/9	1/8	-	3/1	-	-	-	1/2		1/4	
Song sparrow R, G arrive	1	1/3 3/27	11	8	4	1/18	1/1 2/15	1/29		1/27	2/18	1/3				
Titmouse R, G	3/1	3/2	3/7	2/8	3/16	1/16	3/15	3/31	2/31	3/12	3/3	2/1	3/9	2/25	2	1/2
White-breasted arrive	8	26	15	15	10	8	4	-	1/1 2/5	1/1 2/13	1/1	1/3	7th	2/24	2	1
White-throated arrive	1/11	1/20	1/7	1/13	1/12	1/19	2/15	1/26	1/7 4/20	1/21	1/19	1/4 5/26	1/10	1/16	1/12	1/12
Total gang	15	16	16	15	16	15	15	11	14	15	13	13	14	12	13	12
Population	37	51	43	33	41	26	33	21	48	28	45	53	31	35	41	32

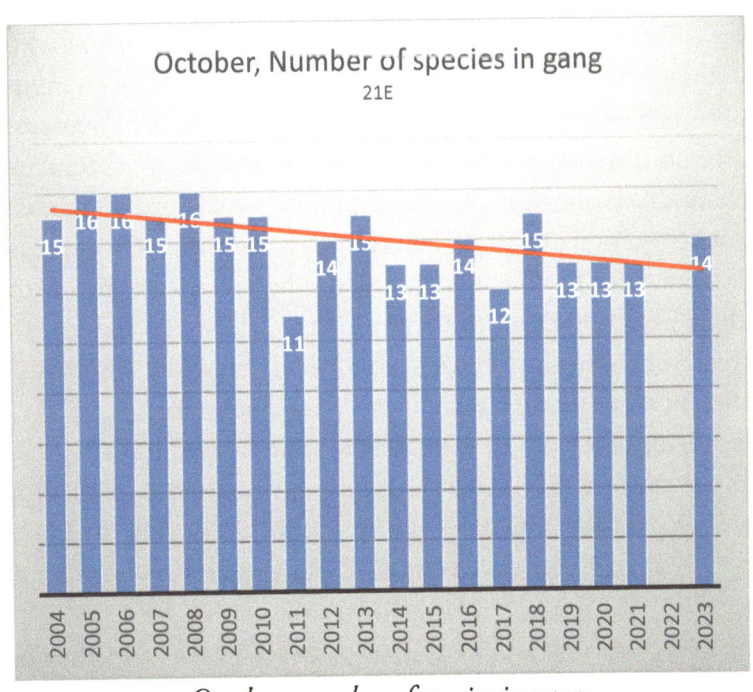

October, number of species in gang

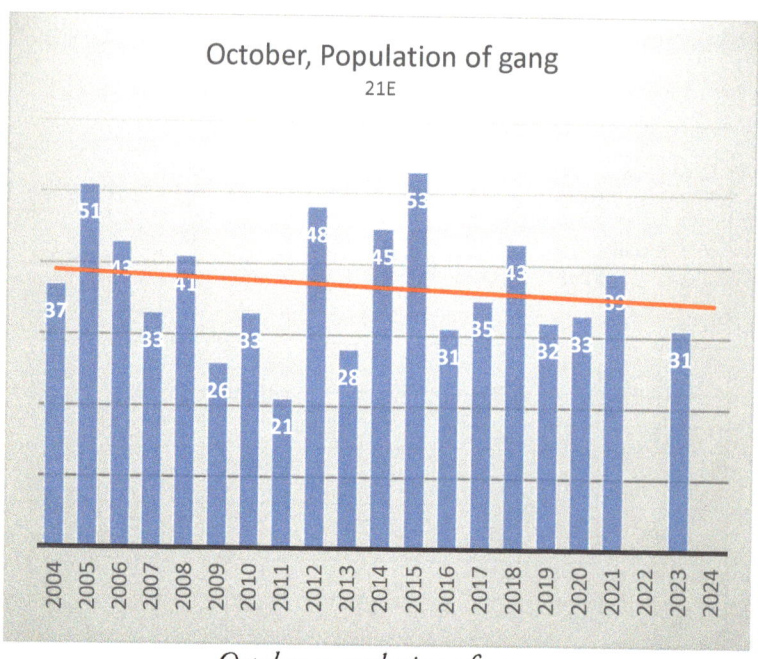

October, population of gang

In October the population of the gang looks like a 10-year cycle, the 2011 dip is showing. The average trendline is about 38 birds.

Over the period, the quantity of jays stays at two, except in 2012 it went to five for the first week.

Carolina wren arrives the first or second week. In 2010, 2011, and 2012 it arrived the third and fourth week (the dip?). Some years it shows up with a mate by the third week. In 2018 it oversummered; in 2019 into 2020 they stayed all year.

Many years there is only a male downy; now and then a mate is seen. In 2015 there were three fledglings! The cardinals bring one to two fledglings the first or second week. Chickadee fledges the first or fourth week. Titmice are consistent, bringing one fledge, usually the third week.

With the climate change, I would think that the goldfinch would be changing later, but the data shows the opposite. From 2004 to 2011 the change was the last week of September or the first week of October. From 2011 on the change seems to be the third week of September.

House finch fluctuates from one to six. Cases of conjunctivitis appears the second to fourth week.

House sparrow goes from none to 10. When it is high, they are very aggressive at the feeders.

Mourning dove is from three to nine, depending on the presence of the sharpie. The sharpie did not arrive until the third week. On the 17th there were nine doves, by the 31st there were only four.

Red-breasted nuthatch arrived the first or second week from 2005 to 2010. There were none in 2011, one in 2012, then none until 2016 which arrived in September.

On a nice afternoon, four flocks of 20 robins fly over in five minutes.

Migrating species are the ones I hope to see. Many times only see a species once or twice over the decade. Most come as pass-through or arrive for the winter during the last week with a possible count of 46. The first week is 29, second week is 27, and the third is only 16, then goes back to 25 in the fourth week.

Flickers arrive the first or second week quite often as a pair. In 2014 they had a fledge. In 2018 they arrived in August.

Song sparrow arrival is a moving target from the first to fourth week of October to the first week in September. They were no-shows in 2012, 2016, and 2017. As pointed out in September the cause for no-shows may be a result of being a surrogate mother to the cowbird. So, what is the proper trendline? When I remove the zeros from the data, an altogether different trendline appears.

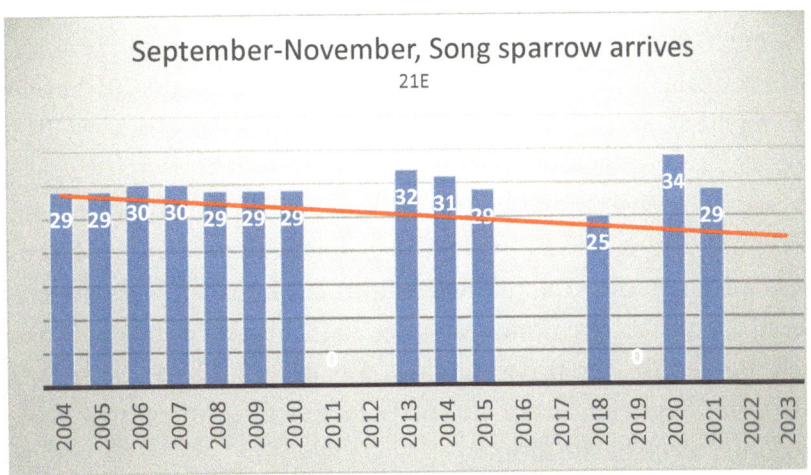

Weeks song sparrow arrives

Chart looks consistent from the overall period for the white-breasted nuthatch. The 2005-2018 period shows how the arrival went from the fourth week of October to second week of September. Then in 2018 it over-summered! They were here all year in 2019.

White-throated sparrow arrival has been from the first to the fourth week of the month, mostly the third week. The arrival depends on the northern weather.

Keep up the hummer feeders through October, that is when the young hummers may still be migrating. The yellow and yellow-throated warblers pass through the first week. Rarely, a flock of cedar waxwing would stop to devour berries on a tree in the first week. Look for the rose-breasted grosbeak in the first week. Flycatcher and fox sparrow arrive any week. Towhee may bed and breakfast for a couple of days the first and second week.

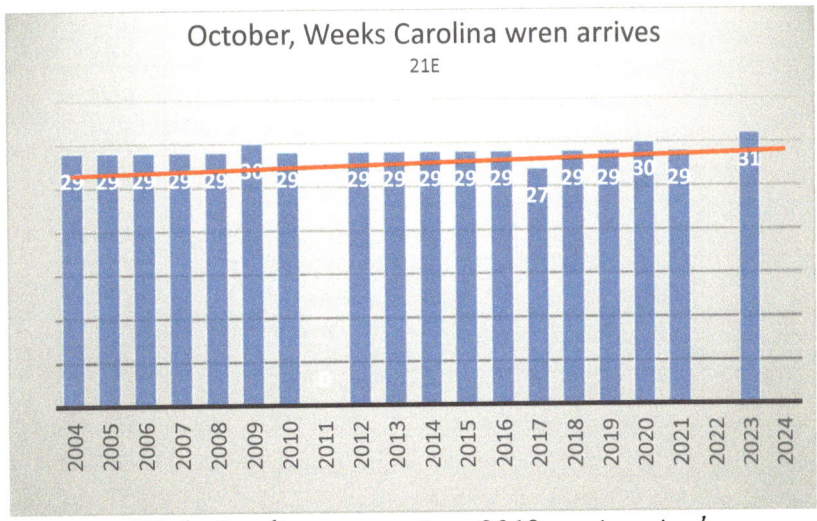

Weeks Carolina wren arrives, 2019 a pair arrived

OCTOBER

Magnolia warbler, northern waterthrush (2004), pine siskin, and purple finch pass through the last week. The catbird was leaving the last week of September, now it is the second week of October. I may spot a ruby-crowned kinglet and a white-crowned sparrow the third or fourth week.

I thought the junco arrival was very punctual during the fourth week, until I made the plot. The trend is moving into the first week of November. The "2012 dip" is evident, but in reverse of the Carolina wren. The junco is arriving from the north and the wren from the south.

Flocks of 20 to 100 robins used to pass over the fourth week until 2010 and then the number dropped to one or maybe 20 in the first week. On the other hand, flocks of 10 to 20 robins have been overwintering.

Canada geese used to arrive at the nearby cornfields by the hundreds until 2011, then it dwindled to 30 or 50. In 2016 it was 20 to 30. The cornfields are being replaced by houses. Large flocks of snow geese would go over the first week until 2011, now it is in November. I used to see flocks by the hundreds of blackbirds; none seen after 2008. Flocks of starlings in large numbers still pass over. Through the decade, the broad-winged, Cooper's, and red-tailed hawks soar through from the second to the fourth week. Sharp-shinned hawks arrive for a winter stay in the third week. Turkey vultures soar any week of the month.

| October, Migrating species from 2004-2018 Year/week = last year observed in the 2000's and week ||||||||||
Species	Yr./week	Week of October observed			Species	Yr./week	Week of October observed		
Bat	05/2	1	2		Ruby-crowned kinglet	15/4				4
Towhee	16/2		2	3	Eastern phoebe	16/2	1	2	3	4
Blackbird flocks	08/1	1	2		4	Starling flocks	17/1	1	2	3	4
Broad-winged hawk	15/1	1	2		4	Northern waterthrush	04/4				4
Catbird, leave	17/1	1	2	3	4	Magnolia warbler	17/1	1	2		4
Cooper's hawk	13/4	1	2	3	4	Wood thrush	17/1	1			4
Hairy woodpecker	16/2	1	2	3	4	Flicker	17/1	1	2		4
Canada geese 20-30	17/1	1	2	3	4	Mallard ducks	16/3	1		3	
Grackle flocks 200	15/1	1	2	3	4	Pine warbler	15/2		2		
Ovenbird	05/2		2		Snow geese flock	10/1	1			
Cowbird flock	16/2	1	2	3	Red-headed woodpecker	07/2		2		
Sharp-shinned hawk	17/1	1	2	3	4	Pine siskin	12/4				4
Pine warbler	15/2		2		Robin	17/3	1	2	3	4
House wren if it has been warm	04/2		2		Yellow warbler	14/2	1	2		
Pewee	14/1	1		Mockingbird	16/2		2		4
Great-horned owl	17/1	1			4	Yellow-throated warbler	08/1	1			
Fox sparrow	15/4				4	Red-bellied woodpecker	16/2	1	2	3	4
Chipper sparrow	13/4	1	2	3	4	Rose-breasted grosbeak	14/1	1			
Red-winged blackbird	17/1	1			4	Red-breasted nuthatch	06/2		2	3	
Red-tailed hawk	16/4	1			4	Purple finch	16/2	1	2	3	4
Cedar waxwing	15/1	1				Winter wren	18/2	1	2		
Cedar waxwing						Total species	46	30	28	16	25

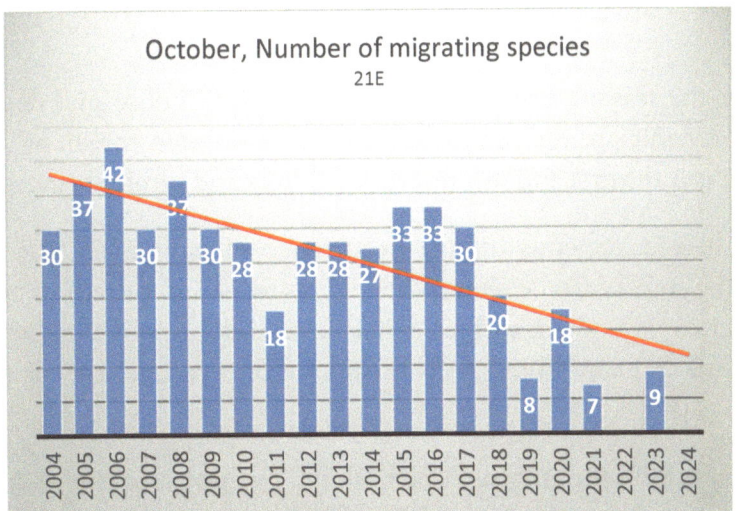

October number of migrating species
Note that the 2011 "dip" shows up in migration

The data for September and October of the "possible" daytime observed species shows how the magnitude of migration is highest the fourth week of September and decreases in October. I stress "possible" because many of the species have not been observed during the same year in years.

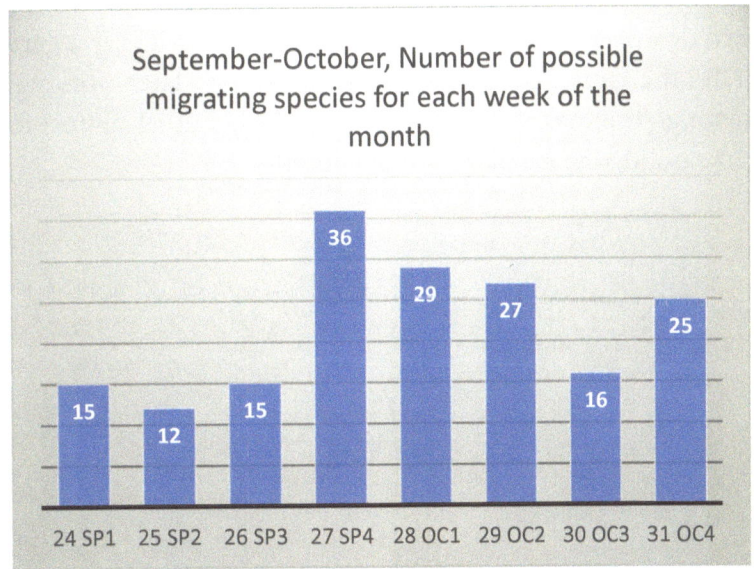

Possible number of migrating species

Quantity of Species

Quantity of species that I have recorded in September is 58, many I have only seen one or two times. For October, the number increases to 67. In either case that is quite impressive. For a given year the quantity goes from a low 18 (2011) to a high of 42 (2006). The first week is when most of the resident species are seen (23). Then it drops to 16 or 18 for the following three weeks; still substantial numbers.

Fledglings

Though it is late in the season, cardinal, chickadee, flicker, titmouse, and downy bring fledglings to the feeders. Most notable are the cardinals, which bring one to three any week of the month. Downy bring two to three the first and second week. Flickers and titmice bring one the second week.

Insects: Activity of insects depends on the occurrence of days over 60F. Then bumblebees, dragonflies, and yellow jackets can be seen. When it gets to 70F, hundreds of boxelder bugs can be seen sunning themselves on a warm surface. Gnats and stink bugs like the warmth. Crickets will be singing until the end of the month, especially on a warm evening. I even spotted hordes of ants on the sidewalks.

Pond: When it is above 60F, the frogs will be basking, but not croaking. I can still see tadpoles scurrying along the shallow water of the pond. During the fourth week, I put the net over the pond. At dawn, as it gets lighter, I see the white outline of the bottom of a stretched-out frog against the rock wall of the pond. The frog tried to jump out, but instead got trapped in the netting that covers the pond. It was still alive but could not move; snipped it lose with a pair of scissors. It was barely breathing. As it put it back into the pond, air bubbles came from its mouth; it was belly up. I thought it was a goner. But no, it slowly turned right side up. After a few minutes, it swam away.

In 2018 there an unanticipated arrival. It was a sunny afternoon. I was cleaning the outside area by the shed and looked down at a small pile of coal. I looked twice to see a garter snake looking at me. I looked back, it did the same and slithered away. Two days later working around the pond I heard a splash, sure enough the snake was in the pond. October 2019, almost to the day, I discover two garter snakes in the pond, one smaller than the other.

Two garter snakes

Snake taking a frog

A couple of days later I find the snake had a frog by its leg. They were both just there. The frog made no attempt to move! I watched for five minutes and then went about my business. When I looked back both were gone. Now I know why the frogs are disappearing.

Flora

Perennials: Not all perennials seem to be affected by climate change. Roses may still have a few flowers in the first week. A few Stella de Oro lily may bloom in the first week.

Purple hardy aster comes into full bloom and the fall sedum turns dark maroon during the first week across the entire decade. The globe thistle has a second bloom after being deadheaded in the third week. Goldenrod wanes in the third week. The yellow iris and Montauk daisy may have second blooms the third and fourth week. Red hibiscus may still be in flower.

Pond: Cattail stalks turn yellow during the fourth week. Pickerel weed has white flowers the first week, the stalks turn brown by the last week.

Trees: A neighbor's maple starts to turn red weeks ahead of the surrounding maples. It starts turning at the top then blushes downward. Its trend shows it is turning five weeks earlier, opposite of most other maples. Silver maples, sugar maples, and dogwoods are peaking a week later. Another neighbor's maple turns red mostly in November. There is not enough data to develop a trendline. The poplar turns bright yellow the beginning of the first week in November.

Serviceberry tree turns orange. Trend shows it is turning earlier.

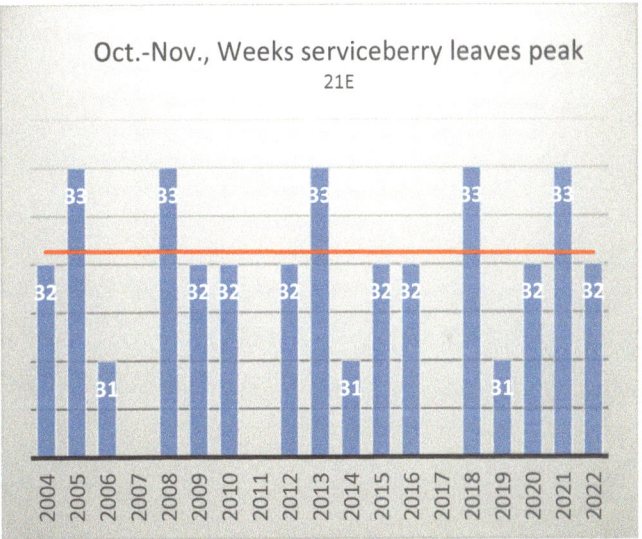

Serviceberry

Pin oak is one of the last to peak color. Trend shows a week later for the period. The last tree to peak in my yard is the black oak; it normally turns brown.

Other than the deciduous trees, depending on the weather spruce cones will form and pine needles start to fall during the first week.

October 6, 2016 – spruce cones

So, ends October.

Index of November

November Overall	284
Weather	295
First Week	297
Second Week	306
Third Week	308
Fourth Week	317
Fall Leaves	319
Table Colors of Leaves	320
Percent Leaf Drop Versus Days	321
Review of Data	326
Weather	326
Fauna	327
Birds	327
Gang	329
Table of the Gang	330
Review of Individual Species	332
Fledglings	336
Insects	336
Pond	336
Flora	336
Perennials	336
Vegetables	336
So, ends November	336

November

Week number for the year of 48 weeks					
Relative Month	Month/Week	First	Second	Third	Fourth
Previous	October	29	30	31	32
Present	**November**	**33**	**34**	**35**	**36**
Following	December	37	38	39	40

Weather

The average high temperature on the 1st is 58F and the low is 36F.

The highest temperature was on the 1st in 1958 at 81F.

The average high temperature on the 30th is 47F and the low is 29F.

The lowest temperature was on the 30th in 1929 at 10F.

Reviewing the high and low temperatures from 1927 to 2018, the decades with the highest percentage (37%) of high temperatures was 1930's-1940's. The 2000-2010's had 17 percent. The decades with highest percentage (37%) of lowest temperatures was 1930's-1940's. The 2000's-2010's had 17 percent.

The amount of daylight is 10 hours.

A full moon is called the Full Beaver Moon.

Overall Summary of November

As noted in the write up of October of how the peak color date of the trees has moved to a later date by a week in the past 10 years, the same changes have occurred in the last bloom of perennials, apple picking, bark peeling from the large birch tree, mushrooms developing in the lawn, and asparagus ferns turning brown. Except for cabbage-types of vegetables, gardening is done. Brussels sprouts are not picked until after the first hard frost because they produce the best flavor after being hit with a frost.

On the 13th in 2016 there was a "supermoon." The moon was the closest it has been to Earth in 70 years.

November is the month that many trees come to peak color and have a complete drop of leaves. Maples and poplar trees drop their seedpods, which get tracked into the house.

Six species of birds still bring fledglings to the feeders. The data shows the reduction of the possible number of migrating species noted during the first week to be 18 to the last week of four. Why I say "possible," because many years some species are not observed from one year(s) to another and arrival of the species may be erratic depending on the weather to the north.

The number of species in the resident gang is five to 17 for the month. These numbers do not seem to be affected by the amount of rain nor change in temperature. The number of possible migrating species drops when the temperatures are unusually higher than normal; the amount of rain does not appear to affect the number.

Weather

Passing clouds at moon rise

Moonbeam

The temperature has high 60's and 70's at the beginning of the month. Some years it is warmer than normal by six to 14 degrees all month! In 2005 we even had the 3Hs (hazy-hot-humid) on the 29th! Now and then the temperature drops to 3 to 5F. On the 11th in 2011 it dropped to 20F but felt like 13F. It was the coldest reading for that day since 1976. Before 2005 we used to receive three to five inches of snow during the third or fourth week. Since then we only had one-and-a-half inches of snow once in seven years, except in 2011 we received six inches and in 2018 there was surprise of eight inches that snarled commuter traffic from a normal 45 minutes to seven to eight hours; a really bad situation. Snow has been replaced with frost, freeze, and fog. During the month the winds come in from the west at 30 to 50 mph. In 2007 it was unusually warm and muggy, and we had a tornado warning. The main change since 2005 is the lack of rain. The exceptions were 2011 and 2018 where we got too much rain. The second week of 2019 the temperatures dropped from a daytime of 50F to a system-shocking nighttime of 20F. It then stayed unseasonably cold for two weeks.

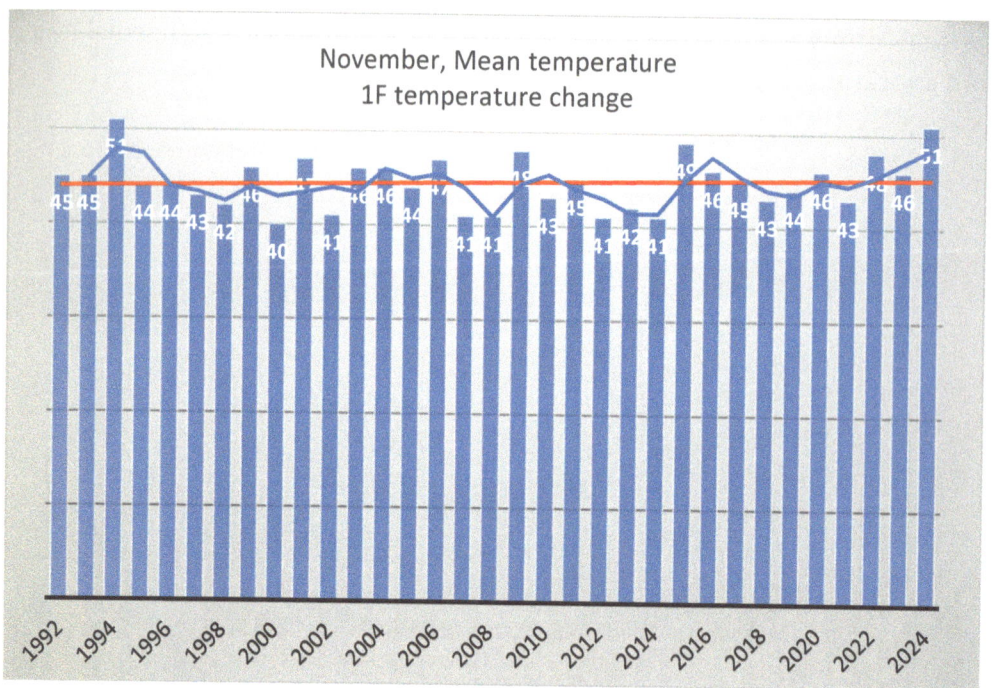

November, mean temperature

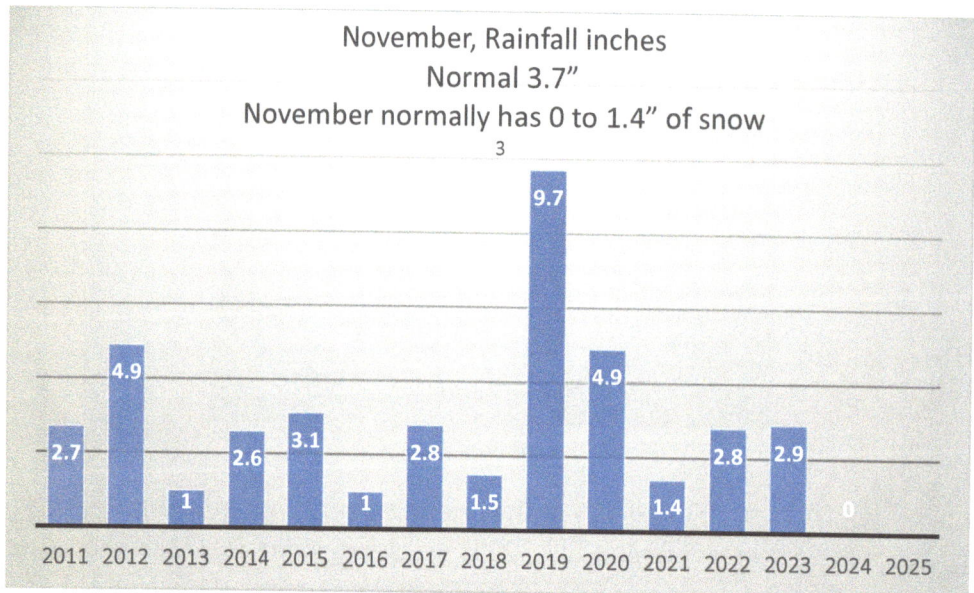

November, rainfall inches

Dawn, there is frost on the lawn and a film of ice on the birdbaths. I watch it disappear as the sun rises and swipes across the lawn. On average the first frost occurs the second week of November. The trendline shows the frost is occurring earlier from the third week of November to the first week, opposite from the climate change and when the leaves peak!

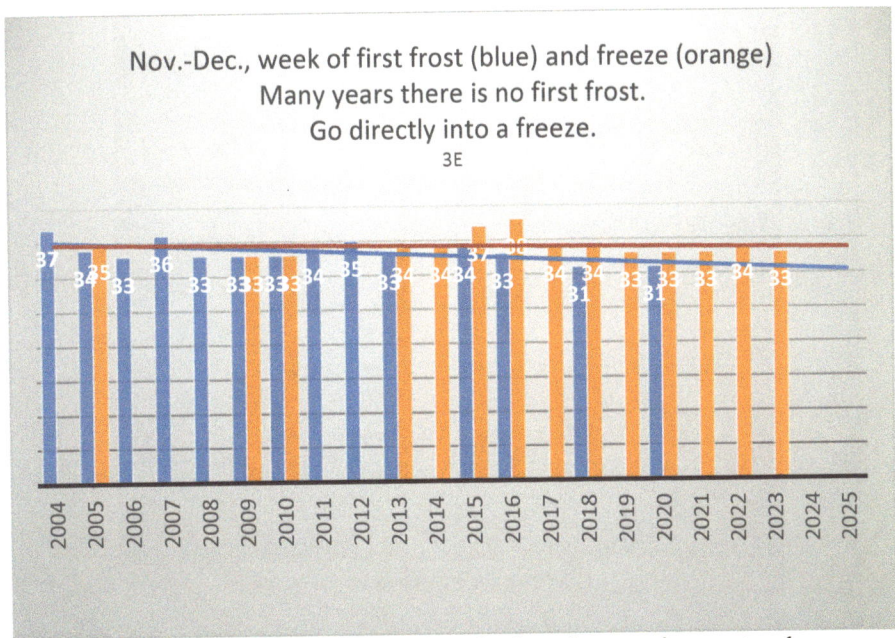

First frost is occurring earlier but the first freeze is happening later

Today was the last "Pollen Count" to be shown in the newspaper. It has been replaced by the amount of "Snowfall."

Hurricanes are still a threat; even if one does not make landfall there are usually high winds during the first two weeks. A hurricane is to hit Cape Cod. At 8:10 a.m. it is light. A northeast wind is picking up from the effect of the hurricane. Few birds until 10:30 a.m., then the gang appears; there are 10 house sparrows. When I go out to take a walk, I hear the tips of the oak leaves rattle like they are on five legs as the wind blows them across the macadam.

Then there is the thick morning and evening fogs. Lots of above and below normal temperatures, dampness which leads to decomposition (rot) of organic material.

November is the month that the reduction of daylight really starts to affect the wildlife. While us mortals keep our hours of breakfast, lunch, and dinner pretty much the same all year because we can produce light whenever we want, the daylight hours for the wildlife changes with the season. In June they have 15.5 hours but in December they only have 9.5 hours. The wildlife must obtain their daily nourishment for themselves and fledglings in the much shorter available daylight of the winter. During the summer the weather report in the newspaper had information on "pollen, humidity and what it felt like." On November 1st it is replaced with "wind chill and snowfall." A true sign that winter is coming.

First Week

It is the first Sunday of November that we move the clocks to Standard Time.

<u>Animals</u>: Rabbits are seen in the evening during the first week. In the morning, I would smell a skunk that visited the feeder during the night. That has not happened since 2008. Voles are scampering around the rock wall all month.

Out for an evening meal

Over the years, nooks and crannies the size of walnuts developed in the rock wall around the pond. During the summer chipmunks occupy crannies of the wall. In October they move out of the wall to their ground tunnels. Then the voles move into the wall. Come snow, the voles make tunnels in the lawn from the wall to the base of the birdfeeders to get the spillover seeds. Spot a nocturnal vole as it is collecting leaves and grass; then pulls them into the nooks and crannies among the rocks. In the evening spot a rabbit in the back garden.

You must look hard, there is a vole to the left of the apples

At 6:30 a.m., 50F, clear and light, one squirrel chases another on the hill. I wish I knew if they were just two siblings having fun or two young lovers getting ready to make another brew of apple eaters. A squirrel is carrying a mouthful of cattail stalks up a spruce tree. The average lifespan of a squirrel is six years.

Making a nest with cattail stalks

<u>Birds</u>: The first week flocks of geese may still be seen going south, Cooper's hawk may go through, owls are hooting. The red-bellied woodpecker and red-breasted nuthatch used to pass through to go south, since 2015 both are overwintering.

At 4:30 p.m. a flock of 20 migrating robins go south. Four crows are hopscotching from one tree to another, cawing to each other as they go.

Three starlings are trying to get into three separate birdhouses. From 8:00 a.m. to 12:00 noon the gang is here. At 4:15 p.m., partial gang is at the feeders for dinner. Dusk, 4:50 p.m., it is that time of the year when you hear small birds chirping around the bushes and lawn before settling in.

Because of change to Eastern Standard Time the gang is here at 6:45 instead of 7:45 a.m. The birds don't know the difference. The cardinals and song sparrows are eating at 6:45 a.m. Most of the gang is here plus a red-bellied woodpecker and a rabbit. I hear a lone robin singing! At 7:00 a.m. two bright cardinals are chasing each other and then a third comes in to make the chase more exciting. Who wins? Are they males rivals or two just having fun? A ruby-crowned kinglet is here and so are 10 juncos. Put broken eggshells on the feeder; a blue jay dives down, looks around and takes a nut and flies off. It yells with the nut in its beak. I had not even gotten to the backdoor; it must have been watching me. I look up to see a broad-winged hawk circling in the clear sky. A great horned owl flies over the house going south at 7:30 a.m. This is a surprise because I normally would expect to see or hear it at 7:30 p.m.! A squirrel climbs up the oak tree; gets too close to the apartment house full of house sparrows. They quickly come out and bombard him. He scurries past them to his nest. Thirty crows fly over in a loose flock formation at 8:30 a.m.

Another day, a female house sparrow got caught in the net over the pond. It had probably been there for 10 minutes before I noticed it; its head and one wing are sticking out one side of the net and the rest on the other side. I gently released the net from it, holding and supporting the sparrow in my left hand. She just sat there calmly while I got her head and wing free. She shook herself to get her feathers back together. She looked at me for a minute, not frightened, and then flew up to the birch tree with the other sparrows.

Two solitary vireos go through at 1:00 p.m. I have not observed that species since. Dusk, two loose flocks of 15 robins are going west.

Clear and light, 6:30 a.m., 41F. Two flocks of 12 crows go west at 7:10 a.m. The gang has been active for seven minutes, suddenly a sharpie lands in the dogwood and all activity vanishes. Blue jays yell at the sharpie. Annoyed, after a few minutes it takes off. The blue jays stop yelling, in two minutes cardinals are back at the feeder. Twenty more crows go over.

At 5:00 p.m. six song sparrows are at the feeder; during the summer there was only one. A goldfinch goes for split peanuts in the swinging basket, this is a first.

To prevent the hawks from getting to the birds under the window feeder I install a three-foot-tall wire fence in front of the kitchen rhododendrons.

Fence to keep out hawks

Empty out the wren house, the internal nesting material consists of small, rough oak twigs. No grasses or other nice cuddly stuff. What an incentive for a fledgling to want to leave the nest.

House wren nest

In nearby Bethlehem, between 3:00 and 3:30 p.m. at least 200 crows are flying to the rookery located by the Monocacy Creek by Moravian College. The rookery has been there for at least five years. This returning of the crows to the rookery occurs every evening about an hour before sunset.

November 7, 2006 and 2013, a white-crowned sparrow stops for a visit.

2013 – white-crowned sparrow

At 8:30 a.m. a song sparrow is under the pond net getting a drink when out of nowhere a sharp-shinned comes down to take the sparrow, however, the hawk did see the net. It bounces off the net. Underneath, the poor sparrow is panicking and flying all around; it flies out the west end. Meanwhile the hawk gathers its wits and pursues the chase. The hawk is too late, and the sparrow lives for another day. As this drama was happening all the other small birds scattered, except a white-bellied nuthatch that happened to be eating peanut butter from the hanging log. It literally froze in position on the log. Two minutes after the hawk was gone the white-bellied was still frozen in place. I looked to the dogwood that was adjacent to the pond and there perched on an upper limb was the hawk. This was on the opposite side of the log that the nuthatch was on. I don't know how it sensed the presence of the hawk. I went out and walked right next to the frozen nuthatch and chased the hawk. I walked back again past the nuthatch. The only thing that moved was its eye following me. After I passed it flew to another tree. When I went back into the house the other birds returned.

Red-bellied woodpecker opening the house hole

Fruit: Come late afternoon it gets cool and damp; it is time to go in and make an apple pie from my Winesap apples; add raisins and sherry. Mmm good!

Pie from Winesap apples

Insects: Some afternoons the temperature rises to 68F. When that happens, swarms of small insects fly about. Boxelder bugs bask in the sun on the bricks in front of the house. They are not on the east side or back which are in shade. They are a half-inch long and are bright red when they fly. When I touch them, they fall to the ground. They are not harmful or bite but would be a bit disturbing to someone coming to the front door. At 4:00 p.m., dusk, the crickets are still singing. No stink bugs are present.

Lawn: The weather can also be warm and humid which is conductive to mushrooms popping up on the hill and front lawn. Lawn is lush with no weeds. I cut the lawn low so the leaves will blow off. Normally the last cut is the second week of November but when it is abnormally warm and rainy the last week in November, the last lawn cut goes to the first week of December; then put down lime.

Perennials: Fern fronds are yellow and light brown. Some gladiolas are still blooming. White iris is still out and roses still have a few little buds. If there has not been a frost, marigolds are still hanging on.

After a November warm spell, the reblooming white and yellow iris have flowers and the large white azalea has a few flowers. Cut perennials short enough to be able to rake over them. The mums have finished blooming; cut them back to six inches. Do the same with rosemary and tarragon and cover them and the mums with lots of straw; I've had an 80 percent survival the next spring. It truly is fall because it is raining orange leaves from the silver maples. Leave ornamental grass seed heads on the stalks until spring. Straw flowers are still out. Verbena is still on the go. Trumpet vine is turning a little yellow. The goldenrod is still in full bloom. The white, pink, and red phlox still have a few flowers. Lungwort looks good.

Lungwort

Pull out the annuals. After a hard frost dig up the blackened dahlias; cut them six inches above the root, turn them over so the liquid can drain out of the stem. After a couple of days, I will put them in a cardboard box with layers of newspaper. The box will go in a dark place in the cellar.

Dahlia before the frost

Easter roses

<u>Pond</u>: Cattail stalks are almost completely light yellow.

<u>Shrubs</u>: The robins, mockingbirds, starlings, and squirrels are still plucking the berries of the dogwoods and winterberries. Lilac leaves are dark green and bright red.

The rhododendron and azalea all have buds for next year. Give the azalea a light trimming. The japonica clusters have formed in the first week. With a warm October the clusters form during the fourth week. Another sign of climate change!

Japonica with fall clusters

Rhododendron leaves with midge

Some of the new growth leaves of the rhododendron are curled up along their edge. Penn State Extension says that this is from a midge, a small wasp is eating the rhododendron leaves. I should spray with Isotox in the spring. As a precaution, cut off the leaves during December and dispose of them.

Many years after a short freeze in October the purple flowered rhodie by the pond would produce a few flowers. In 2015 it became quite warm for many weeks after a freeze. The rhodie was almost in full blossom. The bush had no flowers the next June.

November 8, 2015 – rhodie in blossom

PJM leaves are turning maroon

<u>Trees</u>: This is the time of the year when layers of bark come off the big birch; I would think this would happen in spring!

In the 1970's through the 1990's leaves were on the ground the first week of November, then it was the second week, now in 2019 it is the fourth week going into December. Back in the 1970's it took a couple of hours to rake the leaves, now 50 years later it takes seven hours. There are two reasons. One, those 10-foot trees are now 60- to 80-feet tall. (It is estimated that a mature oak has 700,000 leaves.) Two, I am much older. In 2018, I decided not to rake the leaves off the garden, just let them act as mulch and groundcover. This is the natural way and the birds and critter will like it. Many garden books say to prevent overwintering of insects, clean off all the leaves. Some will say use them as a mulch. Others say beware, that if there are too many leaves, the mice will make a nest around the trunk and eat the bark during the cold winter. This will girdle the trunk, which will eventually kill the shrub. So, I choose to let the leaves on the ground. The white-throated, song sparrows, juncos, and squirrels all like to scratch those leaves during the fall and winter to go after insects, seeds, and nuts for sustenance. Under those leaves they also find moisture. In early spring pass through migrating species like to scratch those leaves before continuing to the north.

Brown thrasher scratching the oak leaves

However, come spring when the winds blow the leaves they continuously go into the pond and on the lawn. So, this time when I rake them onto an 8- by 10-foot tarp instead of dragging them to the circle for the township to pick them up, I drag them up the hill to the compost pile. Do observe that the birds in the winter and spring do scratch the garden area much more than when it was cleaned off. At dusk the fallen leaves give off their own distinct fragrance.

Leaf pick-up: The last township leaf pick-up depends on the weather of the season. One season I lost track of time for the last pick up of the leaves by the township. I used the light of a full moon at 6:00 p.m. to sweep and rake the leaves to the curb for the next day. When it is warm the leaves do not come down until the second week of November; they extended pick-up to the first week of December. Climate change?

Township leaf collection

Vegetables: Asparagus has small red berries and the ferns are light green or brown.

During first three weeks of November is when many of the trees reach peak color and drop their leaves. In 2005 the weather station says that the first week is the best time for leaf color. Now peak is the third or fourth week in 2019. The reason I say "many," is that the pin oak, Barnet pear, and some apples keep their leaves until late December or sometimes straight into March. The first week of November little brown crosses are falling from the birch trees.

The buckthorn tree is light green. Blueberry bushes are all orange-red. Maple tree B is getting a touch of red around all the leaves. Maple C is at peak. Overnight the silver maple by the steps has changed 100 percent to orange-yellow-red. The three big maples have dropped 100 percent of their leaves. When an oak leaf lands, it always lands stem first! Leaves are all over the ground. Rake them onto an 8- by 10-foot tarp which I drag to the circle.

Waiting to be raked up

Leaves over the net

The net is over the pond to keep the leaves out but it is raised enough at each end so the birds and squirrels can get under it to get water.

Second Week

Weather: Venus-Jupiter-moon all lined up at 6:30 p.m. facing southeast. With the proper line up this combination could be the making of the Christmas Star! When there are water crystals in the cool night sky a big halo may appear around a full moon.

Animals: Squirrels are still running around with walnuts in their jaws to find a burial spot.

Birds: At 2:00 p.m. a flock of 20 of what could have been Cape May warblers were flying very fast around the third tier. They did not land for more than a couple of seconds, then flew up and re-landed. This went on for five minutes before they took off, not to return. The resident birds did not seem to take notice of what was going on. I think these were the same birds that were at the fall sedum the day before!

The sun is setting; crows are on the top leaders of the spruce trees. A turkey vulture soars overhead. Three blue jays take cracked eggshells and peanuts. A jay drops directly down from the peanut globe perch to catch a nut in midair. A red-bellied woodpecker chases a blue jay off the main feeder. Nuthatch and chickadee are tucking seeds under lifted bark of the silver maples. House sparrows are active at the houses. Three goldfinches are on the thistle; they are also chasing each other. A male cardinal does not like the male purple finch that has landed next to him and chases him away. Find a headless goldfinch, which is a sign that a screech owl is in the area. Red-breasted nuthatch gets a whole peanut and pecks and pecks at it as he rotates it in his claws like a cob of corn. House finches are in poplar branches eating seedpods. Five flocks of 20 each of Canada geese fly over.

Canada geese

House finch (top of the picture) eat poplar seedpods

<u>Insects</u>: Spot a snout butterfly whose wings look like chewed leaves. If it is warm a yellow sulphur may be is flying about.

Perennials: Pine needles are hanging from the dogwood branches on the hill. A few white irises in the lower garden have new buds. The autumn clematis is still bright green. Cattail stalks are yellow and broken. After a frost all the leaves have fallen off the trumpet vine overnight. Pampas grass is in full tassel. Cut back the globe thistle, loosestrife, fall sedum, and brown fern fronds.

Shrubs: The big 60-year-old all white azalea has 10 percent yellow leaves; these azaleas do not keep all their leaves. Most of the leaves will turn yellow. It is good to add sulfur at the root line. The burning bush is red. It is 37F, rhodie leaves are horizontal. The japonica by the gate has deep purple clusters.

Trees: Since 2005 the dogwood tree berries in the backyard ripen from October into November. Climate change?

Vegetables: Brussels sprouts are coming along fine, cut off two four-foot-long stems for Thanksgiving. A lettuce called Zen comes back in October it is like bok-choy. Pick it and sauté in butter. Cover rhubarb and asparagus with dry cow manure and straw. On the 15th in 2009 and 2015 Swiss chard still has full leaves. Broccoli is still blossoming. Cover beets with leaves. Buy a bale of straw to cover the strawberries and perennials after the ground is frozen. One year I forgot to cover the bale with heavy plastic, and we got a hard rain a couple of times before a freeze. When I decided to put the straw on the plants, had one big heavy ice cube. Used a pitchfork to rip it apart; not easy.

Third Week

When I was first taking notes of the plants, I was not aware of the climate change until the early 2000's. Now reviewing the data for past 18 years the change in plant occurrences within the cycle is quite apparent. Ferns that change from green to brown in the first week now change in the third week. The same has happened with the dropping of pine needles and pine and spruce cones. Species of yellow iris that rebloom, rebloomed the first week of November in 2004 are now reblooming the first week of December in 2015.

Weather: Temperature for the third week can be a yo-yo. Dense fog can quickly form in the evening or be there in the early morning. On the 19th in 2008 the daytime high was 34F, 15 degrees below normal. Winds were nine mph with gusts up to 24 mph from the northwest. A year later, October of 2009, temperatures for the nation were below normal; for November the national and local temperatures are above normal. Mold is moderate.

Normally we do not get snow in November; the record in 1962 was 3.2 inches. In 2018 the weather station predicts one to three inches of snow starting at 10:30 a.m., which comes and goes. Then at 12:00 noon it starts to snow hard at two inches or more per hour. Schools let out, then businesses were quick to follow. In a short time, there were traffic jams all over, not only on local roads but also the major highways. It took people seven to 12 hours to get home that normally would be 45 minutes. People abandoned their cars. Many ran out of gas. The plows could not do their job because of all the jammed cars and trucks. When it stopped there was eight inches of snow, the second highest for a November. The record was in 1938 when 15 inches fell on two consecutive days. Traffic did not start to move again until the next morning. People were not prepared. The finger pointed to a very bad call from the weather departments.

From my standpoint, the leaf piles in the curb were now buried under not only the fallen snow but buried further from the plowed snow. Most of the oak leaves stayed on the trees during the storm. Then it did snow oak leaves, which gave an unusual sight of the brown on white.

Pin oak leaves on snow

<u>Animals</u>: At 6:30 a.m. it is light enough to see a rabbit. The squirrels always arrive at least 15 minutes after daylight. There are three families (drays) of squirrels; one from the west, one from the north, and one from the south. They arrive separately and at different times. If they do arrive together, they usually chase each other until only one family is left or each family stays at a different end of the yard.

I see you and you see me

A squirrel rocks its body up and down and moves its head up and down as it gets ready to attempt to jump on the feeder cable. The squirrel jumps upward and over five feet horizontally from the top of the seven-foot-tall pole to the green cable from which the peanut butter logs hang. It balances itself, then goes upside down and goes paw over paw to the feeders. Hey, congratulations, it deserves the feed. After that I removed pole.

Made it

I cut a 10-inch diameter plastic bowl to three inches high. I now have rat deflector like they use on the ropes of docked ships. I mount the deflector through the cable. When the squirrel comes, it gets on the awning rail and looks at the deflector, then at me, then the deflector. It decides to leave. The squirrel comes back and looks again from the rail. It decides not to make the jump. I won this round.

A squirrel jumps onto the cable holding the birdfeeders; I put grease on the cable. That stops the jumping.

A squirrel jumps two feet high onto the edge of an 18-inch diameter disk that is screwed horizontally to the hanging safflower feeder. The squirrel's weight tilts the whole feeder. It slides off, hits the ground, and goes away. It is not to be undone. Two days later it jumps to the top of the feeder and then slips down to the disk and the feeder without tilting the disk. It had time to think it over or got coached. All that effort and then it found out the feeder was filled with safflower which squirrels don't care for.

A squirrel tries to jump from one tree to another, misses and falls 20 feet. It just gets up and runs to the pond to get a drink. They have resilient bodies.

Squirrel getting a drink

One squirrel chases another one up and down and around the spruce trees; their tails are corkscrewing as they run. Another buries a peanut where I just planted the tulip bulbs. Then another comes with a walnut and buries it in the same spot! During the winter apples and acorns are put out for the squirrels.

I have five metal pails with lids where I store the birdfeed at the back of the house. One is 18 inches tall and the others are 12 inches tall. Each one is filled with different types of feed. Usually I go out at sunrise to replenish the feed. I noticed that the lid was off one of the cans. Looking into the can with sunflower seeds I saw that something had a feast. Being up in years, I said to myself, I must have forgotten to put on the lid. A week later I found the lid of the split peanut can off. A squirrel ran off as I approached the can. I looked in and there was a panicked field mouse running around in circles trying to get out. I tried to scoop it out, but it was too fast. Finally, I tipped the can so it could jump out. I must be slipping about putting the lids back on. Then one afternoon sitting in the family room I noticed movement outside by the cans. There was a squirrel on its outstretched hind legs with the tip of its nose against the rim of the lid pushing it upward. Mystery solved. Now I make sure the locking handle is in place before going back in the house.

Feed cans with locked handles

Planted the hawthorn tree for berries for the birds. The berries ripen to an orange-red the second week of November. The third week the squirrels discover them and literally go to all lengths to get the farthest berry on a branch.

Going down

Going up
stretched out squirrel to get that last red berry

Birds: Some species come to the feeders in the twilight of dawn, then others after sunrise. It is not always in the same order. Many times, the cardinals, mourning doves, Carolina wrens, and white-throated sparrows are here at twilight. Then the chickadee arrives, followed by blue jays, red-bellied woodpecker, song sparrow, downy woodpecker, house finches, female house sparrows (the male is a late sleeper), white-bellied nuthatch, junco, and goldfinch. By 7:30 a.m. the red-breasted nuthatch and starlings arrive. At 8:30 a.m. the gang is still here. Two blue jays are taking the eggshells. Sometimes the junco does not arrive until 9:30 a.m. This is a typical population of the gang for the third week of November.

The crotch between upright branches of the maple tree is hollowed out and has become a reservoir for water. This is bad for the tree because it is a prime rotting spot but good for woodpeckers to take a drink.

Hundreds of overwintering Canada geese are flying over Dorney Park at 9:30 a.m. A flock of 30 crows in their usual loose formation go east.

Driving to the shopping center I was lucky to a catch a glimpse of a bird larger than a red-tailed or a broad-winged hawk soaring above the highway. I learned that a bald eagle had been seen in a nearby park.

At 1:00 p.m., 69F, a turkey vulture is soaring overhead. Starlings are whistling from the poplar tree. Their whistling sound is just like ours, but they do not pucker up their lips! A song sparrow is jumping at itself in the mirror below the raised feeder. As a red-breasted nuthatch picks upside down at the split peanuts feeder it keeps flapping its wings to stabilize itself. Titmice eat the punks on a cattail. A big dove flies up into an evergreen. A minute later it flies out of the evergreen with a hawk right behind it. From 6:15 to 6:30 p.m. a great horned owl is hooting from the back maple.

The blue jay population has increased, most of the juncos have arrived, the hairy woodpecker has left, the house finch population has increased. The red-bellied woodpecker leaves but comes back in December. Why? The sharp-shinned arrives to give havoc. The titmouse population has doubled. The crickets have hibernated. I don't know where they go but there is no more chirping. The boxelder bug population is reducing but not enough. I still spot turkey vultures soaring. From 9:00 to 11:00 a.m. exercising Canada geese fly. The gang comes back for brunch. From 3:00 to 4:00 p.m. a sharpie comes through. I find the feathers of a cardinal that a hawk had snatched on the patio. I keep tracking little feathers into the house for a week!

It's 6:30 a.m., 31F. A nor'easter is coming tonight. No birds until 7:00 a.m. and then only a few. At 4:28 p.m., almost dark, very overcast, very damp, still only a few birds come to the feeder. They know a storm is coming.

"What do they eat?" In the late fall and winter, as natural seed gets harder to come by, the birds come to the feeders. Some eat multiple choices of feed; others strictly stay with one type.

Bread is not good for the digestive system of most wild species because of the additives and yeast. No bacon grease, it contains compounds that are not good for the birds.

Eggshells are a good supply of calcium. They should be put in an oven at 240F for 20 minutes, not a microwave, or your house will smell like rotted eggs. Crumble the cooled shells before putting them out on a feeder. When put on the ground the shells may attract raccoons. Blue jay and titmouse go for the eggshells.

Eggshells

Crumbled up

No grapes or orange slices in the fall, they may go moldy. Millet is for ground feeding species like the sparrow and junco.

Millet

Mixed seed

Some birds do not eat milo; the seeds will produce unwanted weeds. Mixed seed arc eaten by every species. Except for an occasional downy or chickadee, peanut butter usually stays untouched in the cold weather.

Blue jays, titmice, squirrels, and raccoons like whole peanuts. Almost all bird species plus the squirrels like the split peanuts.

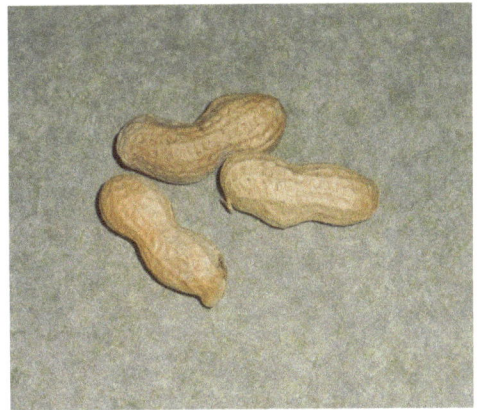

Whole peanut

Split peanut

There are no takers for caramel popcorn.

Safflower is enjoyed by the mourning dove (can eat 100 per minute), house finch, Carolina wren, chickadee, titmouse, and cardinal; squirrels do not eat the seeds unless it is very cold.

Safflower

Suet cage

Suet cage with mixed suet is enjoyed by woodpeckers, titmice, starlings, and nuthatches. No takers for pure suet in the winter.

Black sunflower seeds are higher in protein than the stripped. The cards, finches, and squirrels like them both.

Black sunflower seed

Black striped sunflower seed

Goldfinch and chickadee eat the thistle.

TRENDS DUE TO CLIMATE CHANGE

Thistle

I put a thermostatically controlled water heater in a birdbath. Almost all the birds, chipmunks, and squirrels come for a drink.

Water heater and small rock for birds to stand on

Fruit: Cut back the raspberry brambles.

Insects: If it is warm small swarms of mosquitoes are in the backyard.

Lawn: Over the next two weeks spread the lawn with 10-25-10 and do a soil test to determine if lime is needed.

Perennials: At 61F, the iris behind the pond has more buds, two are in bloom. If there is a freeze the buds and blooms will turn to mush.

Shrubs: Mound topsoil, not mulch, around the trunk of the roses; mulch encourages pests to girdle the trunks. Some of the leaves on the mountain laurel on the west end are turning yellow.

Trees: From the high wind, the oak has lost all its leaves. In the evening I smell fresh dampness from the newly fallen leaves.

As I open a tarp to rake the leaves onto, I find stink bugs in the folded laps. They had already nestled in for the winter. Many years, five to 10 percent of the leaves remain on the tree until the beginning of March. When that happens, you can hear the sound of winter as the leaves rattle during a night gale.

The poplar tree releases a twirling cloud of hundreds of seedpods during the second to third week. In 2014 the pod dropping now starts in the first week of December. False start, in 2019 the drop was back during the second week. The pods have a bad habit of landing by the front and back doorway. Even though I wipe my feet again and again, I still track them into the house and even find them in the bedroom. I should invent a vacuum cleaner in the floor by the threshold to suck the leaves off my shoes.

Poplar tree seedpods (samaras)

The hawthorn tree leaves usually turn yellow this week.

Vegetables: Dig up the whole Brussels sprouts plant. The stem is four- to six-feet tall. Spray the plant with water at a strong nozzle setting to wash off aphids. Put the plant in the garage so I can pick the sprouts as desired.

Aphids on Brussels sprouts

November 2015 – Brussels sprouts

Fourth Week

<u>Weather</u>: I get out the snowblower and give it a few yanks and with a chug-chug the 60-year-old starts up. Good for another season. I put one ounce of gasoline stabilizer in the tank of the lawn mower and let it run for five minutes. Then top off the tank, put a piece of plastic wrap over the opening of the tank, screw on the top and put the mower away for winter storage.

Two things happen this week. One, we are likely to get a frost and, two, the hurricane season ends.

<u>Animals</u>: One afternoon a pair of squirrels were having a good time making whoopee on a branch of the west dogwood.

Whoopee

<u>Birds</u>: On the 29th we have a heavy morning frost. When I get the paper, the frozen grass bends and crunches underfoot, the birdbaths are frozen, the pond has a light coating of ice. As the sun rises four crows are going west and a sharpie sits on the lower maple branch. When it is just becoming light, a song sparrow and cardinal silhouette can be made out. It is light, 7:10 a.m., juncos, blue jays, and a squirrel are at the feeders. I hear a crow cawing from the poplar lower branch. I go out to investigate; the crow does not fly away as I approach. I do not find the cause for it being upset. Hundreds of local geese go over at 9:00 a.m. A cardinal chases the house sparrows from the feeder. Noon, 54F, a pair of broad-winged hawks soar in the thermals. At 5:00 p.m., dusk, 20 local ducks circle and circle around the neighborhood. The fourth week it is 26F. Between 6:00 and 6:30 p.m. I hear a great horned owl in the back trees. A full moon is peeking through the tall evergreens as it sits behind the house on the hill. On the 23rd at 7:00 p.m., I spot a great blue heron flying toward its nest.

Another day, 6:00 a.m., 41F, light rain. It has been mild for over a week. Carolina wren, house sparrow, and the nuthatches are here in the twilight. Light, 7:10 a.m., the song sparrows have come. At 7:30 a.m. chickadee, blue jay, cardinal, house finch, white-throated sparrow, and dove arrive. A red-bellied woodpecker chases a house sparrow off the peanut flake tube. They are all gone by 8:00 a.m. Fifty geese fly over. At 9:20 a.m. there are three pair of house sparrows, a pair of house finch, goldfinch, red-breasted nuthatch, fox sparrow, and a brown thrasher. Quite a combination. The fox sparrow will stay on for a few weeks. The thrasher will only be here a few days. I feel lucky to see it at all.

Fox sparrow

Carolina wren

I look out to see a broad-winged hawk sitting in the branch of the dogwood by the pond. I take my camera to get a close-up. I get within 10 feet; it looks at me and I at it. It does not move. I take pictures and go back into the house. I look out again, it is still on the branch looking down into the pond. It then drops down to the edge of the pond, walks onto the netting, bends down, and starts pecking. Feathers fly into the air. I then realized why it had not moved when I went out. It had dropped or its kill had landed in the net and it wanted it. I let it consume its meal.

It wants its meal

A trait I have seen with the sharpies is they will go to the ground level branches of the hemlock hedge to hide or rest. An unwary bird or small mammal will suddenly be ambushed by the hawk from the undergrowth.

On the 30th at 7:00 a.m. 10 species are here. At 4:00 p.m. a tight flock of 100 starlings are going east. A loose flock of 100 robins go west. Flocks of 30 local geese go to the field in the east.

Lawn: Do not put down weed killer, it is ineffective this time of the year, can still put down fertilizer. Make the last lawn cutting of the year. This ritual has gone from the first week of November to the fourth week. In 2010 and 2017 the last cut was done during the first week of December. The last week can be warm and humid, still conducive to mushroom growth in the lawn.

Perennials: All the leaves are off the trumpet vine. Goldenrod still has whitish-gray flowers. Butterfly weed forms pods. Put urea (nitrogen) on the flower gardens. Dig up summer bulbs. Augmented soil to the garden while replanting daffodil bulbs. Put dahlia tubers in a box with peat moss and store in the cellar. In 2010 and 2015 after few warm days of 60 to 70F, daffodils were up a quarter-of-an-inch.

Pond: Take the net off the pond. If I leave it on, heavy snow will rip it. I used to remove the net on the fourth week; with the climate change, removal may be the second week of December. Removal takes longer than putting it on because there are a lot of leaves that have formed collection pockets in the net. I use a leaf blower to get rid of the pockets. After laying the net on the ground, little wet branches get caught in the netting. I learned not to wear anything with buttons because they manage to get tangled in the netting as I roll it up. It takes an hour-and-a-half to get the net into a big plastic bag. If there is a storm after the removal, I use a net to skim off the leaves.

Shrubs: The upper shrubs feel the brunt of the winter winds. Spray the leafy shrubs with winterizer wax. This spray seals the pores of the leaves. I do not want to seal the leaves too early. I wait for a warm day in November or December to spray. In late November the burning bushes may still be brilliant red.

Fall Leaves

Leaves are a continuation of October. Of course, we picture fall as the leaves turning to a kaleidoscope of color. Living in the northeast we look forward to seeing the change of colors of the leaves from the various greens to the orange-red-yellow-brown-maroon colors. We ask when they will change? How long will the colors last? When will the leaves drop? When will the last one drop?

Another question is why do they change? It is a combination of the amount of sun during the day, the temperature of the day, and amount of precipitation in the ground. In September the newspapers show a map where and when the peak color of foliage will be by the days of the months. The changes start in New England and goes through upper New York State and progresses south into Pennsylvania. The map shows Allentown's peak as late October. Part of this progress from north to south is due to the tilt of the earth on its axis from going from summer to fall. The tilt results in less daylight in the north as the months progress. Most of us see the change as the green changing to vibrant colors but experts take the romance out of the changeover. The experts say "No." The green does not change to a color, rather the bright color(s) was already there and as the photosynthesis chlorophyll factory of the tree shuts down for the season, the green is no longer produced and the underneath color appears.

At the base of a leaf is an abscission layer (separation cells). During the summer small tubes in this layer return water and food back into the tree. With reduction of sunlight and heat the layer swells to form a cork-like material. Glucose and waste are trapped in the leaf. Without water the chlorophyll (green color) disappears.

There are many factors effecting color. Summer rains mean healthy fall trees and bright colors. A hot-dry summer results in less color. Leaves may wither and drop. Brightest colors are inspired by sunny fall days and cool nights below 45F (but not freezing) because leaves cannot use all the sugar produced during the day.

Unseasonable October warmth slows colors, so they last longer. Dry soil brings early changes. A damp cold October explodes the colors. Cloudy October days and warm nights equals drab colors. High nighttime temperatures slow the change. Heavy frost brings out color, but colors will not last long. Heavy rain and Indian summer extend color.

Fall Colors of Leaves of Trees and Shrubs							
Type	Color	Type	Color	Type	Color	Type	Color
Apple	Dark green	Dogwood	Maroon	Pear	Maroon-green	Sugar maple	Yellow
Azalea	Dark green	Japanese maple	Bright red	Pin oak	Reddish-maroon	Serviceberry	Reddish-orange
Blueberry	Red	Lilac	Light green	PJM rhododendron	Dark maroon	Tulip tree or Yellow poplar	Yellow turns brown
Burning bush	Red	Mollis azalea	Pinkish-green	Red maple	Yellowish-brown		
Cattail	Light brown	Paper birch	Yellow	Rhododendron	Hunters green		

The next question is "When do plants change color?" Well, fortunately for us they do not all change at once, so depending on the species the phenomenon can last for a week or a couple of weeks. There is a neighbor's tree that changes completely red almost two weeks before mine. Then there is another early changer that is different in that the whole west side changes red and then the whole east side changes yellow. It then drops its leaves before most other trees have even started to change. Or there are the walnuts that have a complete drop of leaves by early October. The barren branches against a setting sun sure makes it look like fall is here.

The time of the change-over not only changes with each type of tree but also changes within each sub-type. For example, the basic maple has sub-types of swamp, sugar, red and Japanese.

As stated in the October review, peak color will occur at a later date if there is an abnormally high temperature and dry spell in September. What is normal? I took what records I had over a 10-year-span and looked at when the process started and ended and took an average.

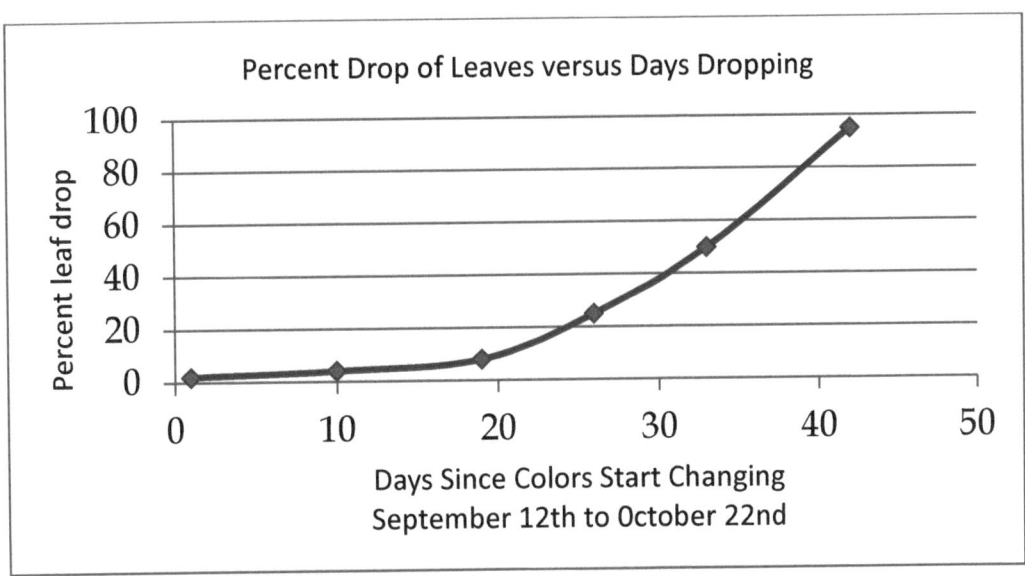

The chart shows the change from September through October. With the climate change, it is still the same principle, but may go into December.

Now that the tree is at peak color, a homeowner may want to know when the leaves will fall and when that last leaf will come down. Again, this depends on the type of tree. Say that 1993, 2004, and 2006 were normal temperature-precipitation periods and length of time that the leaves stayed on the trees, the hot, dry years were usually 10 days longer. The 2007 and 2017 Octobers were the warmest, driest on record, which pushed the process another 10 days beyond the 2005 season.

The color change occurs with the fall's reduction of daily sunlight. The plant's ability to produce chlorophyll (green) is reduced; likewise, water and nutrients can no longer go into the leaves. With the lack of the chlorophyll, the bright colors of autumn are revealed. As that happens the plant goes into its winter state of being. Photosynthesis decreases and instead of producing chlorophyll, the plant cells produce an antifreeze of sugars and amino acids, which go into the root system. The shutdown of the leaves causes two different cells to form where the stem of the leaf joins the branch. On the stem's side the cell is soft; on the branch side the cell is in the form of a waxy impermeable seal. With the shutdown of the chlorophyll factory the veins which are holding the leaf to the branch become weak allowing the leaf to break off and fall. When this happens, a scar is formed on the branch on which a bud for next year's growth is formed.

From what I said here and what I reported above are contradictory. For in one case I said that with the reduction of sunlight the water can no longer get to the leaves which shuts down the chlorophyll factory and the cells to form. Yet, when we have a dry, hot September there is already a lack of water, so you would think that the leaves would fall early. Maybe during the hot Septembers, the trees semi-close their photosynthesis to retain water in the upper systems of tree. Then it takes longer for their factory to shut down in October and hence the leaves stay on longer. Or due to lack of water the leaves are lighter, so they do not have the weight to readily break.

I have recorded the date when a tree was at its peak color, then recorded the date when 10 percent of the leaves had dropped off, then did it for 20, 40, 60, 80, 90, and finally 100 or very close to 100 percent. Again, like the occurrence of peak color, the days to drop depends on the climate conditions for the year.

Of course, it is very subjective of when you say the amount of dropped leaves is at the various percentiles. I have done this for the maples, dogwood, birch, popular, fruit, oak, and serviceberry tree. In almost all instances, the graph of percent of dropped leaves to the days is a straight line. The slope or rate of change the leaves fall varies with the tree type.

Let's have some fun. Pick out a couple of trees, preferably those that you rake the leaves. Record the day that you judge that 10 percent of the leaves have dropped. Then do it again when 30, 40, and 60 percent drop has occurred. Now make a graph of the days on a horizontal line and percentage from zero to 100 on a vertical line. Plot the data you have taken. You should find that when you connect the points, it will be a straight line. When you extend the line to 100 percent, it will give a good insight on what day all the leaves of a tree have dropped. You will also find that the plot or line will have a different day of 100 percent drop for the various tree species.

The pin oak is one of the exceptions, it sometimes holds about 10 percent of its leaves until March of the next year.

Plot of when all the leaves will fall

The way the leaves fall is different from type to type. Some leaves like the poplar are heavier and come straight down; others like the sugar maple dry out a little and then slowly float down. When most of the red maple releases, it is like a rain of leaves and even the birds must dodge them as they drop down. They land on the pampas grass reeds, then bounce off one reed to another until they have landed.

Some maples start to drop their leaves from the bottom of the tree and the tree gets bare in an upward process. Other maples go the other way around and start dropping from the top. With this type of tree, it is something to watch because at the top one leaf lets go, and on the way down it hits another leaf. That leaf releases another leaf. Those two hit two more, so now the four hit others. In a short time, you are observing a cascade of leaves. In one or two days the entire tree will have a 90 to 100 percent drop.

The Japanese maple gets bright red and then dull red and then suddenly a key is turned and the whole tree drops its leaves all at once. Birch leaves are small and light and drift lazily to the ground.

I had a couple of weeping willows. They grow tall and fast. Their branches form archways that sway gracefully as a wind blows through them. However, when their long slender leaves fall it is not just leaves that fall but their whip-like end branches that are two to three feet long also drop off. These whip-like branches are difficult to rake. They roll up in a ball and get caught in the teeth of the rake.

The pin oak leaf is something different. It is curled around the edges. Its stem gets stiff. When it is released, it does not come straight down. It sails and turns; it goes down and up and over and under; it lands close by or lands far away. How can this be? Well, I will tell you what is happening. When I was a small boy my father bought me a book, "The Night Before Christmas." The book had illustrations of Santa landing on the roof, coming down the chimney, decorating the tree, and with a finger to his nose up the chimney he goes. However, at the top and bottom of each page was a border of elves with sleigh reins guiding an oak leaf down to earth. To this day when I see an oak leaf being blown over the roof, I know who is guiding that leaf to its landing.

Over the years I found that evergreens are not completely evergreen. One spring, I bought a healthy five-foot sugar pine to put in the garden in the back west-corner of the house. During the summer it grew eight inches, but then in early October it started to lose its needles. A few releases at first and then they all came down rapidly. I thought the tree was dying. Since we had been in a drought, I watered it heavily. Finally, I called the nursery and told them what was going on. To my chagrin there was laughter at the other end. The man apologized and told me that this pine sheds some of its needles in the fall just like a maple. Over the next eight years my little tree grew to 40 feet; it shed its needles profusely each fall into the roof gutters. I also found the branches to be brittle when we had a heavy snow. Between the gutters getting full and the possibility of the tree snapping and landing on the roof, I had to cut it down. I used the needles to mulch the blueberry bushes. The cones of the pines, spruce, and hemlocks fall throughout October. The cones close tight in the cool weather but will open when you bring them inside for decorations.

Most leaves and needles that drop from the evergreens are soft. The one to watch out for is the holly trees because their fallen leaves get brown and hard. When you rake them or pick them up, they bite back. I use rubber coated gloves.

Raking the poplar and some of the maple leaves can be laborious when they are wet but others like the oak can even be fun because they are light and pile easily. In 1995 the 8- by 10-foot smooth-coated blue tarps became available; what a God send. I lay it down on the lawn and rake the leaves on it, pick up the side to roll all the leave in the center, then just pick up two corners at one end and slide the tarp across the lawn out to the circle and just turn the tarp upside down to empty it out.

The raking all seems nice and a fresh-air thing to do on a crisp autumn day or evening, however, even on the most seasoned or callused hands, raking invariably produces blisters between your thumb and forefinger. I have tried different types of gloves, different material gloves, put band-aids on my hand before raking, two layers of gloves, held my hand in different positions on the handle, and different types of handles. I almost gave up and said, "Let the blisters come," until one day I put on a pair of those plastic surgical gloves under a pair of work gloves. As I raked a complete wet layer of perspiration developed between my skin and the surgical glove. The result was there were no blisters. I have now been doing this successfully for many years. However, I used the same pair two or three times, not realizing that a virus had developed in the sweating gloves. A couple months later I got a wart on the back of my left hand. My doctor told me where it came from. The next fall I only used a pair of gloves once and then got rid of them.

The last stage of the fall leaves is getting rid of them. The circle being an asphalt surface, no burning was allowed. For the first few years there were some empty lots behind the house so guess where the leaves went. After houses were built on the lots, I went to plan B. The trees were still relatively small, so I made a burning drum. I got a 55-gallon drum, put holes in along the bottom edge and raised the drum on cinder blocks. It was raised to prevent the grass from catching fire and to give it a good draft. I put the drum up on the hill. I used this for quite a few years. The burning leaves gave out their distinctive fall fragrance. I had an iron grate I could put on the top of the barrel to prevent burning cinders from blowing out.

As years went on, we were not allowed to do that because of air pollution and city codes. That was okay because there were too many leaves to burn. The township now wanted us to pile the leaves in the street. Then they would come with a front-end loader and scoop the leaves into a large dump trunk. After a couple of years, the front-end loader was replaced with a large vacuum cleaner built into a special machine that was pulled by the dump trunk. Even though it had a canopy over the backend, if the leaves were dry, large clouds of chopped leaf dust would rise above the truck and spread everywhere. Talk about air pollution! Everything was not good with piling of leaves along the curbs. Occasionally you read about someone parking their car over a pile of leaves and the heat from the car's catalytic converter would ignite the leaves and goodbye car.

When I visited my sister in November of 2005, her gardener came to rake the leaves for the last pick up. He had a team of four men. They cleaned up her 60- by 100-foot yard in less than 30 minutes. When they left, there were 16 big plastic bags of leaves on the front curb. No more burning. Everybody had these anti-environmental plastic bags of leaves. To make matters worse, the garbagemen would only take 10 bags per household per time. Just think of some dumpsite with all these thousands of unopened plastic bags filled with compost pile material!

When I was kid on Long Island in the 1940, we had swamp or Norway maples lining the streets. They dropped big heavy leaves; we would pile them in rows along the curbs and burn them. There would be fires up and down the streets. That is when the fragrance of the leaves really permeated the air. There was no such thing as air pollution control. We would take potatoes and put them in the middle of the piles of leaves before we started the fire. I don't know why but we called these baked potatoes Mickey's, "No, it was not after Mickey Mouse." In the 1940's there was no aluminum foil. After the fire would go out, we searched for the Mickey's. They were usually black on the outside. The baked skin would form a thick, crunchy, blackened crust and the tuber inside would be steamy white. We would take them in the house and put lots of butter, salt, and pepper on the tuber. The whole thing was good. The thick crust was like a special candy and the tuber was delicious in its own special way in that it absorbed the flavor of the woodsy leaves. Mom would say that that the black charcoal of the crust was good to clean your teeth.

November 22, 2007
first year, black oak, pin oak in background

2019 – black oak, 12 years later

Still raking oak leaves on the 30th. Today is the last day for the township to pick up leaves. Japanese maple turns bright red. Some years its leaves hang on until spring.

The number of acorns on an oak varies. It may be three to five years before the first crop. Then it may take another seven years before the next crop. The result of a crop is usually of what happened three years before! A large crop is called a "mast." Then the squirrels really go into a planting spree.

November 22nd   *Leaves still hanging on in December*
Japanese maple

Why do oak and Japanese maple leaves hang on? It is called "Marcescence." Marcescence is the retention of dead plant organs that normally are shed.

Ten to 25 percent of leaves fail to drop in the fall because the abscission layer fails to completely develop. These leaves have a tough fiber called lignin. Lignin decomposes slowly giving extra bond of the leaves to the branch. They will finally drop off during the winter or spring.

A tree's colors are not always the same, even during the same season. Following is the color variation of the silver maple by the steps.

October 24-30, 1992 and 1993 is bright orange.

October 1999 is red.

October 28-31, 2004 is red.

October 28, 2005 is bright orange; two days later the leaves are orange-red.

October 27, 2006 it starts with yellow-red color changes, November 6th it changes to orange.

October 24, 2008 the tree is bright red, October 28th the color is yellow-orange, November 1st it turns to peak orange.

October 24, 2010 it is all red.

November 5, 2016 it is all red.

Vegetables: May still picking broccoli, Swiss chard, chives, thyme, and rosemary. In 2015 it was so warm during November and December that I did not do winter garden preparation until December 27th.

Review of Data

Weather: On the 25th in 2004, 6:50 a.m., with a warm 63F, it is still dark and cloudy; a storm is coming. As the cold front pushes through, the barometer drops to a low of 29.4 inches. Liquid is coming out the top of the New England barometer. Sunrise is at 7:00 a.m. with no bird activity; 8:00 a.m. little activity. At 9:00 a.m. blue jays are calling loudly; all the other birds have disappeared. At 9:15 a.m. the sky gets black and dumps heavy rain for 10 minutes. The temperature drops from 64F to 56F in 10 minutes. That's it! The month received 134 percent of normal precipitation. The 23rd in 2005, 44F, heavy wind and rain overnight, received 1.23 inches rain. Yo-yo temperatures.

On the 29th in 2005, a record 69F, 73 percent humidity, 30 mph winds, 23 degrees above normal, the low was 57F, 28 degrees above normal.

In 2006 it was unseasonably warm; 12 days have been 20 degrees above normal. All this is a good condition for an asthma attack. Fortunately, I don't have that problem. Some yellow bell flowers are out. November has been 4.5 degrees above normal.

In 2007 on the 23rd we have a high of 38F, 9 degrees below normal, and a low of 18F, 12 degrees below normal. A year later, almost to the day, we have a repeat performance. In 2008 on the 22nd we have a high of 32F, 16 degrees below normal, and a low of 20F, 11 degrees below normal. Then a year later the performance swings the other way. In 2009 the whole month has higher temperatures than normal. The bird activity for the last two weeks was low.

In 2008, the hurricane season has ended, but for the first time, six consecutive storms hit the mainland. They say it is due to the El Niño. Even though hurricane season has ended we still can get high winds and heavy rain.

November 2015 was one of the warmest on record. The weather people predicted that we will not have a mild winter. Last year's weather was 180 degrees from their predictions. They are predicting 40 inches of snow for this season. At 3:30 p.m. the southwestern sky had clouds that looked like a hand with spread out fingers swirling through gray paint. Reality, I only needed to use my new snowblower once and that was for a three-inch snow!

Fauna

<u>Birds</u>: The data on the quantity and date observed on the birds is from 2004 to 2024. For a period of 37 years, I have data for the species and quantity observed at and around the feeders. The amount of data makes a big difference in conclusions of what is happening to the population and when. A graph of the number of species in November from 2004 to 2021 shows the quantity going up with an average of 25 per year with a deceasing trend the last five years.

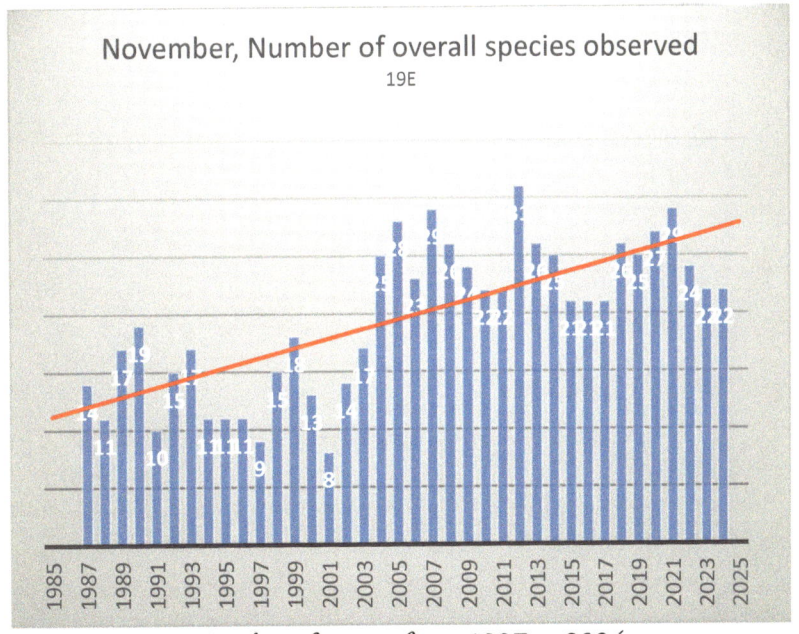

Number of species from 1987 to 2024

The population of the species observed is almost cyclic from 30s to 70s. The changes may be due to climate change, environmental changes of the surrounding area I live in, environmental changes I made to my property, and environmental changes due to my trees and shrubs growing and dying over the 50-year period.

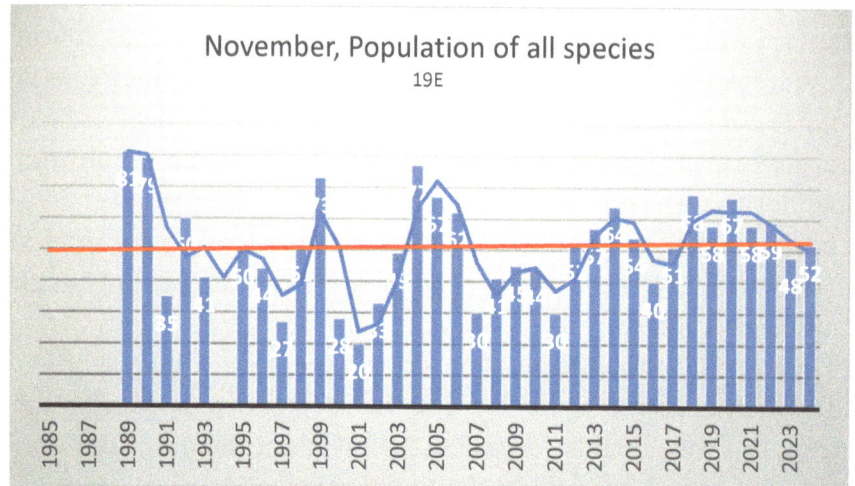

Population of overall species from 1989 to 2023

The main contributors to the high populations in the 1990's are the house sparrow, house mourning dove.

Graphs of their populations shows their changes.

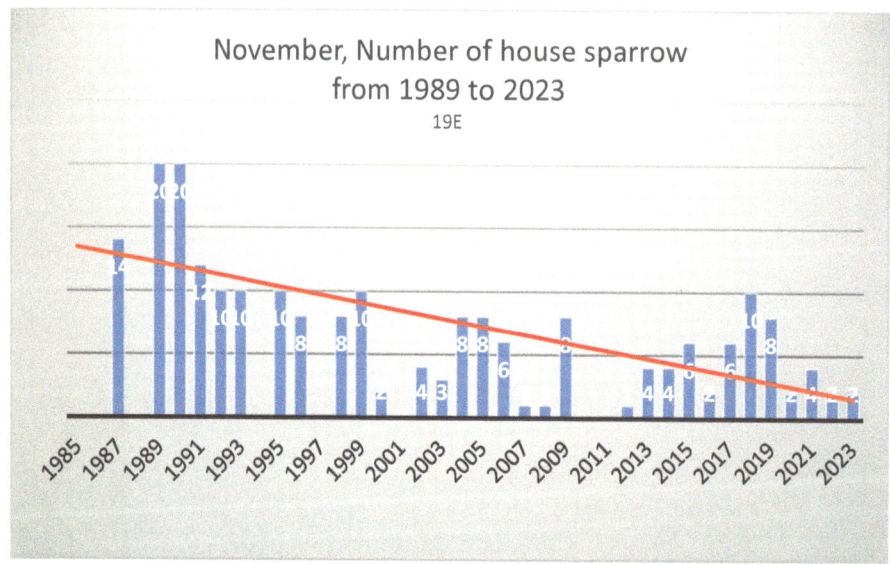

Population of house sparrow from 1989 to 2023

Cornell's site "All About Birds" shows a house sparrow decline similar to mine. Investigators found that the decline might be due to excess pollutant gases like carbon dioxide and possibly electronic wave admission.

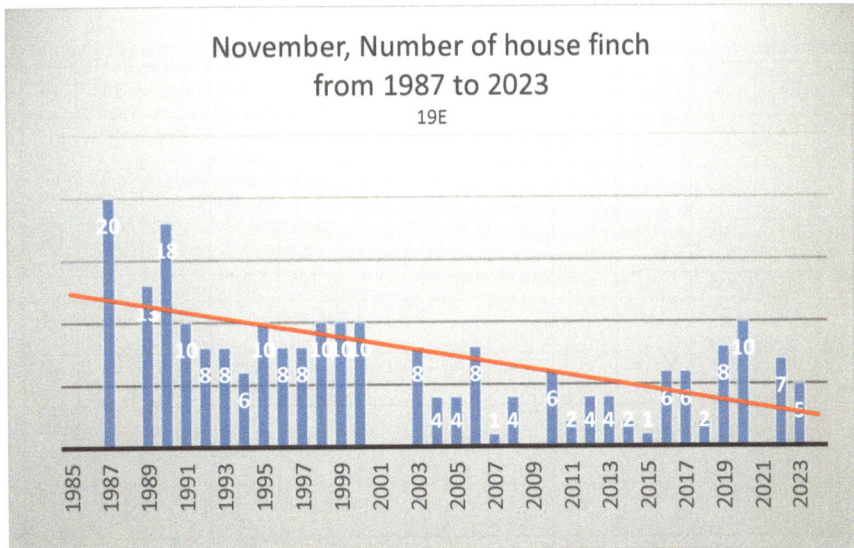

Population of house finch from 1987 to 2023

The decline of house finches is from parasitic diseases and conjunctivitis.

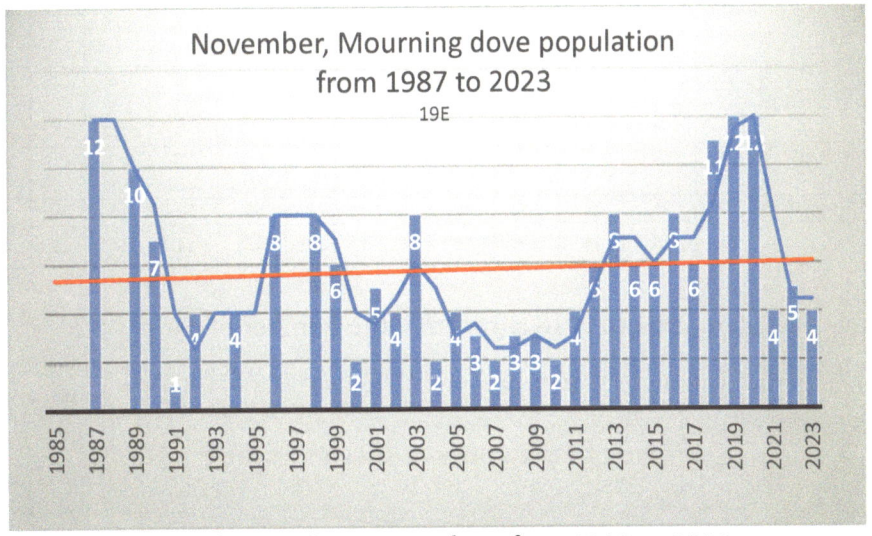

Population of mourning doves from 1989 to 2023

Decline of doves may be due to my cutting lower branches of evergreens and shrubs that the doves used for nesting, also killing from sharp-shinned and Cooper's hawks.

Other contributors to the high population of 1989 were the short visit of 10 cedar waxwing and the passing of a flock of 30 starlings.

Gang: The quantity species of the gang over 37 years has increased from 11 to between 13 to 15; some years it drops to seven. Some species are not observed some years, other species are overwintering or staying longer before migrating. The increase in the gang species from 2004 are the Carolina wren, downy woodpecker, goldfinch, song sparrow, and red- and white-breasted nuthatch. This may be due to the fact that the migration line has moved north hundreds of miles over the past six years.

NOVEMBER

Gang Members and Quantity for November																
R=resident, G=gang member, arrives=arrives from migration, y=young, t=teen. Date without a number is a single bird spotted. Quan/date																
Year 2000 Species/Quan.	04	05	06	07	08	09	10	11	12	13	14	15	16	17	18	19
Blue jay R, G still here	2/3	2/7	2/5	3/4	4/4	3/4	3/11	2/7 3/19	2/2	2/1 4/2	1/2 4/2	3/3	2/7	2/4	3/2	4/8
Cardinal R, G largest #	1+2y	1+1y	4+1y	1+2t	1+2t 8/17	1/3	2/1 3/26	2/8	6+2y/2	2/8 6/20	6+2	2/1	2/6	1/4	3/2	5/3
Carolina wren R,G arrives	2/7	1/1	2/24 5	1/2	1/1	2	2/7	1/6 2/14	2/9	1/4	1/2	3	1/5	1/1	1/1	2/11
Chickadee R, G	3/6	3/1	2/18	1/2	3/2	2/21	2/8	2/5	3y/2	1/1	2/2	2/1	2/6	2/9	2/10	2/4
Downy wood R, G first sighting	2/8	1/3	1/18	1/19	1/3	1/3	1/1	1/6	1/1 2/2		2/2	2/1	2/6	1/5		2/28
Goldfinch R, G still here	5/7 3/16	3/8 1/27	2/9	5/812		4/4 1/30	1/13	1/14	1/2 2/26	1/9 2/15	3/2 4/25	1/8	1/7	5/4	1/11	
					30											
Hairy wood															1/1	1/17
House finch R, G conjunctivitis	3/7 26	4/1 3	8/24	4y/11 18			6/19	2/5	4/11	4/2	2/1	1/1	6/7	6/23		8/25
House spr. R, G active	4/8 26	8/10 9	6/5	1/5	1/3	8/5		1/6	4/2 13	4/1	12/3		6/23	10/11	8/1	
Junco arrives	1/6 4/26	1/2 9/18	10/3 9/18	2/7	1/5 2/30	1/2 4/21	1/5 3/13	1/14 2/18	1/2	5/11 8/19	1/1 1/20	2/5 5/9	1/3	4/14	2/2	1/2
Mourning dove R, G largest number	2/6	4/1	2/9 3/25	2/ 19	3/10	3/5	2/7	3/6 4/14	6/5	4/1 11/23	16/30	4/1 6/14	3/6 8/7	6/11	11/11	12/3
Red-breast nut arrives	2/6	1/1	1/5		1/8	1/1			2/2	1/8			1/5	1/25	2/8	
Song sparrow R, G arrives	1/16	3/5	6/3	1/1	1/2	1/5	1/1	1/18	1/1 2/8	1/4 2/20	2/20	2/7				
Titmouse R, G	4/8	3/5	3/5		6/16	4/25	2/14	2/5	2/2	3/11	2/2	3/26	1/6	2/11	1/1	
White-breast nut arrives	1/7	1/1	1/5	1/10	1/8	1/1	2/14		1/2 2/11	1/2	2/1	1/4 2/27	1/5	1/4	2/1	2/8
White-throated arrives	1/6	1/4	1/3	1/1	1/8	1/1	2/1	1/6 2/10	1/1 2/2	1/2 2/4	2/2 2/20	4/1 6/9	1/6	3/16	2/1 4/13	2/4
Population	37	46	52	26	34	37	30	24	39	50	50	46	29	36	42	49
Total gang	15	15	15	13	14	14	12	12	15	15	14	14	13	14	13	12

The following chart from 2004 to 2018 shows that the number of gang species stays between 13 to 15 but increased in 2019.

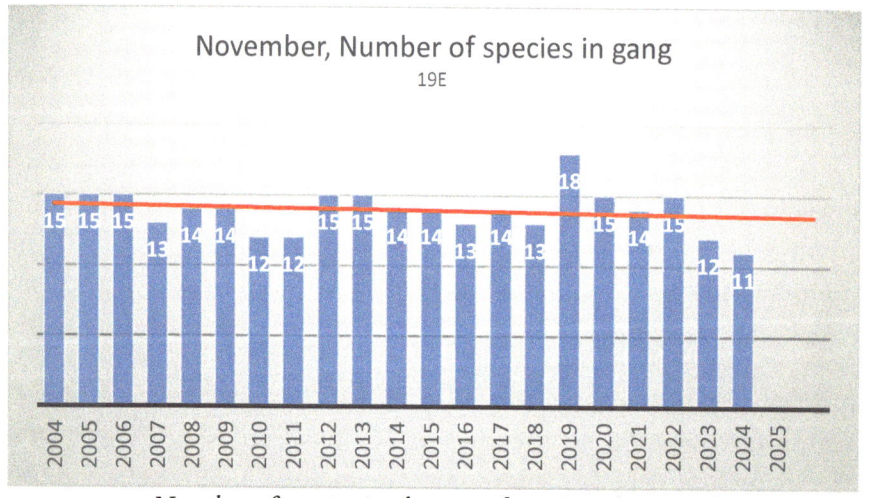

Number of species in the gang from 2004 to 2024

Again, the picture changes when the data goes back to 1987, the trend is upward.

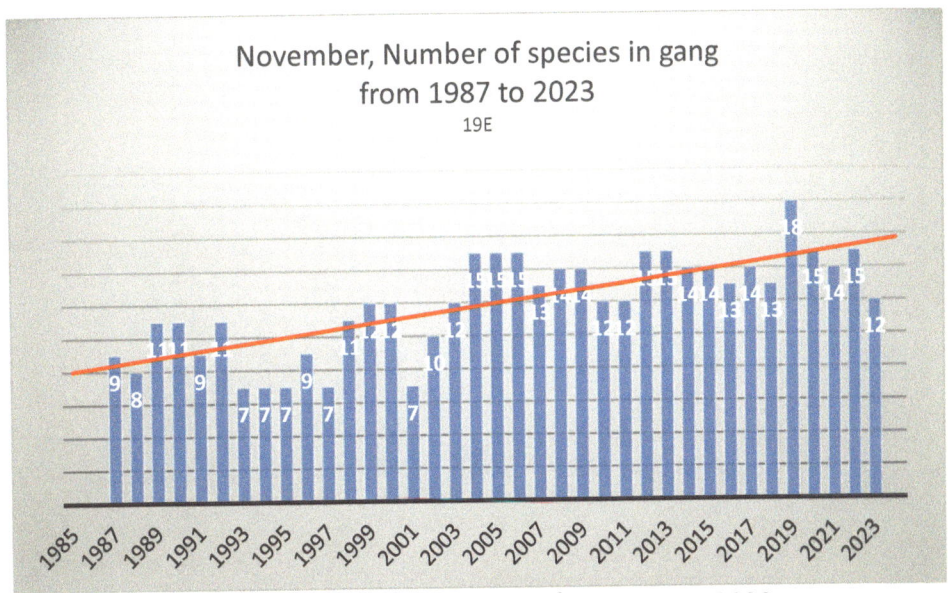
Number of species in the gang from 1987 to 2023

Population from 1987 to 2017 is cyclic but the trendline is slightly downward from 40 birds. The population that the yard sustains goes up and down like a yo-yo, 40 to 50 to 25 to 30 to 40 to 50 to 30. It seems to depend on a combination of the rain-to-date and temperature of the month. The up spikes of population are due to house sparrow, house finch, mourning dove, and arrival of juncos. In the 1990's there was a sharp decrease in house sparrows and house finches due to diseases and man-produced pollutants. Both species are not native to the east coast. The dove and junco are part of the native species.

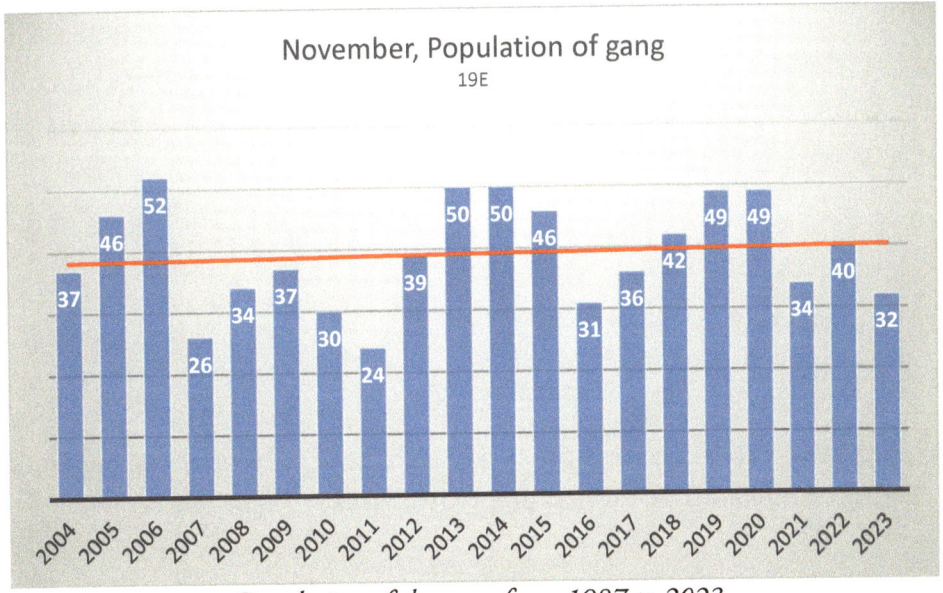
Population of the gang from 1987 to 2023

Removing these two species from the count shows that there has been a steady increase in population from 20 to 40. The 2012 dip that has been prominent in gang population since March shows up again.

NOVEMBER

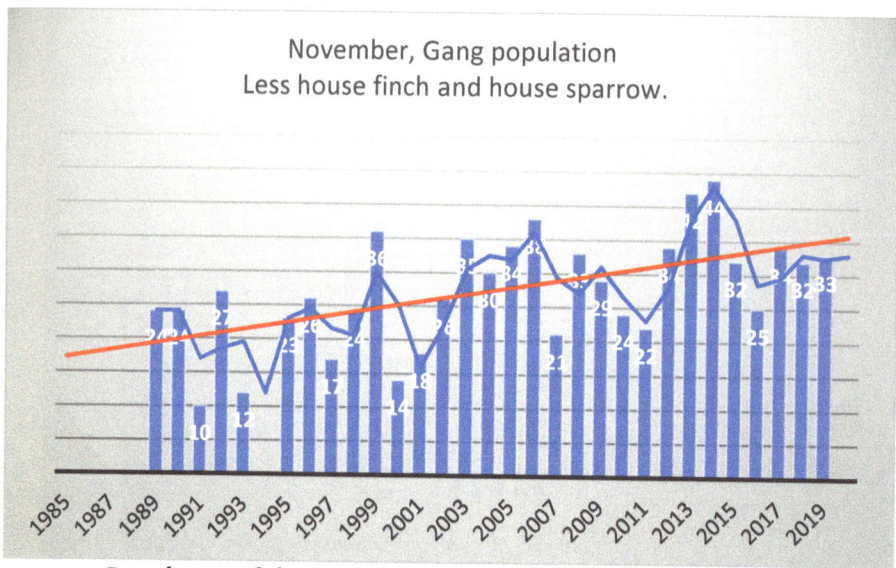

Population of the gang less the house sparrow and house finch

Review of Individual Species

Some years blue jays bring one to two fledglings to the feeders.

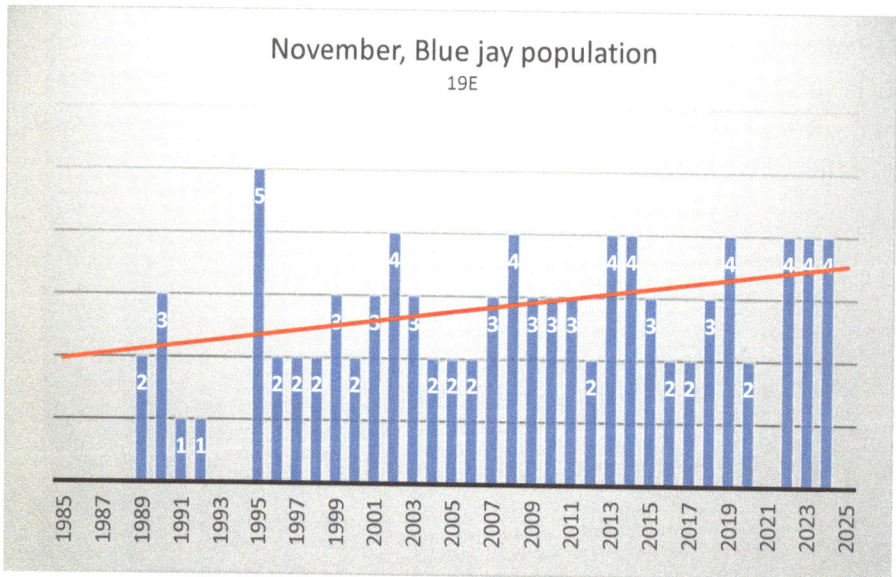

Population of blue jay

Cardinal became very prolific, going from two to eight until 2016 when the population dropped back to two. Young cardinals arrive the first and fourth weeks.

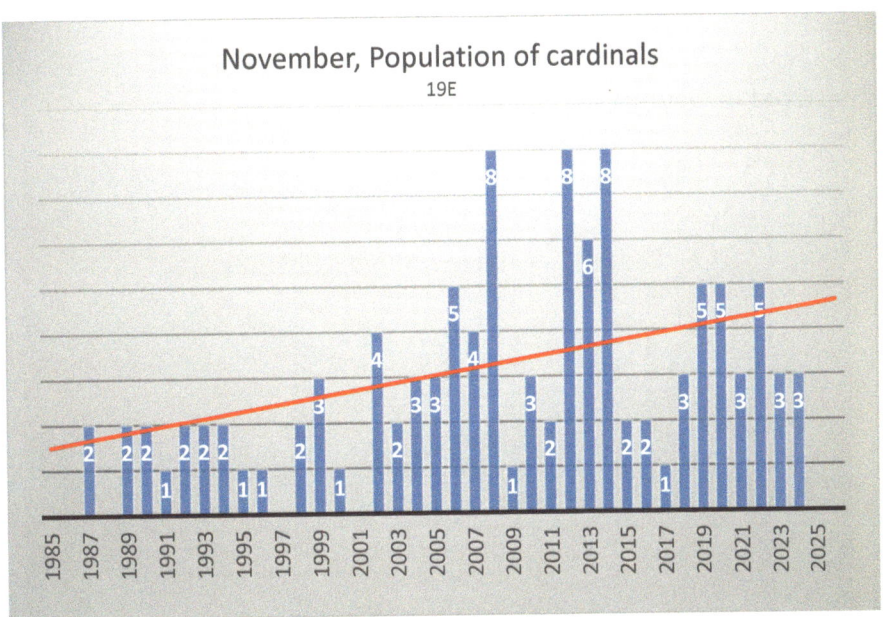

Population of cardinals

Chickadee population is usually just a pair, however, every two to three years they have two to four fledglings. Like the cardinal, the population dropped!

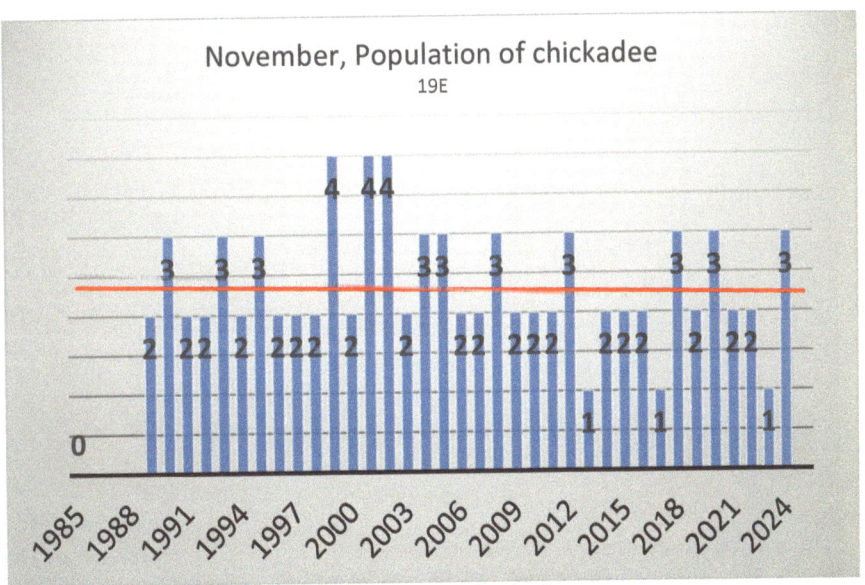

Population of chickadees

Since 2014 house finch population went from eight to none back to six.

Winter Arrivals

Junco mostly arrived in October.

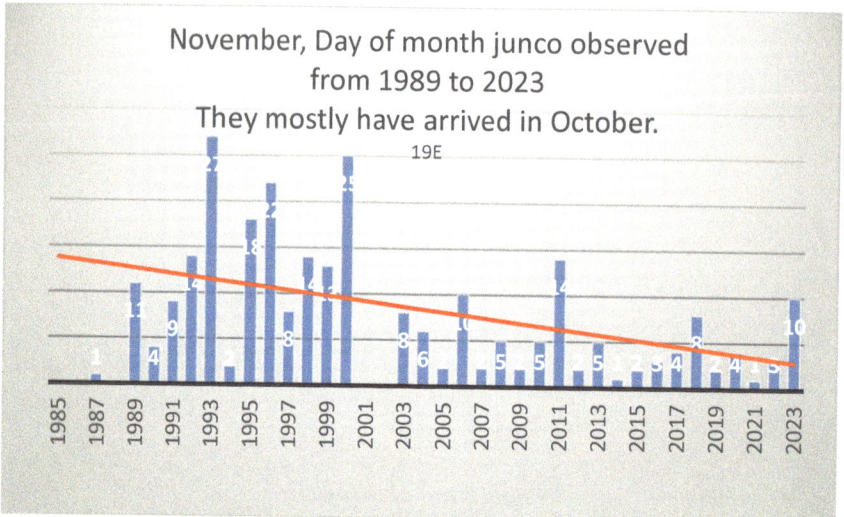

Day of the month that junco observed in November

White-throated sparrow arrived in October from 1989 to 2023.

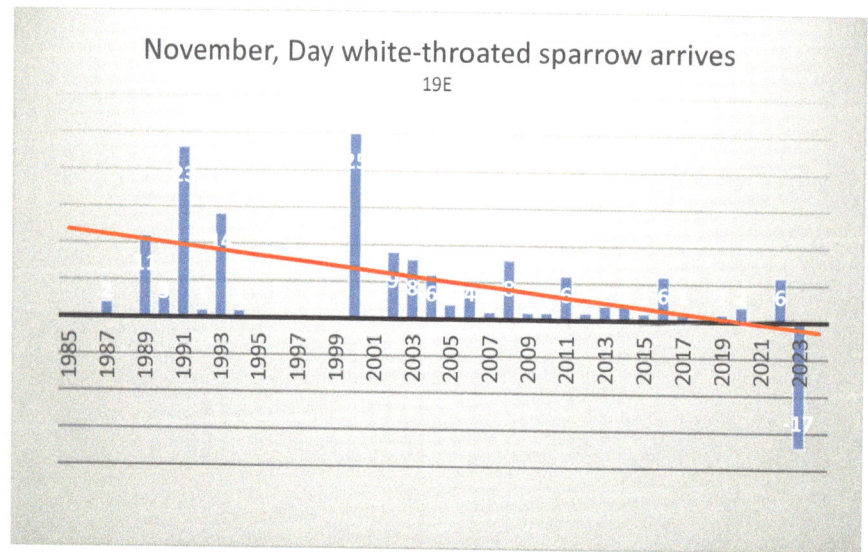

Day of month white-throated sparrow observed

    I have no substantial reason why these species are arriving earlier. It could be that it is getting colder earlier in the north but then I heard the geese are migrating later because the northern lakes are freezing later. The trend is two to three weeks earlier.

    The Savannah sparrow looks similar to a song sparrow but has stronger breast strips. It was first observed the second week of December of 2008. Then it went to the first week of November. After 2011 for two years it arrived the second week, then back to the first week of December. Each season I have only seen a single Savannah sparrow.

Savannah sparrow

Purple finch was first observed the first week of November in 2012, some years it arrives in December. Usually observed as a pair. Hairy woodpecker and red-bellied woodpecker since 2004, they have been on-again, off-again year-round residents.

Sharp-shinned hawk returns the first to fourth week for its winter stay to mainly play havoc with the doves. It does catch a cardinal now and then.

Driving home, I spot an object in the middle of the road. Approaching closer, it is a broad-winged hawk with its head down pulling on a kill. As I pass slowly within three feet, it defiantly looks up directly at my eyes as to say, "What can I do for you?" From my mirror I saw it continued to pluck at its kill.

Titmice goes from three to six week in November.

Turkey vultures soar in the brisk November cloudless sky.

Looking at the trends of the migratory and now and then migratory species, November quantities of all species went from 28 to 19. From 2004 to 2018, 11 species that I have only observed once during the decade are brown thrush, Cape May warbler, cowbird, great horned owl, great blue heron, pine siskin, screech owl, snow goose, solitary vireo, and tree sparrow. The great owl went through in 2022.

Species I have only seen twice in November are the house wren and red-headed woodpecker.

Many years the fox sparrow arrived during the fourth week but in 2011 it arrived the first week, which corresponds with the "dip." In 2018 one arrived on the 16th after a snowstorm. One year it arrived in September. 2024 it arrived December first.

In reference to migration and overwintering, during the first week the broad-winged hawk arrives and stays for a couple of weeks. The Carolina wren also arrived the first week of November until 2018 when it arrived in October.

At the beginning of the decade there were flocks of 40 to 100 crows and starlings, now I only see two to three birds of each species. This may be a result of the West Nile virus. Yet in Bethlehem, Pennsylvania, an hour before the November sunset, hundreds fly back to the rookery.

Those species which have been present over most of the period give me another mystery. Some show an increase in population others a decrease, some are arriving earlier others later.

The returning flocks of Canada geese for winter residency comes in two waves. The first wave in the first week and may be 10 to 40 but the last week the second wave goes from 20 to hundreds. Pass-over of the flocks has gone from the fourth week to second week.

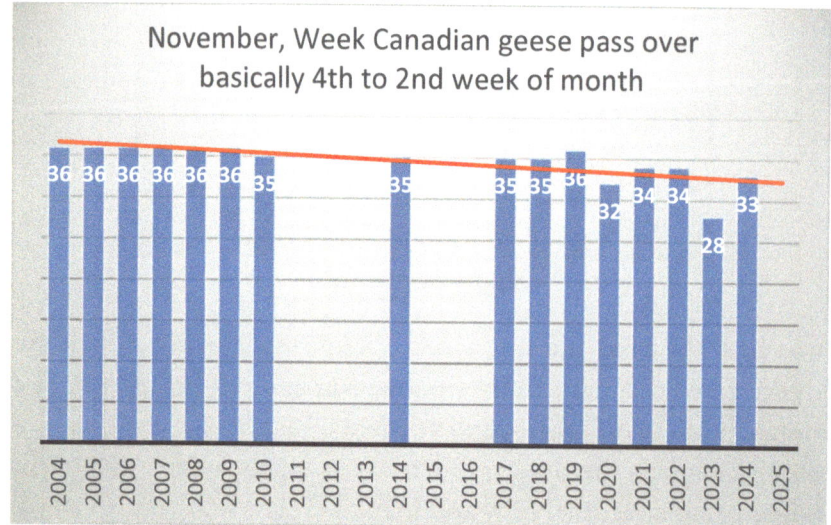

Day Canada geese fly over

I hear the mating call of the great horned owl hooting the third and fourth week.

A mockingbird and one or two robins go through the first week. During the past four years, flocks of 20 to 25 robins arrive the first week to overwinter. The song sparrow arrives the first week. Purple finch passes through the fourth week.

Fledglings: Some species still bring fledglings (blue jay, cardinal, house finch, and mourning dove) to the feeders. Sharpie also bring their fledglings to the feeder area and not for seed. Blue jay brings one to two fledglings to the feeders in the fourth week. Cardinal brings young the first and fourth week.

Insects: May still see a yellow and a white sulphur flit through during the third week. The crickets have hibernated. I don't know what they do but there is no more chirping. The boxelder bug population is reducing but not enough as far as I'm concerned. When the temperature goes above 50F, boxelder bugs like to sunbathe on a light-colored surface. Bumblebees will be out looking for flowers.

Pond: When the temperature goes to 70F, I see frogs sitting in the rocks.

Flora

Perennials: All the trumpet vine leaves fall during the first to third week. Yellow iris bloom for second time during the second week.

Vegetables: Cut back asparagus ferns during the first week.

So, ends November.

Index of December

December Overall	338
Weather	338
First Week	342
Second Week	344
Third Week	346
Fourth Week	348
Growth Rate and Lifespan of Trees and Shrubs	352
Review of Data	352
Fauna	352
Animals	352
Birds	355
Table of the Gang	350
Flock Migration	358
Small Bird Migration	360
Pecking Order	362
Fledglings	363
Flora	363
Trees	363
So, ends December	363

December

Week number for the year of 48 weeks					
Relative Month	Month/Week	First	Second	Third	Fourth
Previous	November	33	34	35	36
Present	**December**	**37**	**38**	**39**	**40**
Following	January	41	42	43	44

Overall Summary of December

The trend is for a drier and warmer December that brings out a groundhog and a chipmunk. Seedpods are still falling from the poplar tree. Depending on the climate conditions in November, the pin oak may still have one to 10 percent of its leaves hanging on. The average species quantity of the gang is 14 to 16.

Weather

The average high temperature on the 1st is 46F, the low is 29F.

The highest temperature was on the 1st in 2006, on the 4th in 1990, and the 29th in 1989 at 72F.

The average high temperature on the 31st is 36F and the low is 20F.

The lowest temperature was on the 28th in 1950 at 8F.

Reviewing the high and low temperatures from 1927 to 2018, the decade with the highest percentage (45%) of high temperatures was 2000-2010's. The 1930-1940's had seven percent. The decades with the highest percentage (42%) of lowest temperatures was 1950-1940's, and again the 2000-2010's had 13 percent.

The amount of light on the 1st of the month is 9 hours and 30 minutes, and on the 31st it is 9 hours and 19 minutes. The average snow for the month is four inches.

The full moon is called the Cold Moon.

Weather is the main factor affecting the presence of life in December. December can have heavy rain, fog, sleet, ice, snow, a short heat spell, strong winds, and thunderstorms. It can be a mixture of many of those plus a variation of 30 to 40 degrees in one day. The change in weather hundreds of miles to our north will affect when species will arrive, such as, the geese do not come south until the northern lakes start to freeze or the small birds have a food scarcity.

Frost and freeze affect not only the plants but available feed for the birds and critters. The first frost seems to be occurring earlier, except in 2018 there was no frost, we went directly into a freeze. In 2019 the first heavy frost was the first week of January. The trend for the first freeze is going from November into the first and second week of December. It seems backward to get a freeze before a heavy frost. 2024, no frost.

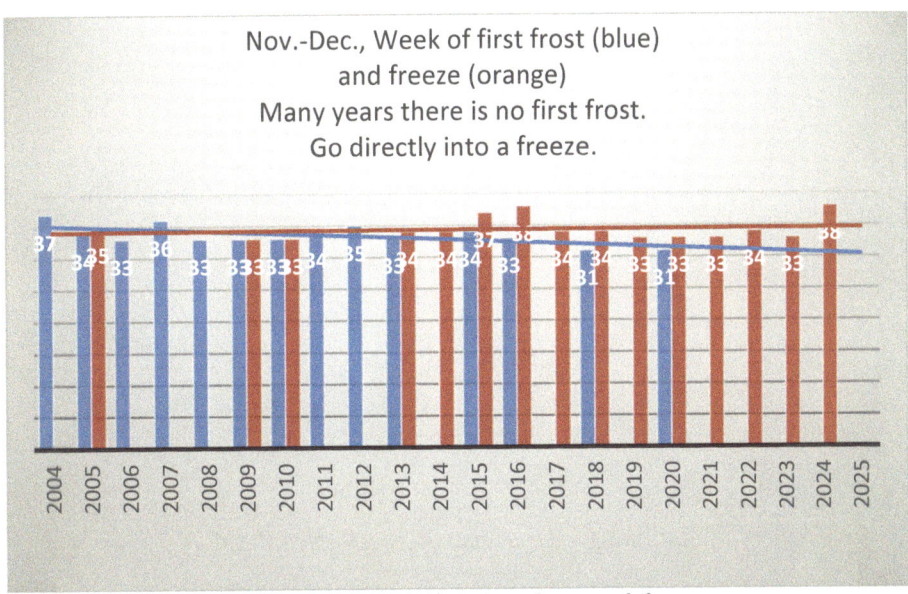

November-December first frost and freeze

Looking at the monthly rainfall for the 15 Decembers show they are becoming drier and drier. Yet the annual total rain for the year is only a slight decline or is it that 2018 and 2019 were much higher than normal.

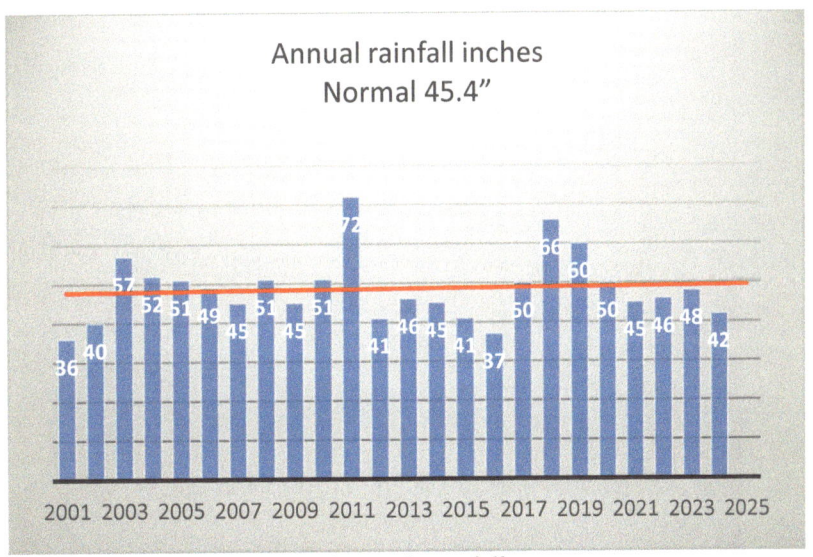

Annual rainfall

Looking at the rainfall in December shows a steady decline or drier month except again in 2018 and 2020.

DECEMBER

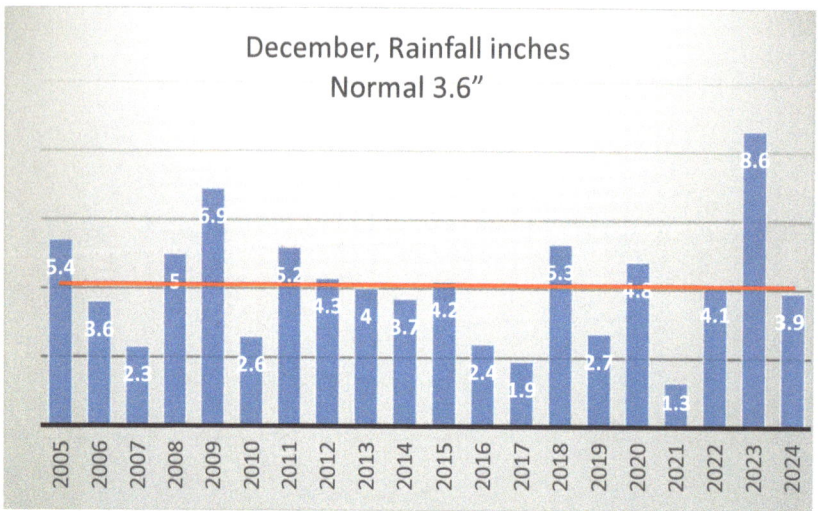

Rainfall for December from 2005 to 2024

Removing the unusual wet 2018, 2020 and 2023 would show drier and drier Decembers which means less possible snow. However, 2022, 2023 and 2024 are all above the average, so the trend is going back to normal!

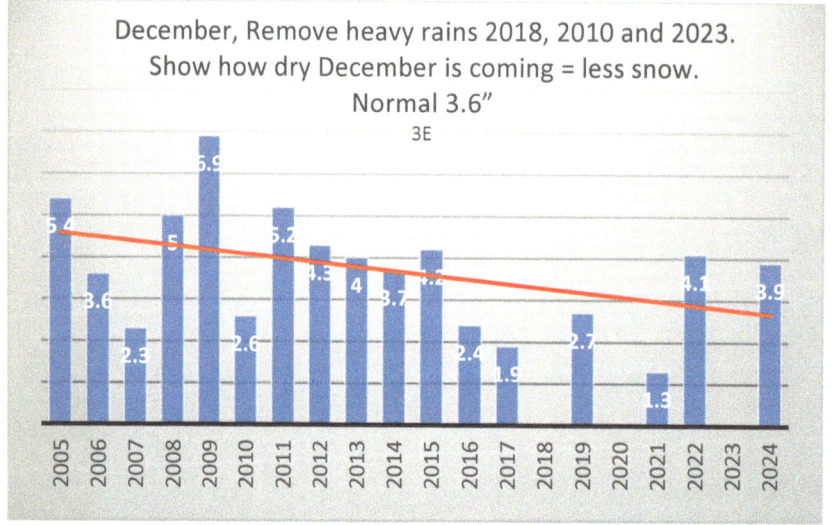

December rainfall less 2018, 2020 and 2023

The average amount of snow for December is four inches. Some years we will get 15 inches, others none. In this area, one inch of rain is equivalent to about 10 inches of snow.

The average temperature is 35F. Again, except when there is a blip like 2015 when it went to 45F. All these factors seem to affect the number of bird species and the population.

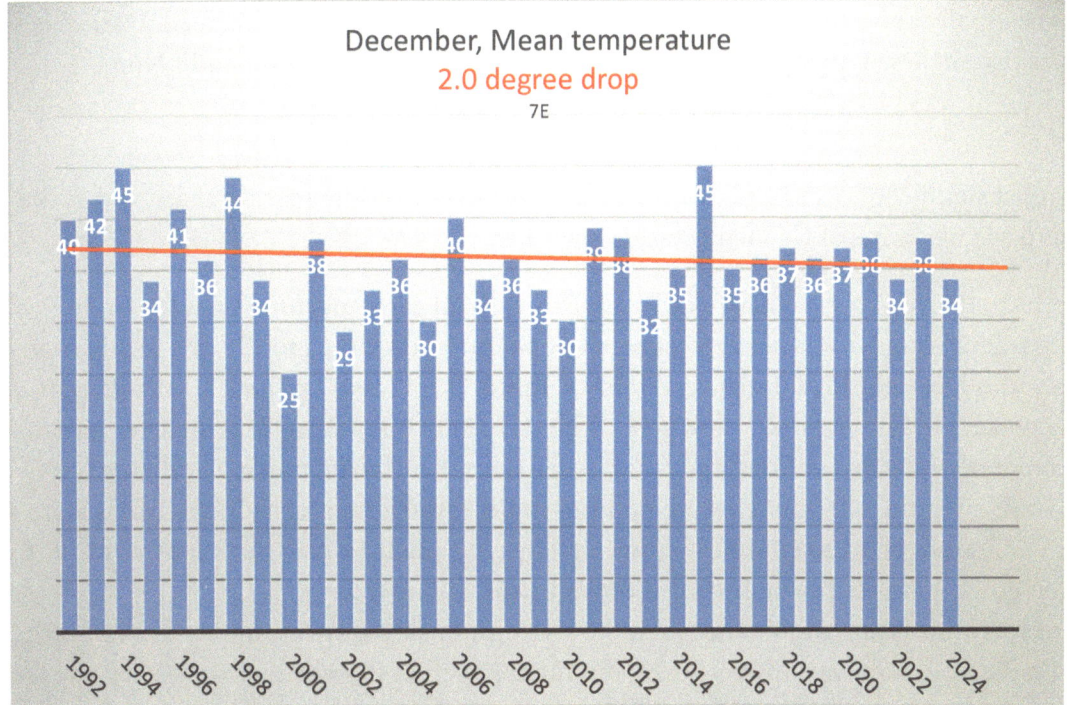
Mean temperatures for December from 1992 to 2024

Brief one- or two-day hot temperatures in the 60's to 70's have occurred at the beginning, middle, and end of the month. In the same manner, cold temperatures of -6F to -15F have occurred through the month. The extreme cold days were in the 1950-1960's. The warm periods have been occurring in the 2010's suggesting climate change.

The average snow for the month is four inches; occasionally we will get six to nine inches or an outstanding storm like 2009 when we received 15 inches. From 2000 forward there has been a drought of snow.

Then we may receive winds of 40 to 50 mph the last two days of the month.

This month you picture snow and keep dreaming of a white Christmas. The odds for a white Christmas are very low, especially with the global change. Instead we get freezing rain-sleet-fog and accidents are prevalent. If we do get some snow, it seems to be followed by a warm spell and by floods.

In 2010 weathermen predicted a cold winter into April, with a total of 28 inches of snow for the season. Wrong, we got 47 inches for the season.

The whole month in 2007 was 10 degrees above normal. December 1st was a new record of 69F. The hottest year on record for the country was 2010. The warmest December on record for the Lehigh Valley was 2015. The temperature was 13.5F above normal for the month. A year later, November 2016 was the warmest on record, followed by warmest December ever! It was 63F on the 24th. In 2017 it hit 71F. Yet the overall mean temperatures for December are going down.

In 2017 oak leaves were still falling the first week; I cut the lawn on the 16th. For the first nine days the temperatures were 50F plus. From the 14th to 24th the frogs were still out. There were a few birds. We had a thunderstorm on the 24th. After the storm it cooled down to 39F. The birds returned but a few days later it warmed up again and the birds disappeared again! Large flocks of snow geese finally appeared on the third week. On the 27th a heavy fog formed. Daffodils sprouts were over two inches. The weather people blame the unusual warmth on an El Nino.

First Week

The most significant occurrences are seeing the squirrels on the mating chase and at the end of the month seeing skeins of Canada geese going north.

Weather: Another yo-yo month. We can have 55F with an early morning light fog or 36F with odd sounding rolling thunder or a heavy frost with temperatures dropping to 15F or a record high of 77F.

Going into December 2006 was very warm. On the 1st it was a record 72F at 4:45 p.m. At 5:30 it was semi-dark at 71.4F with 60 mph gusts. The stars were out with a three-quarter moon. I could see my moon shadow on the lawn and could smell the earth. To the north I could see heavy clouds with lightning within. The clouds were coming south toward me; tornado warnings were issued. We got wind gusts of 30 to 40 mph followed by heavy squalls of rain but that was all. A tornado did touch down 10 miles north. It cooled down to 46F the next day.

There was an unusual occurrence in 2008. I saw an upside-down rainbow to the north. It was unusual not only that it was upside-down, but it happened in the north, usually they appear in the east.

Weather people say that the warm November-December is caused by the large El Nino in the Pacific and a long dip of the jet stream into the middle of the country. This happened in 2013, 2014, 2015, and 2016.

Animals: This is an active week for the squirrels. After eating a piece of apple, one is going bananas running up and down a japonica bush, and then jumps from one dogwood to another. Was there apple jack in that apple? Two more are on the chase.

Squirrels on the chase

The original Gorilla Glue

Another squirrel, which has already been chased, takes daylily and oak leaves up the oak to build her nest. The chases will continue all of December and into January. Squirrels are still finding walnuts to bury. A vole is running among the ins-and-outs of the pond wall.

TRENDS DUE TO CLIMATE CHANGE

<u>Birds</u>: Just after sunrise a flock of 100 crows go north. Then a flock of 250 local geese go west for exercise. Late afternoon, nice and brisk, I look up to see another large flock of crows going north; I look up higher to see a large flock of geese passing over the crows but going east!

Red-bellied and hairy woodpeckers arrive at the suet cakes for breakfast. Within two minutes after I put cracked eggshells on the feeder, blue jays swoop down for them. A one-legged goldfinch pulls out safflower seeds from the feeder. As the gang finishes breakfast, a sharpie comes through, causing a panicking scatter. A downy freezes its body to the split peanut feeder. A red-bellied woodpecker does the same thing. A blue jay yells at a sharpie until it leaves. The sharpie returns and takes a house finch in mid-flight.

Hawk looking for breakfast

Unwanted starlings are awake and coming out of the deserted owl house.

I have placed a mirror vertically on the ground at a slight angle below the kitchen window so I can observe the ground-feeding birds. A song sparrow is aggressively pecking at itself in the mirror to win territorial ground.

Because I do not have enough open space, I do not see mockingbirds most of the year. I did have them when my tree canopy was small (example of environmental change). However, one, sometimes a pair, do come back in December and stay for a few days and then move on. Some years a purple finch passes through.

Robins are still gathering into flocks of 20 to 30 to go south. The white-throated and fox sparrows are scratching the unraked leaves. At 9:00 a.m., clear, I first hear, then look up and see at 2,000 feet a large flock of snow geese going south. First their formation is in a "U" and then it gradually changes to a "V." Later in the afternoon a flock of 1,000 starlings goes west.

It is evening, 20 mallard ducks are still here; they fly around in circles over the circle. The last year ducks were here was December 2016.

Neither birds nor animals like caramel popcorn. Also, during the winter they do not touch the peanut butter, at least the birds at my feeder don't.

The first of the month, 7:30 p.m., it is dark, I am sitting in the kitchen, I catch a sound, hear it again. It is the mating call of the great horned owl. I listen closer, I now hear two calling back and forth. The calling lasted until 10:00 p.m. This happened until the 8th. Then they left. Another year I was awoken by the hooting at 3:30 a.m. I have always heard, but never have seen, the owls. The next evening at 5:00 p.m., it was dusk, as I walked up the back hill, all of a sudden from the shadows of the oak, 10 feet in front of me just at eye level a four-foot pair of wings flew by. The bird landed at the very top of a spruce. The last light of the setting sun silhouetted the outline of the great horned. Quite a sight.

The owls arrived the fourth week in 2005, in 2017 they are arriving the first week!

Flowers: After a freeze put compost from the hill onto the perennial and vegetable gardens.

Insects: After a warm 54F with heavy rain, the next morning the circle was filled with earthworms all facing south. This can also happen after a warm, rainy night during the fourth week. This also happens whenever it is warm all winter.

Worms in the circle

Shrubs: Put oak leaves around the base of shrubs but not against the trunk. Some japonica have maroon clusters, others will turn maroon or pink at the end of the month. Some years the large, white azalea leaves turn yellow-green.

Trees: Seedpods are still falling from the poplar. Spruce trees are full of cones. Pin oak and Winesap apple has 95 percent leaf drop.

Vegetables: Some years December is still warm enough that I am picking small broccoli heads.

Second Week

This week is the last leaf pick up that the township makes. Large flocks of snow geese, Canada geese, and robins are still migrating.

Weather: Normal high temperatures are 40's to 50's. We may get one to three inches of snow; could feel like 13F.

TRENDS DUE TO CLIMATE CHANGE

A swing of weather, we get three inches of snow at night, in the morning it changes to rain. By 6:00 p.m. there is fog followed by high winds. Or temperature will go from 38 to 58F resulting with, again, earthworms in the circle. If there are still robins in the area they do not feed on the worms! Four seagulls fly over from the east, usually a sign that a nor'easter is coming. A couple of days later the storm happened.

Seagulls

The full moon was 17,000 miles closer to earth in 2008, 2017, and 2019; the luminance was 15 percent brighter. This is called a "supermoon." Occurred again December 15, 2024.

A big halo around the full moon usually means a storm is coming. It happened in 2010, but no storm, rain or snow followed.

<u>Animals</u>: A vole is running around the base of the small birch tree. Except for December and January, I normally do see any cats, and then they only drop by for a drink at the pond; they do not go after the birds. One year it was an old, black cat, the next year it was an old calico. Last cat seen was in 2015.

It is a cold, windy 24F; it feels like 9F. Four squirrels sit on branches with bushy tails curled up their backs. I throw out acorns; the acorns are gone in a half an hour.

Brrr! It is cold!

Flowers: In 2015 it was warm for the past week causing the daffodils to emerge two inches on the 12th. Milkweed pods are opening to send out the parachutes of faraway seedings.

Milkweed pod opening

Pond: Take off the net and put winterize in the ponds, it is supposed to keep the water clear; other years it turned brown.

Trees: Seedpods (samaras) are still falling from the poplar. White oaks have 95 to 99 percent leaf drop.

Third Week

In 2010, for the first time in 327 years (1683), the first day of winter and a full eclipse of the moon occurred on the same day. Venus is in the southeast and is high and bright.

Weather: When there is a fog, except for possible Canada geese getting exercise, there is no bird activity.

Animals: Vole goes into the holes of the pond wall. Spot a squirrel eating a piece of orange, again, unusual. To top it off another one was chewing a Reese's Peanut Butter Cup, paper and all. Four squirrels go into the dogwood to eat the remaining berries.

He likes his peanuts one way or another

Birds: There are six robins in the circle in 2010. On the 17th in 2013, 25F, for some reason that only he knows, a song sparrow is singing its mating song! For the past week, a hungry sharpie has been harassing the gang. It misses most of the time. A broad-winged hawk is also on the hunt. The two hawks are chasing each other.

A flock of 200 snow geese go north on the 19th. House sparrows disappear for weeks! Titmice, chickadees, and house sparrows all take a drink from the heated water. Titmouse eats cattail punk. Savannah sparrow arrives. A chickadee is putting seed under japonica bark. Blue jay takes crumbled eggshells. Birds do not eat plain suet or bacon fat. After a snow, both a downy and house finch are eating snow.

When there is a heavy snow, the quantity of birds at the feeders doubles. A male blue jay chases a female cardinal from the split peanut feeder; not very gentlemanly. The jay then eats an orange peel; most of the time no birds touch the orange slices in the winter. A male house finch does not move when a male cardinal aggressively approaches. A female house finch has conjunctivitis. One year a male cardinal also had the disease. Six crows land in a tall arborvitae and eat its cones.

Young hawk

Like last week, hundreds of snow geese go south, 400 local Canada geese go north. Five crows chase a broad-winged hawk from the tall spruce.

A sharpie was sitting on an oak branch when suddenly a squirrel ventures out on the same branch. The hawk raises it wing, the squirrel takes a double take and turns around and runs down the branch.

Ten to 20 ducks circle the circle five to seven times just at dusk, and I mean just at dusk. As dusk gets earlier with the year, the ducks stick to flying just at dusk. I am surprised that they are still here.

Purple finch on top of the main feeder, house finch is at the bottom. It is hard to tell them apart. The purple finch head is squarer. The flicker goes south, no more grubs or ants available. When it is cold and clear turkey vultures still soar. Two male cardinals yelling yeh-yeh-yo-yeh at each other. I do not know how to translate. A male downy and a male house finch are sparing it out on the split peanut cage; house finch wins.

Flowers: To protect the perennials from ground heave I lay straw on the gardens after the ground is frozen. I bought a bale of straw in October from the local nursery and forgot to cover it with a tarp, so it got soaked and sprouted. Then in December when I went to spread it, I had one large, heavy ice cube. I used a pitchfork and spade to pry it apart.

Straw growing well

Spread out

Trees: Large strips of bark are falling off the big birch; the tree is expanding for next year's growth.

Fourth Week

Weather: After a 50F rainy night, I find the circle filled with earthworms the next morning. Another day, 34F goes to 45F, raining with 40 mph winds, the birds are hungry and active. It is cold and damp and the annual rotting of vegetation has set in. There can still be morning fog with geese flying in it. I can wake up to a 19F clear day with a snow-covered yard. It is just twilight, 7:00 a.m., I can see the silhouettes of cardinals and white-throated sparrows at the feeder. At 7:15 a.m. I can see a song sparrow, a pair of juncos, and a titmouse.

The day before Christmas we get a freezing rain; the Interstate is shut down because of all the accidents. The next day the temperature goes way up, resulting in mass runoff and local flooding, then fog that night.

All month in 2014 there were above normal temperatures, plus 4F, then 2015 with a higher temperature of six degrees above normal. In 2015 the whole globe hits a new high mean temperature for the year. This has been the trend for the past four years, a new mean high each year. Don't tell some people that when January and February arrive. The end of December 2017 was one of the coldest on record.

Animals: On the 24th two squirrels are on the chase. A vole goes across the steps. Rather than eat it, a squirrel buries a whole peanut for a future snack.

Birds: At 9:00 a.m. on the 24th, 37F, a skein of 200 Canada geese go north; 9:30 a.m., 100 more; 9:45 a.m. 150 more go north. Male cardinal takes a whole peanut! A chickadee puts its head down to drink the snow. A junco sits at the end of the swinging perch under the globe split peanut feeder. It is waiting for the house sparrow to finish eating. After the sparrow leaves it moves to the middle of the perch to have breakfast. It knew that the aggressive sparrow would not let it interfere. Likewise, a pair of chickadees wait for a white-throated sparrow to finish at the safflower feeder. A pair of juncos goes after a pair of house finch, the house finches win. On the large feeder, there is only room for three or more species to eat at a time. Birds of the same species will chase each other while at that feeder. As a mass, all eyes are on the lookout for a sharpie, which does come, and the mass scatters.

One male cardinal chases another for territorial rights. A red-breasted nuthatch has arrived for the winter. A mockingbird stops for lunch on its annual migration before going on. A house finch with conjunctivitis dumped seed from the platform feeder for two hours!

A gang of at least 20 are eating on a sunny morning. I see a large shadow cross the lawn, the gang scatters. A large broad-winged hawk lands in the tall maple, it looks down and around. After five minutes members of the gang come out to feed. Soon they all are out. The hawk just keeps looking down. He sits there for 45 minutes. The little birds must know or sense that they are too small of a meal for the large hawk. It sits there for two more hours. All of a sudden at a lower level, in comes a sharpie. The underside of its widespread orange wings is facing my window. The gang scatters, some hit my window, others the siding. The sharpie missed and flies off. The broad-winged has been entertained enough, it too flies off. Clear, 2:15 p.m., another broad-winged is soaring majestically. Not much later it is devouring a squirrel.

Just like the morning, but dusk, the local geese have their ritual exercise at 5:00 p.m. I hear and look up into the night sky to see 300 going south. As the snow crunches underfoot, it alerts a broad-winged in a nearby spruce. Look up to see it glide silently overhead to a further tree.

With the light snow, the doves gather at the feeder. It was like an accident ready to happen. There were 20 doves feeding, when in swoops a sharpie for a meal, it got one as the others scattered. A chickadee had just landed on the trunk of a dogwood, when from below under the hemlock hedge shoots up a hidden sharpie to snatch the chickadee.

In the morning blue jays are gathering whole peanuts, red-bellied woodpeckers and young cardinals are getting split peanuts, and a Carolina wren is throwing safflower out of the feeder.

Blue jay with a whole peanut

Red-bellied and young cardinal

Carolina wren throwing out safflower

A flock of 30 robins go west. Dusk, 5:00 p.m., a flock of 50 to 100 local geese are getting exercise. The same flock of ducks that were flying around the first week still circle around and around for 20 minutes then settle nearby. A titmouse attacks the punks on the cattail to use for lining its nest.

I hear a great horned owl at 8:00 p.m., wonderful call to hear on a winter's night. One evening I did hear a screech owl hooting for 15 minutes.

We get a freezing rain, everything is icing up; birds must eat and drink, they slip and slide trying to land on the feeders. House finches and house sparrows are walking on the ice of the pond to get water from the small forming puddles. I break the ice from the feeders and supporting wires and throw feed under the bushes where there is no ice. The fish are slowly trolling around the pond under the ice.

Trolling slowly under the ice

Male house finch in the noon day sun

Some years a great blue heron will stop at the pond and try to get lunch or dinner. Great blues are very leery, when they get the slightest inkling of my presence in the kitchen they take off.

A house finch has conjunctivitis. A crow in a farm field is eating smashed pumpkin. A late migrater is a brown thrasher who stayed for two days. A white nuthatch raises its wing to chase a titmouse from the feeder.

A Cooper's hawk gives a warning shout. It is sunning on a branch of a dogwood when from below a squirrel climbs towards the hawk. Like I have seen birds do to other birds, the hawk raises its wing at the squirrel. It only takes the squirrel a second to accept the warning and quickly retreats. Later I spot six crows chasing the Cooper's hawk from their rookery area.

Cooper's hawk looking for breakfast

<u>Shrubs</u>: Some mountain laurel leaves turn yellow in the fourth week.

Trees: From 1969 to 1985 I kept some records of the newly-planted trees and shrubs. It takes a year for the plants to get established and get over the shock of being transplanted. Then depending on the soil condition, sun location, and amount of rain or snow, the plants grow at an average rate until maturity.

Growth Rate and Lifespan of Trees and Shrubs			
Species	Average Growth Rate Inches per Year	Mature Height Feet	Lifespan Years
Silver maple - hill	2.5 to 3	60	28, disease?
Silver maple - 3rd tier	2 to 2.5	60	50+
Weeping willow - 3rd tier	3 to 5	50 to 60	10, wind
Weeping willow - hill	5 to 7	60	Cut down
Green spruce	1 to 1.5	60 to 70	50+
Blue spruce	2 to 2.5	50+	Cut down
White pine	3 to 4	40+	8, Cut down
Pin oak	6 to 7	60 to 80	45+
Dogwood	1 to 1.5	20 to 30	43+
Canadian hemlock hedge	4 to 6	8 to 10	Cut back to 5' 50+
Arborvitae, tall	6 to 10	40	
Arborvitae, medium	6 to 10	8 to 10	Cut back to 7' 52+

I had no idea how fast or how high some of these species grew resulting in lot of planting and cutting down. Something like, "Too late smart."

Review of Data

Fauna

Animals: Depending on the conditions of the year, the members in a family of squirrels can go from two to five. I only see one rabbit during the month, usually at dusk or twilight. When the temperature is 50F, I may see a groundhog and chipmunks running around. I spot vole tracks in the lawn the fourth week.

Rabbit

Birds: The average number of species observed during December from 1989 to 2003 was 15. In 2004 additional species were observed. The same species are not seen every year but there is an increase. From 2004 to 2019 the average number increased to 24. The population of all species from 2004-2015 was 54. 2020-2024 increased to 60.

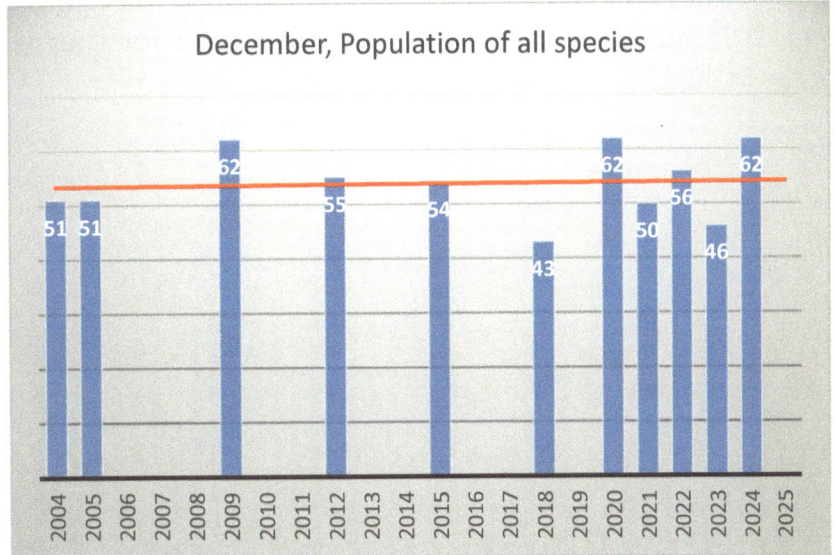

December, number of overall species

From 2004 to 2008, eight gang species were only observed twice over the period.

The average number of species from 2004 to 2024 is a about 24, but when looking back to 1989 there is an altogether different picture which goes from 10 to 25.

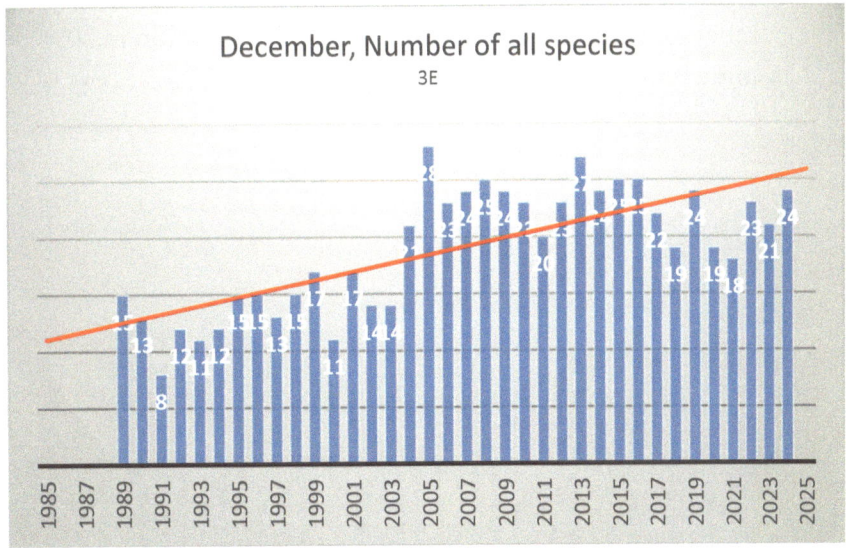

December, number of species observed

The makeup of species other than the gang were the broad-winged hawk, brown thrasher, Cooper's hawk, Mallard duck, fox sparrow, grackle, great blue heron, great horned owl, mockingbird, screech owl, red-headed woodpecker, purple finch, redpoll, red-winged blackbird, flocks of snow geese, and turkey vultures. Except for the flocks of snow geese and ducks, all the other migraters were a single bird of a species. During the third week there are only a few migrations except maybe a brown thrasher or the hoot of an owl. Most passed over the fourth week.

The population within most species in the gang stays the same year after year. There were some exceptions like the mourning dove, house sparrow, and house finch that changed. They doubled to tripled in 2016 and 2017, maybe because the hawk population went from four to one! From 1989 the number of species in the gang has increased from 10 to 15.

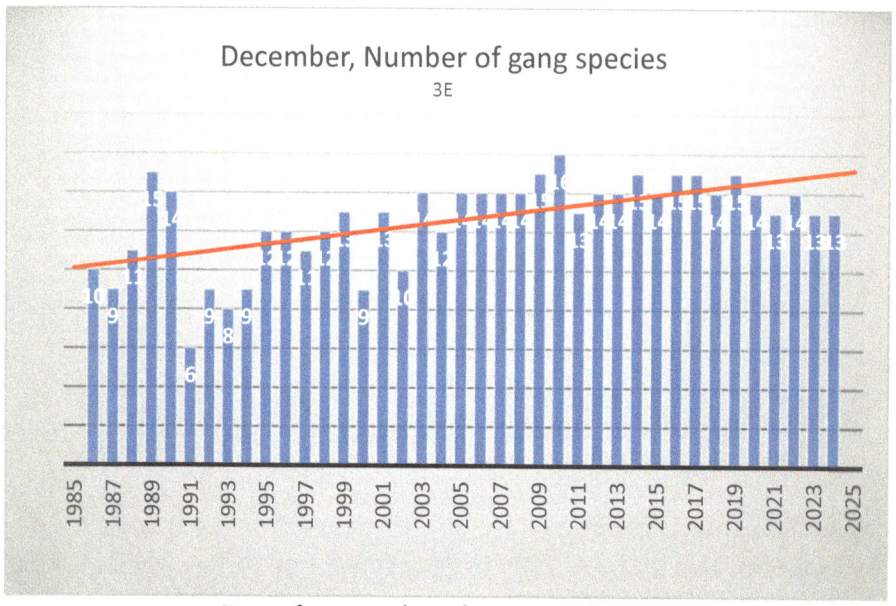

December, number of species in the gang

Even though the number of gang species has increased, the population in 2016 decreased to half it was in 1989 but has come back by 2019.

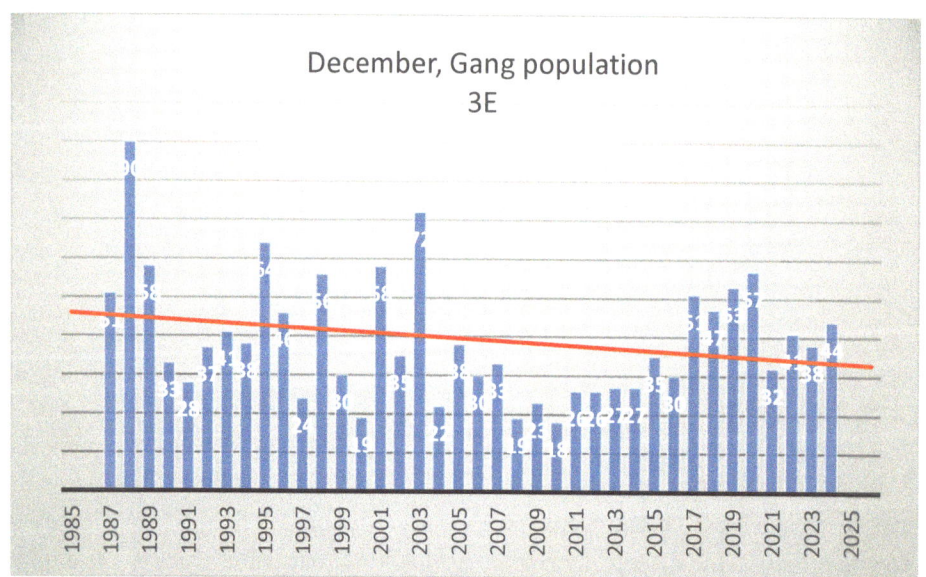

December, population of the gang

Again the 2011-2012 dip appears. Like November, the main reason for the decrease was the population of the house sparrow and house finch.

Table of the Gang

The number of species for the winter gang has been steadily increasing.

Species/Quan.	04	05	06	07	08	09	10	11	12	13	14	15	16	17	18	19
Blue jay R, G	6	3	2	4	3	4	3	1	1	3	3	3	2	3	2/3	4/19
Cardinal R, G	2	4	1	2	8	3	10	2	9	10	6	2	8	5	5/4	7/17
Carolina wren R, G	0	2	1	1	2	2	1	1	2	2	1		2	1		2/2
Chickadee R, G	4	4	3	4	3	2	3	2	2	2	3	2	2	1	3/22	2/7
Downy, R, G	1	1	2	1	1	1	2	2	1	1	1	1	2	1	1/4	2/19
Goldfinch R, G	1	3	2	2	2	1	1	0	0	3	4	1	0	6	1/4	3/7
House finch R, G conjunctivitis	6	8/14	4	3/19	8	4	6	4	4	6	6	8	6	20	3/9	10/18
House sparrow R, G	1	5	4	6	6	2	4	0	1	2	4	4	6	2	10/8	15/21
Junco, R, G	5	7	2	5	6	6	2	3	4	14	2	2	5	6	4/9	5/8
Mourning dove R, G	5	6	4	8	4	3	6	6	5	12	7	9	12	18	4/8	14/8
Red-bellied woodpecker R	1	0	1	1	1	1	1	1	1	1	1	1	1	2	1/4	1/8
Red-breasted nuthatch G	1	1	1	2	1	1	1	0	1	0	0	0	2	1	2/9	1/12
Savannah sparrow G	0	0	0	0	1	1	1	1	1	1	1	0	1	0	0	0
Song spa R, G	2	2	3	2	1	1	1	1	1	1	1	1	1	0	0	1/21
Titmouse R, G	4	4	5	2	3	3	2	2	2	1	2	3	3	2	1/9	2/15
White-breasted nut R, G	2	2	2	2	1	1	1	1	2	1	2	1	1	1	1/5	1/17
White-throat sparrow G	2	2	3	3	8	3	6	3	6	4	3	4	4	5	3/9	3/21
Total Gang	12	14	14	14	14	15	16	13	14	14	15	14	15	15	14	15
Population	22	38	30	33	31	39	49	30	26	27	27	35	30	51	43	53

Over the years during the last week of December the population of the gang drops by 20 to 50 percent, then through January it increases.

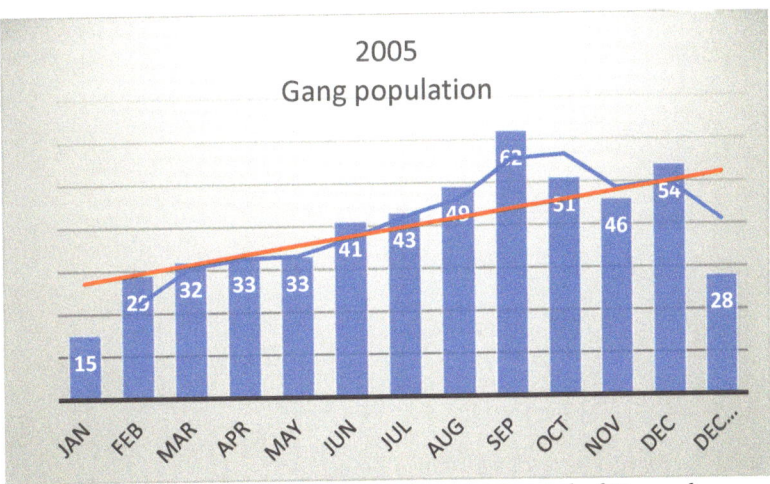

2005 showing drop in population at the end of December

DECEMBER

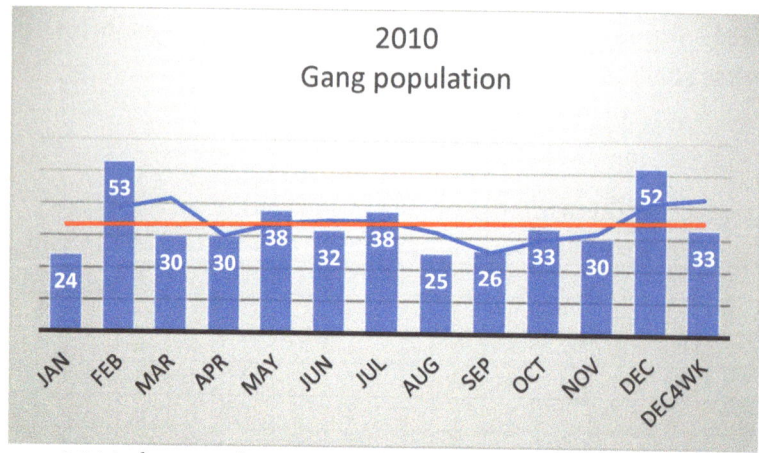

2010 showing drop in population at the end of December

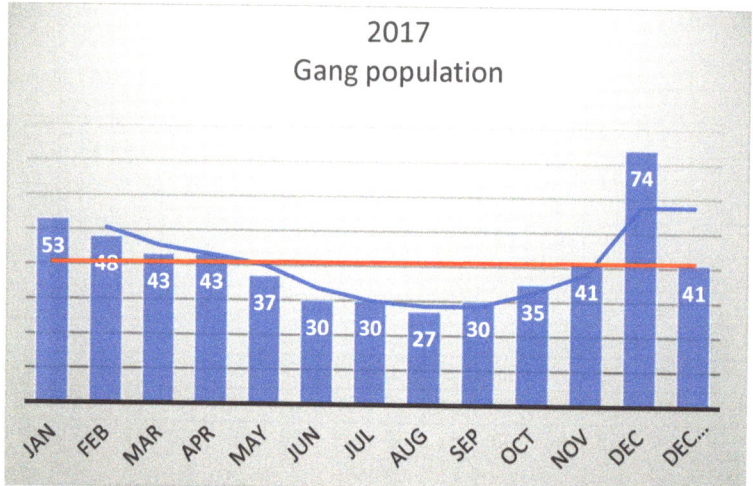

2017 showing drop in population at the end of December

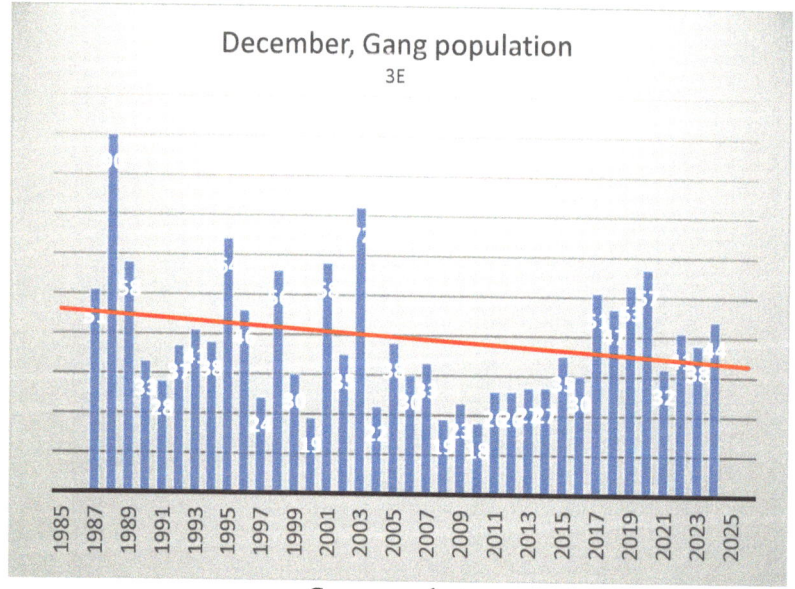

Gang population

In December 2019 the population of the house finch, house sparrow, and mourning dove was exceptionally high. **Removing a quantity of 12 to level the field**, the population of 2019 would be about 4.

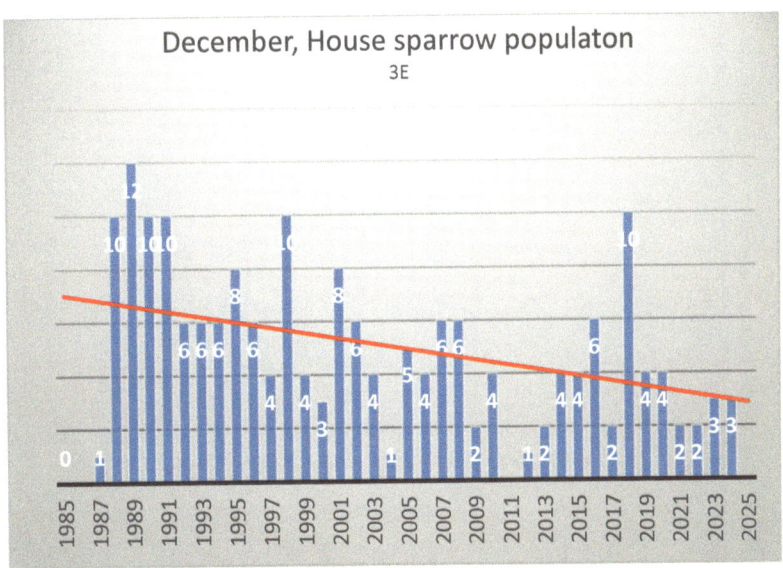

December, house sparrow population

The high population of 10 in 2018 was there for most of the year?

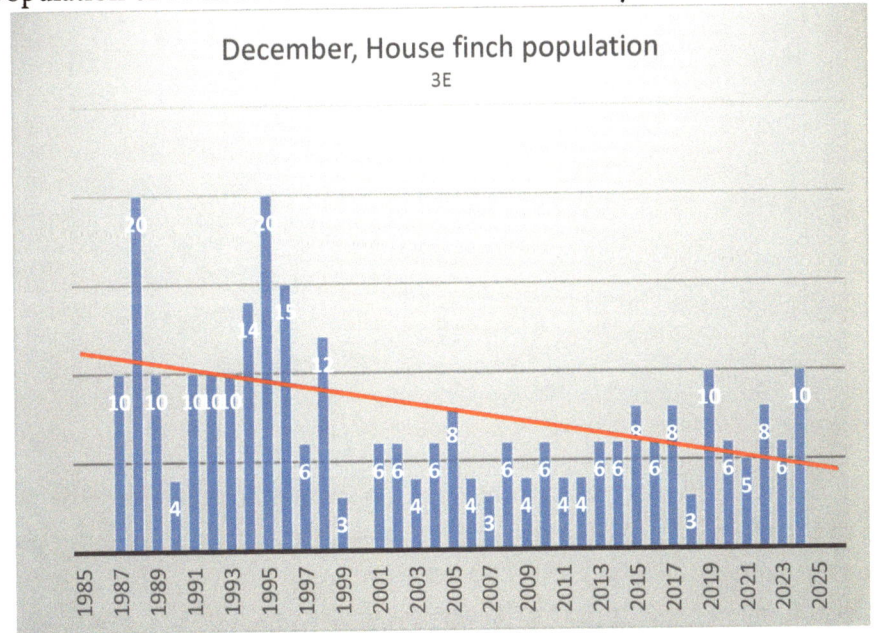

December, house finch population

On the other hand, the house finch population in 2018 was very low?

The dove population peaks then dramatically drops. It could just be a coincidence that there is a 14- to 16-year period between peaks.

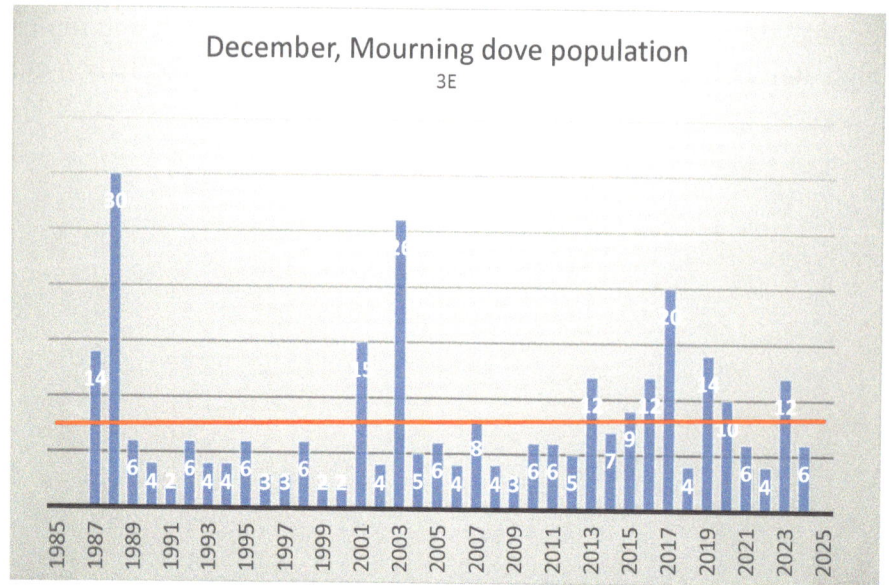
December, mourning dove population

Seventeen of 20 doves observed in December 2017, in 2018 there was only four

<u>Flock Migration</u>: Before 2008 there used to be large flocks of starlings. Then the population went to none or one to three birds. In 2015 a flock of 30 passed through and then the population again dropped to a few. In January 2018 a flock of 20 overwintered.

Locally, flocks of Canada geese appear to be very cyclic. Before 2004 there were large flocks of 300 to 400. Then five cycles from 300 to 20 to 350 to 15 to 350 to none to 200. These cycles were almost in a straight line, each point was taken 12 months apart. The dates of the geese arrival are changing from the first week to the fourth. Flocks were not observed in 2018, 2019 and 2024. The cause is the area's environmental change of large warehouse and apartment-housing developments. West of my area that is still agriculture, large flocks of Canada and snow geese are noted.

TRENDS DUE TO CLIMATE CHANGE

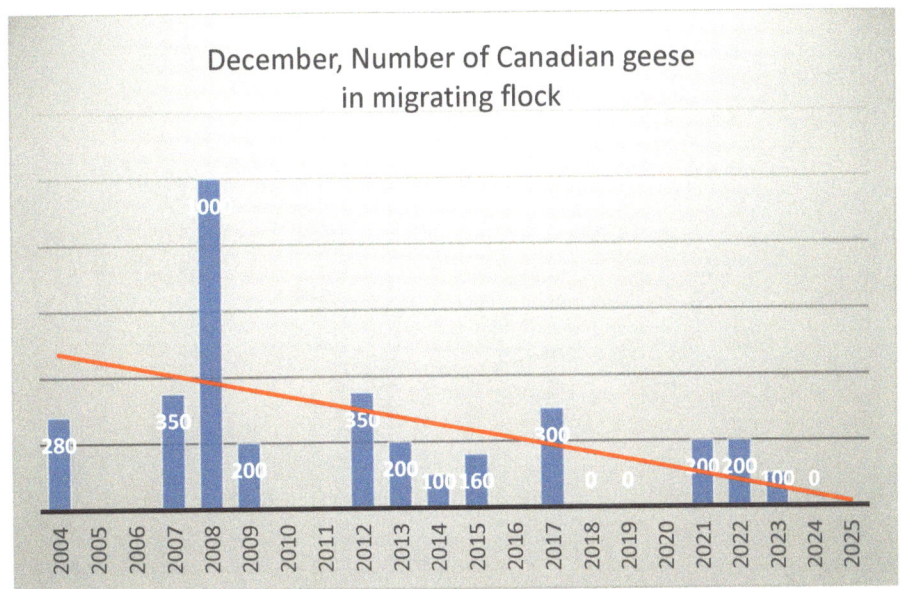

December, flock size of Canada geese

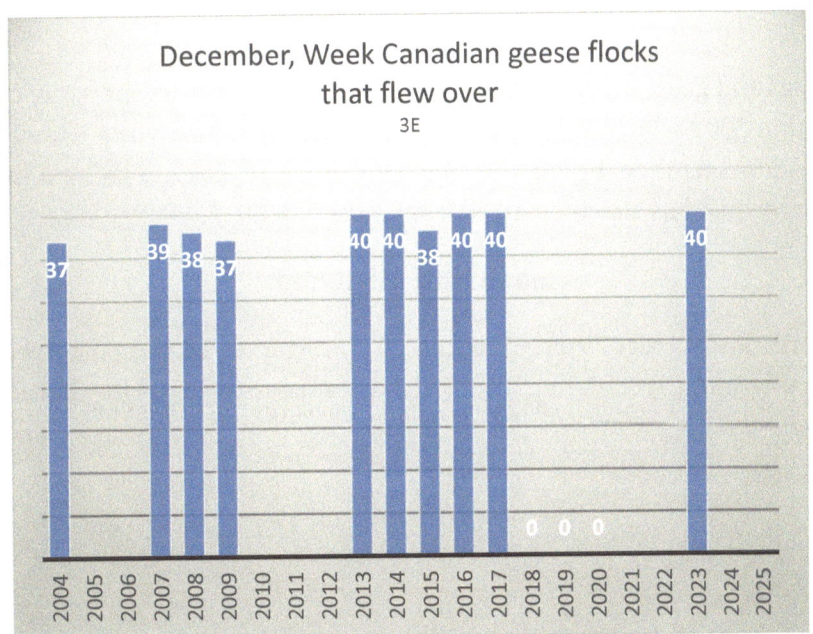

Week number of flocks of Canada geese flew over

There is a contingency of flocks of local Canada geese that reside in the dormant fields for the winter.

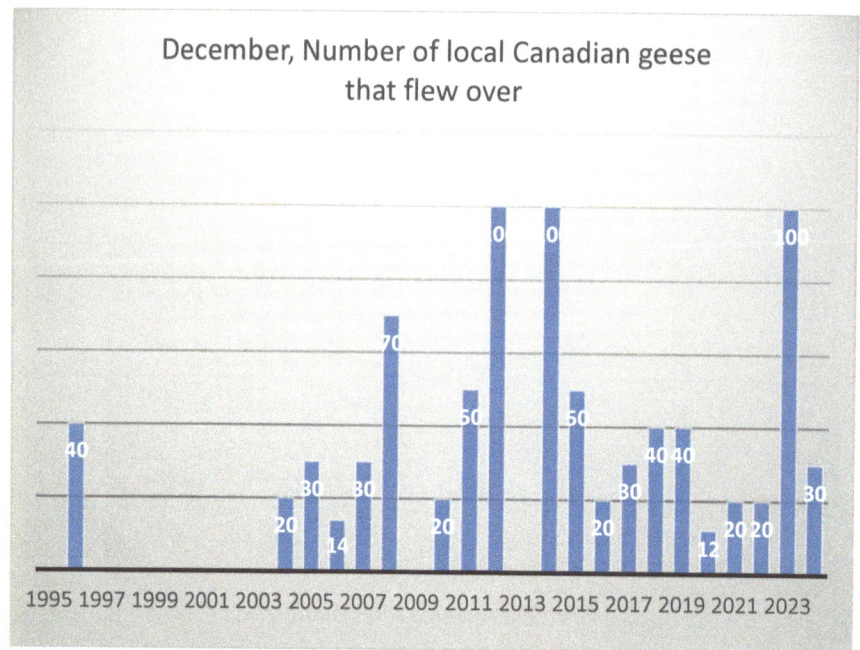

December, size of wintering local Canada geese

Like the starlings and geese, the population of crows almost follows the same cycle. The high of 100 dropped to 10 in 2009 which was probably due to the West Nile virus. However, on the good side, their population seems to be on the upswing to 25 in 2017. The size of the rookery in Bethlehem, Pennsylvania, has also increased dramatically in 2017.

Small Bird Migration

The junco population averages four, except in 2013 it went to 14, then dropped to two the next year. Even in the migrating species the 2011-2012 dip appears. The white-throated population is increasing.

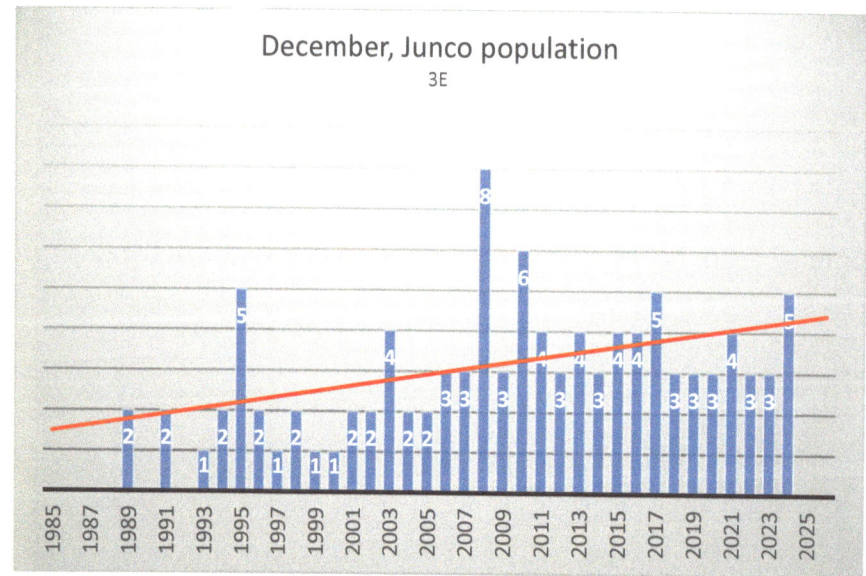

Population of junco for December

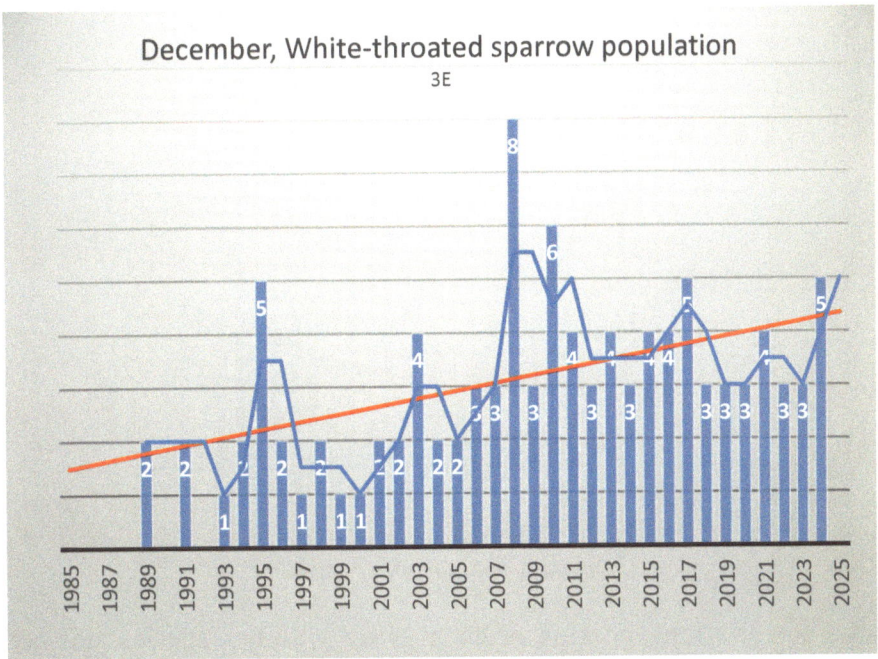

December, population of white-throated sparrow

The blue jay population is increasing in December with fledglings of one to two.

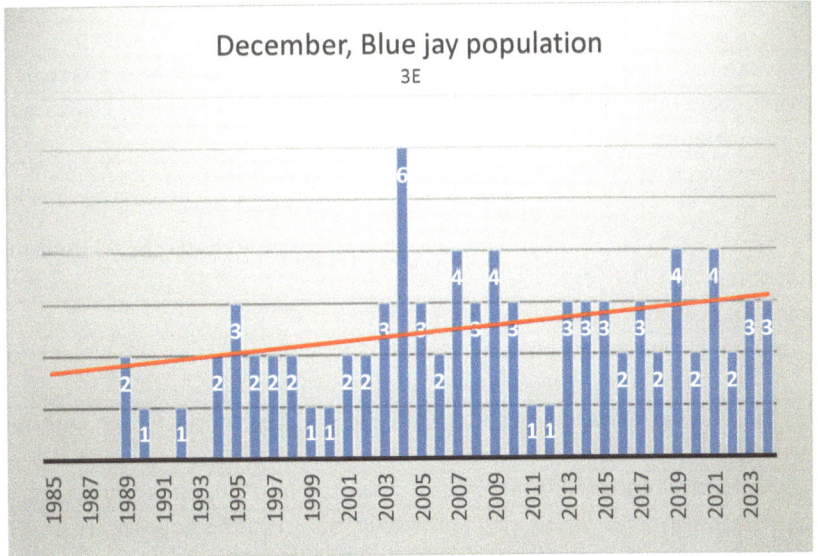

December, population of blue jay

The cardinals are more prolific than the jays by a four-fold since 1989.

DECEMBER

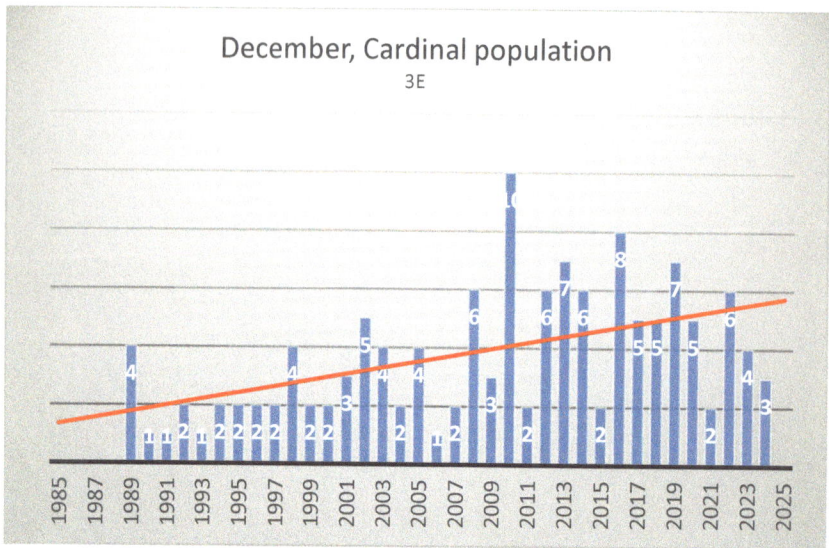

December, population of cardinal

In December a lot of strong pecking order goes on within a species and between species. Interestingly, the pecking order takes a time out among the various species during a snowstorm. I spotted four to five different species on the feeder at the same time during the height of a snowstorm. There must be a mutual understanding.

Pecking Order in December
4th 2004, titmouse chases a red-breasted nuthatch, titmouse chased by chickadee which is replaced by a titmouse.
5th 2006, white-throated chase other species from feeders. Two males and one female chase each other.
10th 2005, jays chasing each other.
11th 2014, house sparrow chases a house finch. Titmouse chases white-throated from feeder.
14th 2005, 100 percent snow, 3F, one cold squirrel. No bird activity. Normally a white-throat would chase a junco from a feeder, today they sit next to each other. However, a song sparrow chases the white-throat. Two jays come in but one chases the other away.
15th 2007, 27F, a junco backs away from a chickadee on the feeder.
19th 2007, two starlings are unsuccessful of getting birdhouse from house sparrows. Grackle chases blue jay and dove from feeder.
19th 2008, 32F, 7:00 a.m., white-throated chases young male cardinal from raised feeder. One junco chases another from feeder. One cardinal chases another from feeder.
20th 2008, 26F, song chases white-throated. Two male cardinals rise four feet into the air going at each other.
21st 2004, 7:00 a.m. just light, two cardinals, two white-throated, a junco, and a house sparrow. At 7:30 a.m., niceties are over; the house sparrow chases the junco.
23rd 2007, 37F, white-throated chases song from millet.
24th 2007, four doves and two jays are on the feeder. A jay tries to chase the bull dove. The bull fluffs up and approaches the jay. The jay flies off.
26th 2004, a junco swinging on perch waiting for house sparrow to finish breakfast of safflower. After the house sparrow leaves the junco goes to the feeder.
28th 2004, song chases junco, then white-throated chases a junco. On the main feeder there is more room, three juncos and a white-throated feed next to each other.
30th 2014, male cardinal chases another for territorial rights.
31st 2009, junco chases titmouse from split peanuts.

Fledglings: Blue jays, cardinals, and titmice are the species that consistently bring one to four fledglings to the feeders in December during the third and fourth weeks.

Flora

Trees: Depending on weather and soil conditions of the previous months, the big birch will shed bark any week during the month.

So, ends December.

Index of January

January Overall	365
Weather	365
First Week	370
Second Week	373
Third Week	375
Fourth Week	381
Blizzard of 2016	383
Other Winter Observations	386
Review of Data	388
Fauna	388
Animals	388
Birds	390
Table of the Gang	393
Individual Species	393
Fledglings	396
Migration	396
Pecking Order	397
Pond	401
Flora	402
Perennials	402
So, ends January	402

January

Week number for the year of 48 weeks					
Relative Month	Month/Week	First	Second	Third	Fourth
Previous	December	37	38	39	40
Present	**January**	**41**	**42**	**43**	**44**
Following	February	45	46	47	48

The average high temperature on the 1st is 35F, the low is 20F.
The average high temperature on the 31st is 36F and the low is 19F.
Extreme daytime high for the month was 72F on the 26th of 1972.
Extreme nighttime low for the month was -5F on the 21st of 1994.
Normal rainfall is 2.37 inches (equivalent to about 24 inches of snow in the Lehigh Valley area).
The amount of light on the 1st of the month is 9 hours, 20 minutes, and on the 31st it is 10 hours, 7 minutes.
The full moon is called the Wolf Moon.

Overall Summary of January

Except for the birds, squirrels, and weather, there is not much outside activity. It usually is cold with wind and snow and an occasional thunderstorm. It can be dank and damp with freezing rain and black ice. There are exceptional Januarys that are warm with no snow, where the grass stays green and the squirrel's mate early. It can be abnormally cold with a blizzard. The number of bird species for the day is 12 to 15.

Weather

January is the coldest month of the year with some bad yo-yo weather. The fourth week is the coldest period.

On the 12th of 1996, 19F, piles of snow in the circle are eight to 10 feet. On the 13th it is 17F in the morning and 35F in the evening; the 14th, 18F in morning. Then on the 18th it warms to 45F resulting in fog. On the 19th it rises to 53F and becomes very foggy; the big thaw begins. Thunder, lightning, flash flood, and rain comes into the cellar. The next morning it is a cold 18F; everything freezes up. The Lehigh River is the highest since 1955. January 20th, 18F, 70,000 homes have been evacuated in Scranton. On the 24th and 27th there is heavy rain. What a mess.

On the 16th of 2009, 6:30 a.m., 9.6F, feels like -4F. The Big Dipper's pan is pointing straight down over the living room. On the 19th and 28th we get freezing fog which is worse than freezing rain. It results in many accidents. At least with the freezing rain you can see where you are going. We have a snowstorm on the 26th with thunder and lightning.

The 3rd of 2014, 16F at 6:30 a.m., goes to 12F at 8:20 a.m., then to 14F at 5:00 p.m. Feels like -1F at 8:30 p.m., the coldest air in five years. On the 5th it was a record -1F with solid ice on the driveway. The next day the temperature goes to 48F with heavy rain, then drops to zero, with the wind (32 mph) it felt like -25F. On the 8th ice jams were reported on the Lehigh and Delaware Rivers. Six days later the temperature goes to 62F. Birds are taking a bath in the water on top of the ice in the pond. Then back to normal January weather on the 12th. On the 22nd we have 1.3F at 6:30 a.m. and feels like -16F with eight inches of snow. At 9:00 a.m. the temperature goes up to 4F. This January we had 12 days in the single digits and received 16.5 inches of snow. This was the coldest January since 1900 and 1978-1979.

January 3, 2015, the snow comes down in big flakes with a yellow cast. Some years there will be 12 inches of snow on the ground, other years the ground will be frozen and suddenly there is a January thaw where the temperature will go up to 60 or 70F (the normal being 35F). Then we get a storm with two inches of rain and gusting wind, resulting in bad flooding because the rain cannot be absorbed into the frozen ground.

New Year's Day, 2016, 39F, Easter rose is blooming. On the 3rd we get a heavy frost, ice forms on the pond. The 4th, 14F; 5th, 10F, coldest day since last March. On the 7th, 44F during the day; drops to 14F at night; on the 10th we have a high of 63F, rain and thunder, next day with the wind it feels like 14F. We get our first trace of snow for the whole season on the 13th. The morning temperature on the 19th is 14F. The ocean is warmer than normal, which will affect the amount of snow from a storm because there is more moisture in the air. On the 23rd, here it comes. We get 30 inches of snow in 24 hours which is a record for Lehigh Valley. It was heavy to dig and blow. The previous record was 24 inches on February 11, 1983. The next day was a clear, crisp 13F. We get a covering of snow on the 31st. Total for the season was 32.3 inches. Last year the total was 19.1 inches for the season. The coldest January in 100 years was 2005. January of 2006 was the warmest in 100 years. The next coldest year was 2014. It went from 48F at 2:00 p.m. to 1F the next morning with 46 mph winds that made it feel like -25F! There were 12 days of single digit temperatures. On the positive side, there were less bugs in the spring.

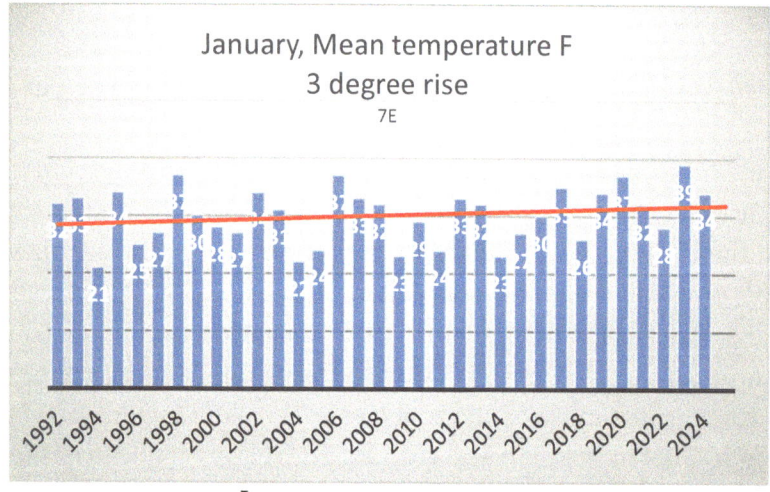

January mean temperature

The graph of the temperature over the period shows a slight downward trend. The average rainfall is 3.0 inches. From year to year it goes from negative 2.5 inches to as high as plus 4.5 inches. The average temperature is 28F. In 2016 and 2017 the average temperature was in the middle 30's.

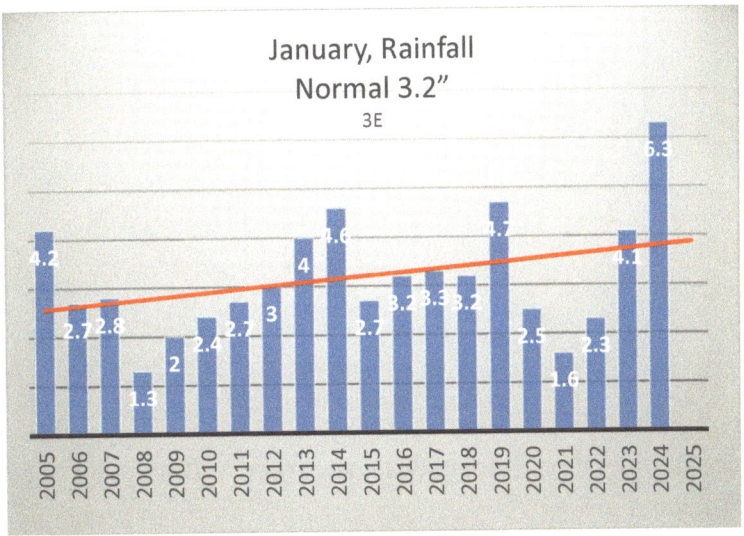

January rainfall

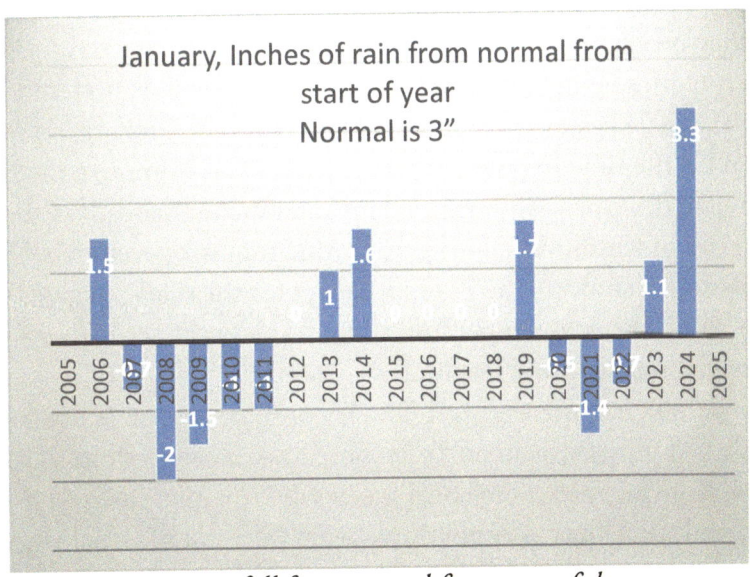

January rainfall from normal from start of the year

You think of January as a month with lots of snow. Many years the first two weeks may be unseasonably warm, followed by cold the last two weeks. For several years, a substantial snowstorm occurred during the third week of the month. Average snow for the month is 11 inches. Now and then we get none or a whopping 27 to 32 inches. Blizzard in 2016 brought a record 30 inches snow in 24 hours.

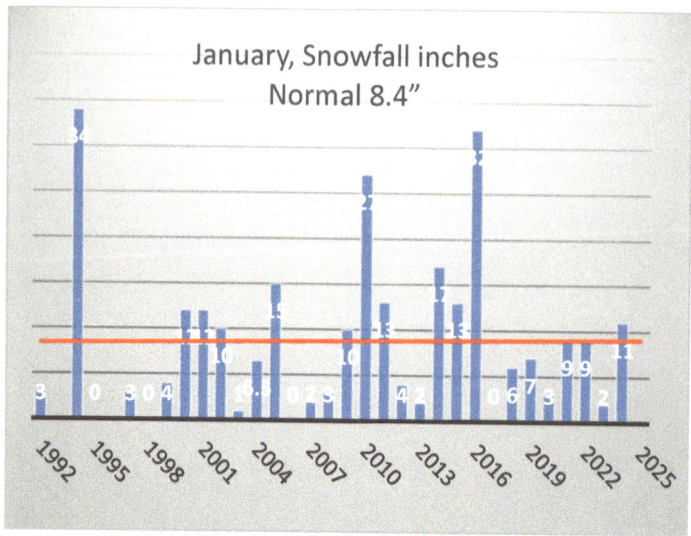
January annual snowfall

A thaw can produce thunderstorms, tornadoes, and mold to develop in heating system air ducts exasperating sinus problems.

If the conditions are right, we may hear thunder during a snowstorm. The main problem is freezing ice which is not only bad for the plants but results in many accidents of people falling and car collisions. Another problem is the ground can be frozen and we have a two-inch rainstorm which is a complete run off, so it does no good to replenish the soil. With temperatures going from the 40's to feel-like -15F in 24 hours, the soil heaves and contracts kicking out the perennials. After the ground has frozen is the time to put straw on the plants to minimize the heaving. This is not always a positive thing to do. One year it warmed up and rained; the seeds in the straw germinated. Come April a good crop of wheat was coming up throughout my perennials.

I prefer a good coating of snow on the ground, except for the roads, to give the plants and lawns a layer of insulation. With the cold temperatures, there is less chance of sap rising into the upper portion of the plants that would freeze resulting in winterkill.

The winter of 2006-2007 had the warmest January on U.S. record. It beat out the 1988 record. El Nino winds had shifted the jet stream north keeping the cold away from the country. As a result, many people suffered from allergies. Some bird species did not migrate from the north. Insects like ticks, boxelder bugs, and stink bugs were plentiful. Bears were not hibernating; gray tree frogs were croaking early. Last fall's grass seed was germinating. Chafer grubs ate the grass roots. Winter wheat became susceptible to mildew and fungus. Daffodils were up four inches. Winter clothes sat on the shelves of the stores. Ice fishing items were not selling as well as snow shovels and rock salt. (A mild El Nino will bring a cold, stormy winter. A strong El Nino brings a mild winter.)

Now here's a good one for those who shovel snow. One foot of snow in the average driveway weighs 2,000 pounds (one ton).

It can be a warm 65F and beautiful with a slight wind. If it is, I go out to prune and pick up fallen branches as well as spend time raking leaves. Then I bring mud into the house. Ugh! All thoughts are on spring fever which should be in March. With the unseasonably warm weather, the front walk gets over shadowed with little ants! The plants and wildlife do not know what to do, nor do the human inhabitants. It can result in more winterkill for the plants and sickness for us. The warm spell causes the daffodil leaves to emerge and grow to three to four inches.

Daffodils up early

You can approximate the outside temperature by the position of the rhododendron leaves. When the January temperatures drop into the teens the leaves point straight down and curl up as small as a pencil. At 38F the rhodie leaves are drooping at a 5:00 o'clock position. At 7:00 a.m., 11.5F; it was 4F overnight, humidity 50 percent. At 7:00 p.m. 1F, the rhodie leaves are really curled up pointing down. You can say, "It is really winter."

Rhododendron leaf position

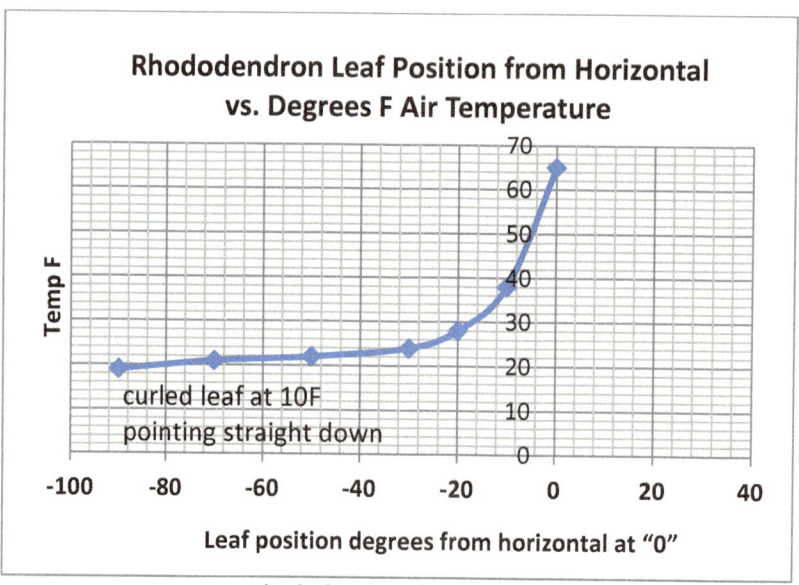

Rhododendron leaf position

First Week

Mercury and Venus are rising in the southeast 10 minutes before the sun rises. As the sun rises it takes half an hour to have enough light to observe what is going on outside, but there is enough light for mourning doves, juncos, and song sparrows to be at the feeders.

<u>Animals</u>: Four squirrels are running around on the hill at 11:00 a.m., two are chasing each other. Over the winter I observe this action quite often. I guess they like to have games for exercise. Yup, 1:00 p.m., the squirrels are still chasing each other. Squirrels are mating on a dogwood branch.

On the chase

It is 70F. I spot a skunk going up the driveway at 9:00 p.m. I find the remains of a cottontail rabbit on the side of the house. Was it the leftovers from a fox hunting?

When it is warm, I may spot a vole scurrying around the rocks of the stone wall. They and their descendants have been here for over 25 years. They bore tunnels under the snow from the stone wall to the bird feeders to get their meal. They kill the grass as they burrow. Come spring I find the tunnels and dead grass.

Birds: Almost light, 6:45 a.m., spot a large white-breasted bird near the top of an 80-foot spruce. I was watching it for a half an hour when eight crows descended on the bird, harassing it. After five minutes the harassed bird flew off; it was a red- tailed hawk with a four-foot wingspan.

Red-tailed hawk

The crows followed it until the hawk was out of their area. Two minutes after the hawk left, the small birds were at the feeders. Three of the crows returned to the oak. I have observed this ritual at least four years where four to eight crows chase either a broad-winged or red-tailed hawk from the area during the first week of January.

At 7:00 a.m. it is overcast except for a slight crack of pink at the horizon. I hear the engine of a diesel locomotive and the rumble of the train's wheels five miles away. The sound of its whistle echoes across the valley. At 7:40 a.m. it is full light; a flock of 50 local Canada geese are flying south at their normal altitudes. The sound of jets or planes does not bother them. Put whole peanuts on the main feeder. Three minutes later three blue jays come screaming in to get them. Some nuts get dropped onto the ground; a waiting squirrel runs across the snow to pick them up. Then a Cooper's hawk came in, causing the gang to scatter. It got no breakfast and left.

For some unknown reason in 2017, 12 blackbirds dropped out of the sky along Route 22.

A white-throated sparrow hops around the edge of the ice in the birdbath trying to find water. A chickadee finds the water along the edge of the pond. A red- breasted nuthatch comes for split peanuts, has a flight problem, and hits the kitchen window; it shakes its head and leaves.

White-throated sparrow

Chickadee

In 2015 at 8:45 a.m. I hear geese; still hear them at 9:00 a.m. I go out to investigate and see 20 "V's" of snow geese. I look again to see "V's" on top of "Vs" on top of "V's." I go up the hill to look south and see the "V's" five miles away flying over Emmaus. There must be 10,000 snow geese going east and northeast. I think they are rising up from their winter layover in the old limestone quarries. It takes 20 minutes for all of them to fly over. It was quite a sight. Later in the afternoon there are 200 Canada geese feeding in the farm field.

At 4:00 p.m. a sharpie swoops toward the kitchen window and grabs a house finch that was eating at the safflower feeder. There are 20 house finches in a front yard maple at 4:30 p.m. A good thing because the sharpie that flew through the backyard did not notice the birds in the front yard. Normally there are only two to four. Dusk, 5:00 p.m., a flock of 30 local geese go north.

With a warm week in January, a starling appears that normally arrives in late February to early March. It is going for suet.

Starling going for suet (picture was taken in spring)

Perennials: It is a record 70F, the ground is not frozen, the perennial flowers, lilies, columbine, and tulips, are showing life. It may be too late; cut back the pampas grass to place over the perennials in the lower garden to act as an insulator. The cattail stalks are still six feet high; punks are puffed up and blowing apart.

Trees: With warmth, trees develop premature buds. If they do not freeze buds will flower in March-April. The grass turns deep green.

Shrubs: Different species of japonica have clusters of light pink, white, and dark maroon flowers. This occurs from the first to third week.

Second Week

Back to reality of winter. It is dark at 7:00 a.m. and 36F. This week in 1953 there was a four-inch ice storm.

Animals: At noon the vole is peaking out of the stone wall. Later in the afternoon a red fox comes down the back steps. It would like the vole.

Red fox

Later spot a red striped cat coming down the back steps; the birds have already scattered. There is a big rabbit munching on something; two squirrels, on the chase. I am awoken at 3:00 a.m. by the sound of the top of the can with peanuts crashing on the patio as a raccoon opens the lid.

Birds: At sunrise, no bird activity; I walk out the front door to see what the world is offering. The sun is rising, I see a sharpie sitting on the branch of a maple. As I go down the driveway to get the paper, it looks down at me and I up at it. It does not move, like "What do you want?" As I go back in the house a flock of 50 local geese fly over. Meanwhile in the backyard the gang is having breakfast.

It is raining, a red-bellied woodpecker flies onto a tube filled with split peanuts. It flies to a branch, tilts its head sideways into rain drops and drinks drop by drop by drop.

Red-bellied woodpecker getting spit peanuts

There is a mockingbird in the front bush. It has been arriving this time of the year but it only stays a week and moves on. Now and then it will come to the backyard for feed!

Mockingbird

Zero degrees, 6:30 a.m., clear, I am awoken by the hoot of an owl in the back tree (and again February 23, 2019 at 4:30 a.m.). At dusk 10 to 15 ducks are making six to eight circles over the yard.

In 2015, 6:00 a.m., we get a freezing rain, only a few birds at the feeders. At 12:30 p.m., 36F, the ice is starting to melt, and the birds are starting to come to the feeders. It is dark, dank, and foggy at 5:00 p.m., I can hear, but don't see, the geese flying over.

Third Week

<u>Weather</u>: The weather can be a yo-yo. At 5:20 a.m., 55F, raining, the temperature goes up to 68F; normal is 34F. At 7:00 a.m., 42F, heavy winds at dawn's early light. Then at 9:30 a.m. it starts to rain, 10:30 a.m. it turns to snow, 11:30 a.m. it stops as rain; no activity. It is clear, 38F, at 5:40 p.m. What a weather day! Rain-snow-rain-clear. The temperature goes from a high of 68F to the 20's overnight. Another year at 6:00 a.m., 24F, clear and dark, there are two planets shining brightly in the southeast. There is just enough light at 7:15 a.m. to see objects on the feeders and on the lawn. As the sun rises it sends a bright beam to the upper branches of the tall evergreens. In January we may not see the sunshine for 15 days. Another year, 60F, we get a heavy thunderstorm between 5:15 and 6:00 a.m. Then at 7:00 a.m., we get a tornado warning. Fortunately, it did not happen.

<u>Animals</u>: Chipmunks and groundhogs are hibernating; skunk tracks and their fragrance are found around the base of the birdfeeders. Chipmunks do not go into a deep hibernation. During a warm spell I see one sunning on top of a snow-free rock. Another one is in on the birdfeeder.

Chipmunk

I hang a safflower tube from the gutter for the kitchen hook that is 18 inches long. Guess what? A squirrel climbs up on the roof and then crawled down on the shank of the hook to the feeder. I don't know how they do it. I hang the feeder from a dogwood branch with a six-foot cord, the squirrel shimmies upside down to the feeder! Then I move the feeder to a branch of the maple tree. I hang it from a 20-foot cord that was eight feet from the trunk. The feeder was also five feet above the ground. No squirrel has gotten to it, but a Carolina wren knocked enough seed down to the ground to keep the squirrels happy. In April, the seeds they missed on the ground will come up as little plants in the flower garden.

Squirrel coming down the rope to the feeder

A squirrel is eating a lump of snow in the dogwood at 8:30 a.m. Two are on the chase. A vole is running behind the PJM on the stone wall.

A person calls the police, thought that a squirrel had gone mad. It turned out it was just trying to get at its mate.

It is cold on the 20th. A field mouse comes out during the day to eat mountain pink foliage and safflower seed. It then comes out at 8:30 p.m. in the evening; the kitchen lights illuminate the feeder so I can see the mouse enjoying the overflow of safflower.

Field mouse

I smell skunk when I open the door in the morning; there are fresh droppings on the lawn. A squirrel has mange on its hind quarter and a female cardinal has it on her head. A rabbit is eating the mountain pink. Neighbors across the circle reported having deer in their backyard in December.

Evening visitor

<u>Birds</u>: I looked out to see two pairs of mallard ducks in the circle! The ducks return in January in quantities from six to 30. They fly overhead around the circle mostly at dusk for 20 minutes. I do not know where they settle for the night. Other years they arrive on the fourth week.

The sun is bright, but it is cold; I hear a high screech above. I am fortunate to see a majestic red-winged hawk soaring below the cumulus clouds.

A turkey vulture soars overhead at noon; not too many years ago I did see them in January. In 2017 they started to overwinter. Their wing positioning and shape results in an erratic flight path. Graceful one moment and the next they are looking like they are going to plummet to earth.

Just light, 7:10 a.m., the gang and a pair of overwintering goldfinches are here until 9:00 a.m. Three starlings go to the small birdhouses. A titmouse takes a whole peanut, puts it between its legs, lands on a branch, and proceeds to twirl the shell like we eat corn-on-the-cob until it can get at the nut. A house finch and a squirrel are eating snow that lies on dogwood branches. Two "V's" of 300 snow geese fly over at 10:30 a.m. Blue jays come for eggshells. Eight blackbirds come for brunch on the main feeder at 10:00 a.m. The flock may include starlings, cowbirds, red-winged blackbirds, and grackles.

The chickadee and white-throated sparrow are in the raised feeder eating millet together, normally the white chases the chick out. The chick goes to the safflower feeder; it takes the seed and goes to a branch to eat it. Two squirrels bury a peanut for the future. Titmice are singing loudly.

JANUARY

A bull mourning dove holds its ground with raised wing when a chipmunk comes to the feeder

Dove keeps itself warm overnight by sitting over the warm water of the deicer

Two blue jays come for peanuts and eggshells. The eggshell gives it calcium. A red-breasted nuthatch takes a split peanut and goes to the tall maple, lands on the trunk, then clambers upside down until it finds a patch of bark sticking out and tucks the flake under it. A titmouse gets to the water, dips its head down and just keeps on drinking. A chickadee gets water from the fountain by dipping down and then raises its head a little above the water to swallow. Two white-breasted nuthatches go for the split peanuts, each at a different feeding tube. A flock of 15 robins land on the front lawn; it is too early for them to be here! A fox sparrow is here for a few days; it runs from under the holly bush, gets seed, and scoots back under.

A house finch has conjunctivitis; this disease can appear in the house finches any time of the year. The disease seems to affect the hearing of the bird as well as its sight. A diseased bird is on the feeder. I walk up to feeder on the opposite side from the bird's diseased eye so it cannot see me. I then talk to the bird, but it does not move until I walk around the feeder and it sees me with its good eye. The disease does not seem to be contagious to other species but now and then I have seen a cardinal with the disease. Unusual, two days later a teen cardinal arrives with the disease. This disease blinds the bird in one eye then the other and then the bird starves to death.

House finch with conjunctivitis. The eye in the reflection is still okay.

At 4:50 p.m. a flock of 100 geese are going west and high above them is another flock of 50 going north! How do you figure, one going north and one going west? Another flock of 30 go west at 5:10 p.m. I guess they need exercise. Dusk, 5:10 p.m., on the ground two white-throated sparrows are bouncing around on the oak leaves, cardinals, house finch, song and white-throated are still at the feeder. At 5:20 p.m., 28.5F, there is no feeder activity. I can make out bird silhouettes in the tree branches.

A red-bellied woodpecker puts a seed in a little cavity on a tree trunk, so it does not have to hold it to eat it. Three titmice are squawking at a big, rust-colored cat and later a squirrel at the edge of the frozen pond. Now the titmice are really yelling. The cat spooks up a rabbit. It chases it but the rabbit leaves it far behind. Now the squirrels yell at the cat.

Thirsty animals

Chickadees and titmice are investigating birdhouses for the coming mating season.

Titmouse

Two unusual observations. I was driving home and saw 20 pigeons sunning on a telephone line. On the next pole over was a hawk looking at the pigeons; neither are moving, just sunning. The pigeons must realize that the hawk had to be in motion to be a threat. The other observation was more unusual. A sharpie is calling from the back maple, yet the small birds at the feeders did not scatter. I slowly opened the back door and looked up to spot another sharpie in the next maple. They were calling back and forth before they observed me and flew off. The male was more interested in other things than the potential food below.

It is 14F, the statue of Buddha is smiling because the sun is out. There is no activity probably because it is too cold. Spot a white-throated sparrow bouncing around among the leaves.

I empty out two birdhouses of last season's nesting material.

The gang of 15 species is here from 7:30 to 11:30 a.m. Just like us, they are stocking up because they know a storm is coming. They are even eating the peanut butter. At 11:30 a.m. it starts to snow; at 2:00 p.m. it is snowing at a good rate. The snow on Buddha makes him look like he is falling asleep. It is getting dark, 4:20 p.m., the snow is stopping; we got two inches. At 4:37 p.m. most birds are gone. It is dark, 4:45 p.m., there are still cardinals, juncos, song sparrows, house finches, and white-throated sparrows. For the day, the high was 20F, 15 degrees below normal; the low was 6F, 12 degrees below normal.

On the 19th of 2008, 7:00 a.m., 30F, overcast, semi-dark, getting cold, the gang is active all day. There are five blue jays. The various species are a little aggressive to the other species as they need food. They know it will be a cold night. A dove lands on the ice of the pond and slides halfway across. A cardinal chases a house sparrow and a white-throated off the main feeder. A Carolina wren is at the safflower feeder; it throws out a lot of seed but only eats a few. A white-throated sparrow and Carolina wren eat snow that sits on the rhodie leaves. At 11:00 a.m. a turkey vulture erratically soars overhead. Overcast and damp, 3:00 p.m., 37F, a vole scurries into the wall of pond. As I go out at 4:00 p.m. to fill the feeder, a broad-winged hawk glides over me and through the dogwoods. Later, just after dusk, I hear a great horned owl giving a mating deep hoot. It is an unsettling call.

Flowers: With a warm January the crocus may be out. The flowers will die with a freeze. In the fall I cover the bulbs with chicken wire to prevent squirrels from digging them up.

Crocus

Shrubs: The arborvitaes have turned a dull green.

Trees: Sheets of four-by-six-inch bark are falling from the white birch tree.

Fourth Week

In the morning on the 28th, 7:15 a.m., the moon is full giving off a bright light on the snow. There is already bird activity at 7:25 a.m. It stops abruptly at 8:30 a.m. when the hawk arrives. At 4:50 p.m. there is no activity as the sun sets, by 5:30 p.m. it is 21F.

On the 29th, 6:30 a.m., 4.6F, clear, and the moon is starting to wane. The rhodie leaves are like pencils. The silhouettes of the bushes and trees can be made out at 6:50 a.m. The tail of a song sparrow is twitching up and down. The white-throated sparrows may look nice, but they are not. They continuously chase house finches from the safflower feeder. The blue jays give a call that they are coming. They are shrewd; they take the peanuts off the ground before the squirrels get them, then they take the ones on the top of the feeder which the squirrels cannot get.

The starlings are going to the owl house and house sparrow houses to take up residency. The starlings try to get into smaller houses but are too big. I had put a ring of epoxy around the entrance of the smaller holes so the starling cannot rip out the wood to make the hole larger.

When there were no small birds present, I put red meat under a feeder. Within ten minutes a young red-tailed hawk lands; it takes 30 minutes to eat all the meat. It was quite a sight to watch how it ate the meat. It took a ten-inch-long piece and put it in its beak and then proceeded to swallow a half-inch at a time, gulping it by the tip of the beak. It swallowed the whole ten inches down its gullet.

At 11:30 a.m., 41F, 30 to 40 mph winds and sunny. Trees swaying, leaves blowing, 20 Canada geese are flying west. Suddenly a big gust comes from the west. The whole flock banks south from the wind like a squadron of fighters.

On the 30th, 7:20 a.m., 24F, a chickadee stays to the side of the feeder until a Carolina wren finishes breakfast. House sparrows are active at the houses. The first downy in weeks is observed.

Just light, 7:12 a.m., 5F, rhodie leaves are curled up like a pencil. Between 10:00 a.m. and noon we are scheduled to get more snow; they were wrong. It starts lightly at 8:45 a.m., by 10:30 a.m. it is coming down at a good rate. Through the snow I observe a squirrel taking nesting material up the evergreen. At noon it feels like -7F. Spot 30 geese flying around in the hard-falling snow. At 5:00 p.m., it is still snowing, the temperature has increased to 11F. So far, we have had six inches. Only see one junco; the others have bedded down. At 6:00 p.m. the temperature goes to 14.5F.

Another year, 6:00 a.m., 31F, a catbird arrives (they usually arrive the fourth week of April) with six robins who are singing spring songs. Starlings are trying to get into the houses; 100 Canada geese and 100 snow geese fly over. There are five titmice and four chickadees. Where has winter gone?

We have an incident, 8:00 a.m., normal feeding activity; white-throated sparrows are foraging underneath the PJM's which are on top of the stone wall of the second level. When bam, a sharpie dives in under the PJM to get a sparrow. The sharpie immediately flies off with its prey. It happened so fast that the other birds did not miss a beat in their feeding. They stay another 10 minutes before flying off.

Another incident. A dove is feeding on the main level of the main feeder. A house sparrow lands on the roof where there is some sunflower seed. The sparrow purposely dumps some seed onto the head of the dove. The dove retaliates by chasing the sparrow off the feeder. A sharpie flies in after the dove and misses. It hangs around three minutes then flies off. Within a minute a chickadee, white-throated, junco, and Carolina wren are back on the feeder as well as a squirrel feeding at the base of the feeder.

At 6:10 a.m., 14.5F, we have had over 10 inches of snow. The squirrels are taking one peanut each. Another squirrel breaks off a maple tree's icicle and chews on it. The wind has come up and the spruce tree branches hang low with the weight of the snow; I can hear the branches rubbing against each other. At noon a house finch and a song sparrow are eating side by side. Usually the finch chases the sparrow. The gang is here, 1:00 p.m., 19F, so are three crows and 10 starlings who eat the cut-up grapes. A sharpie arrives and sits in the oak tree. Bang, all the other birds leave with alarm. Hey, the sharpie is hungry too. It sits there for an hour-and-a-half; a high wind comes up and the hawk leaves. Three minutes later all the small birds are back. The sun is setting, 4:45 p.m., 14.8F, most birds have gone to roost. For the day the high was 16F, 19 degrees below normal; the low was 14F, four degrees below normal.

At the end of the month winter is really settling in. At 6:00 a.m., it is clear, 4F, the stars are out, and a full moon is setting behind the tall evergreens. There is 12 inches of snow on the ground. At 7:00 a.m. there is no activity, a lone cardinal is at the feeder. I may hear courting owls hooting from 10:30 p.m. to as late as 6:00 a.m.

The activity starts at 7:10 a.m. A crow is cawing from the top of the 80-foot poplar tree, another caws from a distance, it is the beginning of morning. Activity is in full swing with 10 species between 7:30 and 9:00 a.m. The quantities of many species have gone from one to two. Two male cardinals are sparing for the same mate. A red-bellied woodpecker comes to the feeder. This would not be the time for its annual arrival; in 2007 they started to overwinter.

I spot a flock of 15 robins under the hemlock hedge. I do not rake the fall leaves from under the hedge in the fall so they can harbor insect and worms near the surface. One possible explanation is these birds were way further north than the normal flocks. This is a by-product of climate change.

Robins under the hemlock hedge

January 23rd – Blizzard of 2016

The weather people have been predicting that we may get a heavy snow on the 23rd. This time they were right on and Allentown was the bull's eye. We got a record 30.2 inches in 24 hours. Everything was shut down, roads were impassible. It took hours to snow blow the heavy snow.

The day before the storm the population make-up of the gang was 27 of birds from a normal of 13. During the storm there were 65 birds. There were three species missing that were present the day before, the mockingbird, red-bellied woodpecker, and titmouse! On the other hand, I got a visit from three species that do not normally come to the feeders, a red-winged blackbird, robin, and a broad-winged hawk. The hawk knew where its next potential meal was gathering; everything scattered when it landed in a nearby tree. The population of some species exploded during the storm; cardinal went from two to 10, house finch from one to 12, and mourning dove from six to 10. They all fed during the height of the storm. The next day was brisk and sunny but only two species came to the feeders! A grand sight was seeing 100 snow geese fly over. I believe it was for exercise not migration.

JANUARY

Quantity of Birds Before-During-After the Blizzard on the 23rd of 2016								
Day	22	23	25*	Day	22	23	25*
Blue jay R, G	2	2	2	Purple finch	1	4	
Broad-winged hawk		1		Red-bellied woodpecker	1		1
Cardinal R, G	2	10		Red-winged blackbird		1	
Chickadee R, G		1	1	Robin		1	
Downy woodpecker R, G	2	1		Savannah sparrow			1
Goldfinch		1		Sharp-shinned hawk			1
House finch R, G	1	12	3	Starling	1	8	2
House sparrow R, G	4	4		Titmouse R, G			1
Junco G	2	4	4	White-throated sparrow	4	5	3
Mockingbird	1		1	Total Species	13	15	12
Mourning dove R, G	6	10	2	Total Number of Birds	27	65	22
*25th on the 24th only one junco, two house finch, and five squirrels returned to the feeders								
				Squirrel	2	1	1

Water pan before and during the storm

White-throated approaching the hole and going in

TRENDS DUE TO CLIMATE CHANGE

Make room – 20 are already here

Surprise visit from a robin

Ah! I needed that!

Snow geese the day after the storm

Other Winter Observations

It is clear and brisk on the 25th, 7:14 a.m., 23F, the sun is coming up, there is 12 inches of snow on the ground. Small birds are on the feeder, there are two doves on the snow. They fly up to the main feeder. There is a sharpie in the oak; it flashes down toward them. The doves sense it is coming; it may be too late. The doves take off and swing sharply to the right; the sharpie is right there and swings right. There is a splash of feathers. The other small birds keep right on eating.

The animals and birds still need water in the winter; the ponds are frozen except for a thin strip of water along the pond's edge. The pond's edge is lined with big rocks that are partially submerged in the water. During the winter, the upper portion of the rocks absorb heat from the sun. There is enough heat to keep the pond's edge ice-free. The water strip is visited by squirrels, chickadees, sparrows, and robins.

At the end of the month the cattail punks puff up before being released. The little parachute seeds are blown away by the westerly winds. The birds take the punk for nesting; the squirrels will take pieces of the stalks up the trees for nesting. The punk is all blown off by the first week of February.

Cattail punk blowing apart

At 10- to 15-year intervals the power company has crews trimming the branches 10 feet away from the line. These guys are real climbers. What I thought was too thin to climb did not phase them. Toppers cut the branches and the lower crew takes them away.

The first day after the power company thinned out branches along the power lines, a squirrel ran along the highway of cables. It stopped and looked around at what was his normal highway stops, baffled as to what has happened. Then he continued down the highway. Some of the spruces are nothing but high stumps.

Trimming out the branches

The big birch on the third level only has one tall trunk that is over 60 feet tall. The right-side trunk is now only 20 feet tall. The north side of the trees next to the cables look like a giant hedge cutter had sheared them off from below to the top of the cables. Surprisingly, with the huge pile of cut branches on top of the shrubs and the dragging of them over the shrubs there is minimum damage. The next day two squirrels are developing a new highway to travel through the trees and cables. After a heavy snow there may be three to four inches of snow on top of the power line. A squirrel puts its nose against the line and runs down its length using it as a snowplow.

<u>Trees</u>: The fourth week is the time to prune the apple trees. The silver maple buds swell in the fourth week.

The day after a storm

Review of Data

Fauna

<u>Animals</u>: When we have a warm spell for a couple of days, it brings animals out of hibernation or their dens. Sadly, in the morning there is roadkill of skunks and raccoons.

Squirrels get nourishment from raisins in cereal, new dogwood buds, and whole peanuts. They bury the whole peanut for a future dinner. I dig many of them up when I turn the garden over in the spring. They not only bury the nuts in the ground but also in the snow. Sometimes they like something sweet; I watch one snap off a sugar maple icicle and chew it like candy. Another time one picked up a piece of ice and sucked on it.

Getting ready to leap

A squirrel tries to jump straight up to the main feeder. It twirls its tail like a propeller as it starts the leap. It did not have enough push to get to the platform. Down it went, shook itself, and ran away.

There is a 5-7-9 rule for placing a feeder out of a squirrel's reach. They cannot jump higher than five feet, and cannot jump more than seven feet horizontally, nor go down more than nine feet vertically. I watched a squirrel jump horizontally for seven feet. I had a feeder hanging from a 10-foot rope. After many failures trying to go down the rope the squirrel simply ate through the rope to drop the feeder to the ground.

Good try but did not make it

The first week of the month squirrels start what I call the "chase." This is when one squirrel chases another. Squirrels run up and down and around a tree trunk at a fast pace. They jump from tree to tree across the yard, jumping horizontally seven feet to get to another tree. This is part of the mating procedure. It is fun to watch. When the one being chased stops, the pursuer stops right behind it. The pursuer does not land on or bite the pursued. Squirrels can come straight down 30 feet along a tree trunk. When the pursued continues, the pursuer quickly follows. This can go on for 10 to 15 minutes and then they just break off. Shortly after they may be found mating in a dogwood tree. The chases continue all January and into early February. In late January, the female takes leaves and stalks into the tall spruce trees. She gives birth to kits at the end of March and into early April. One year there were 13 squirrels running around at the same time. On the 28th of 2012, a squirrel has mange on its hind quarter.

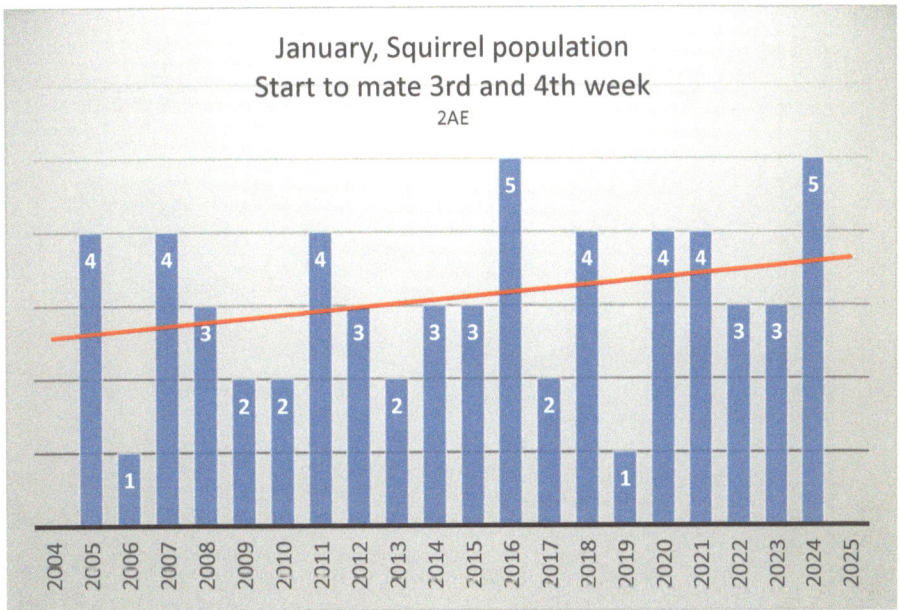

January squirrel population

On a cold, sunny morning, a field mouse or vole eats the leaves off the mountain pink (creeping phlox). They don't bother the pinks when the leaves turn dark green in the spring. The gnawed pinks will recover in the spring to have its flowers. A vole was skating on the ice of the pond; it loses control and goes straight into the rock wall. In a warm spell, I get the fragrance of a skunk when I open the door to get the morning paper. Raccoons come 3:00 to 4:00 a.m. to raid the peanut can. A hungry red fox appears on the hill, the sighting causes all the birds to scatter.

For years there were no deer in the neighborhood. Then in 2008 neighbors across the circle reported having deer in their backyard. Two days later there was a doe and an older fawn in my backyard having a feast on the perennials and shrubs. The doe has an injured front left leg. I was amazed how they would stand lengthwise next to a four-foot fence, then in a flash with no running start, leap over it. They stayed a week before moving on. They were probably driven from the field a half mile west of us as it became a development for 240 homes.

Doe

2011 – going out of the circle

Until 2014 deer would come in the evening to settle in the upper part of the backyard. Since then neighbors have reported seeing a few but I have not.

Birds: The yearly range of the quantity of species of birds for 13 years went from 10 to 30, most years it is 23 to 25. What throws the quantities off is eight species only appeared once or twice during the period. They are the black vulture, brown thrasher, catbird, cowbird, fox sparrow, pine siskin, purple finch, and red-winged blackbird. The quantity of the other species basically stays the same year to year. Two exceptions are the cardinal, whose quantity has gone from two to eight or 10, and the crow quantity went from 10 or 20 to none in 2013 (West Nile virus). By 2018 it was back to 15. The reason species have arrived late, passed through late or not observed some years is probably due to weather. The population trend is about 43 but has decreased to half that number by 2017.

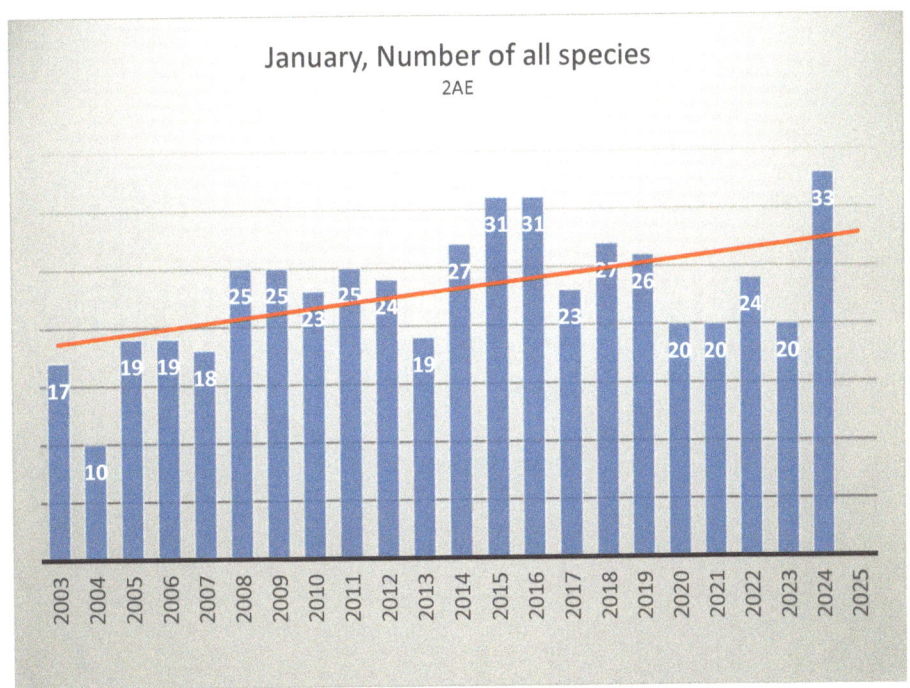
Total number of species observed in January

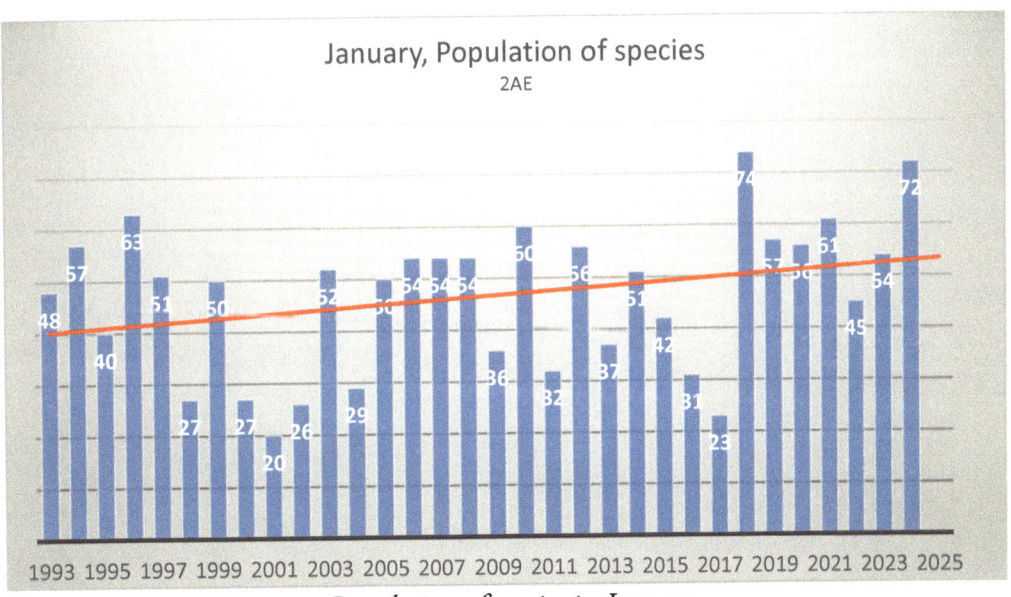
Population of species in January

    The 2018 population was high because of 24 doves, 12 mallard ducks, and 16 house finches added 52 to the count.

    The average quantity of gang species in January is 13. The trend is an increase in species. The population of birds within a species does not fluctuate too much except for the cardinal, house finch, and mourning dove. That may be due to how many hawks are in the area.

JANUARY

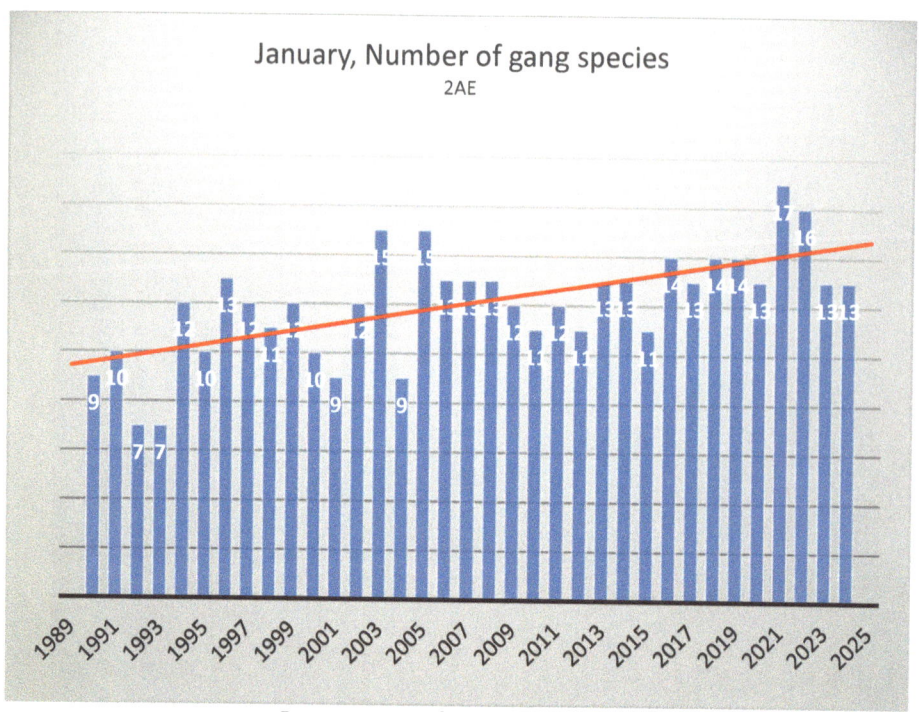

January, number of gang species

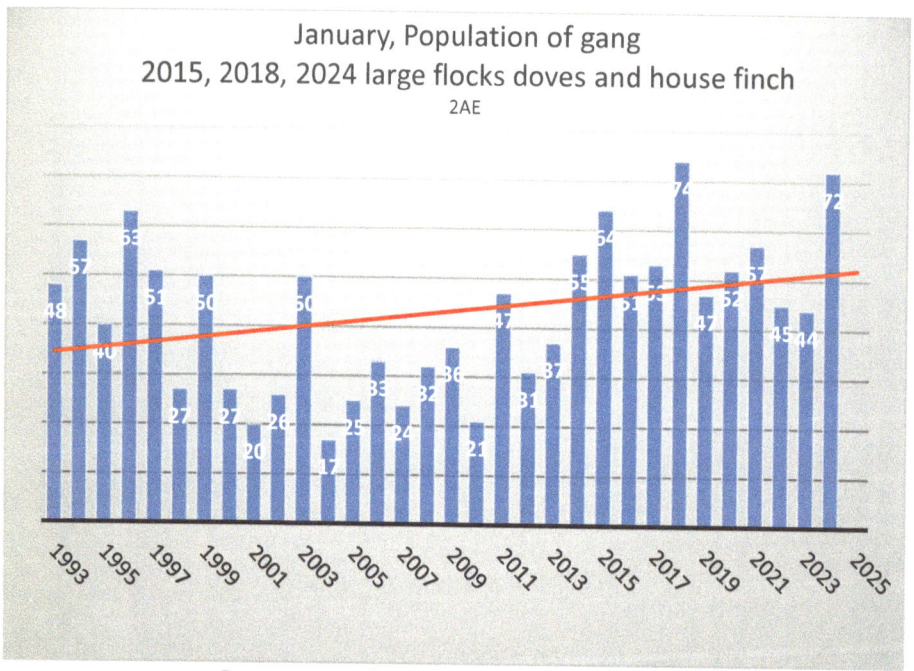

January population of gang (low in 2004)

Table of the Gang

January quantity of birds for the Gang																
Species/Quan.	05	06	07	08	09	10	11	12	13	14	15	16	17	18	19	20
Blue Jay	3	2	2	3	1	2	2	1	4	3	2	2	3	2	3	2
Cardinal	2	2	1	2	6	2	10	2	6	4	8	7	8	8	5	4
Carolina wren																1
Chickadee	3	3	4	3	2	2	2	1	1	1	3	2	3	1	2	2
Downy	1	1		1	1		1	1	2	1	2	1	2	1	1	1
Goldfinch	1			1	1	1	1	1		2	2	2	1	1	4	1
House finch	2	4	5	1	4	2	8	6	6	1	6	12	6	16	8	8
HF with conjunctivitis		1			1					2	6	4		1		
House Sparrow	4		5	4	4	4	4	1		2	6	4	6	4	8	2
Junco	1	5	1	4	4	1	2	4	3	5	5	2	6	7	5	4
Mourning dove	1	2	1	3	1	1	8	4	5	22	12	6	8	24	6	19
Red-bellied wood																2
Red nuthatch	1	1	1	2			1		1				1	0	2	
Savannah sparrow				1	2	1	2	1		1	1	1		1	1	
Song sparrow		3	1	2	1	1	1		1	1	1			1		
Titmouse	3	4	2	1	3	1	1	2	2	2	2	3	1	2	2	
White-breasted nut	1	1		2			1		1	2	2	1	4	1	2	1
White-throated	2	4	1	3	6	2	4	6	4	6	6	4	4	5	3	3
Total gang	13	13	11	14	14	12	15	12	13	14	14	13	12	14	14	13
Population	25	33	24	32	36	21	47	31	37	55	64	51	53	74	47	50

<u>Individual Species</u>:

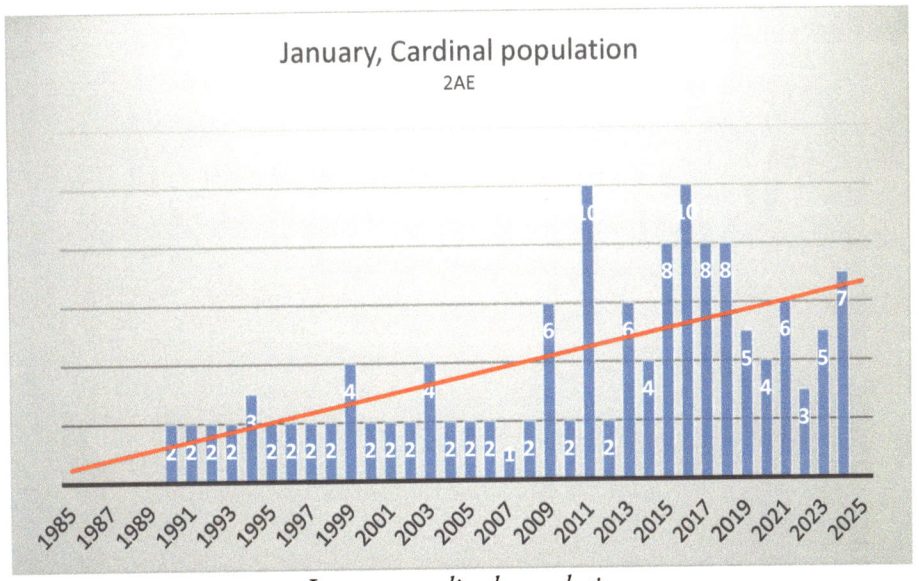

January cardinal population

JANUARY

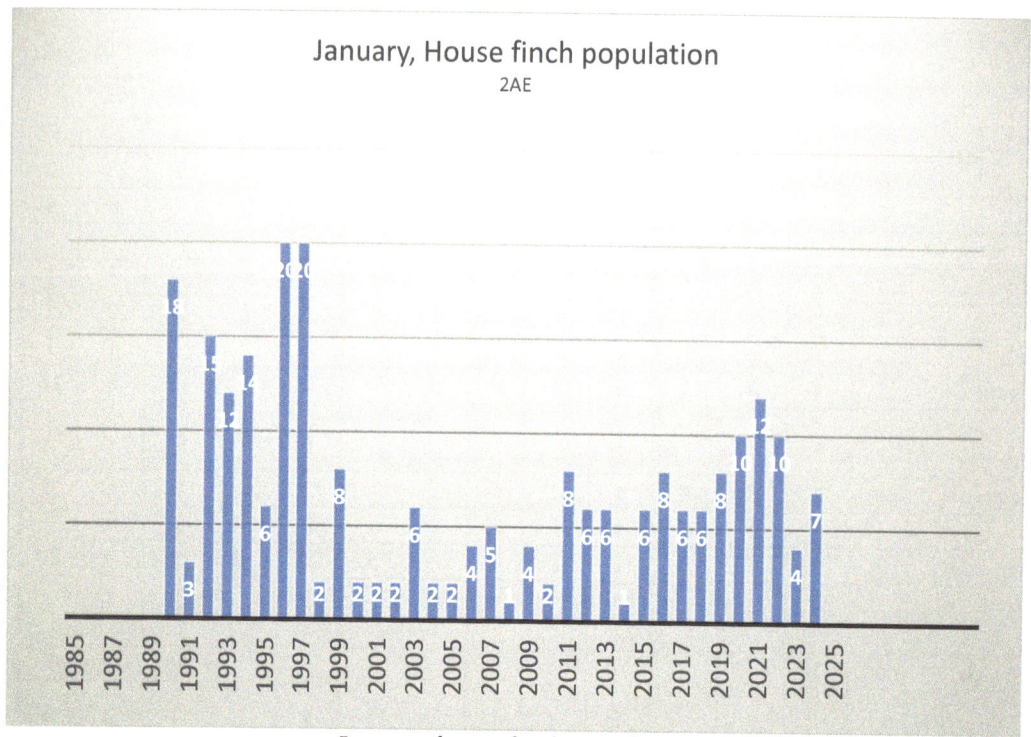

January house finch population

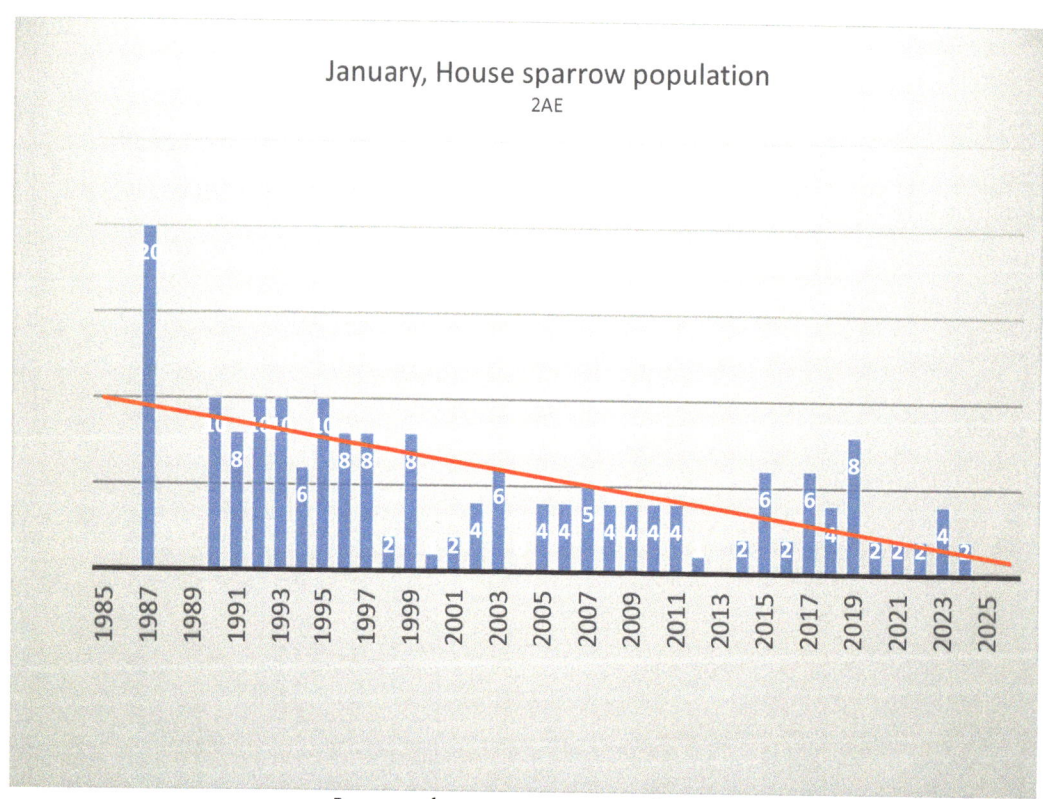

January house sparrow population

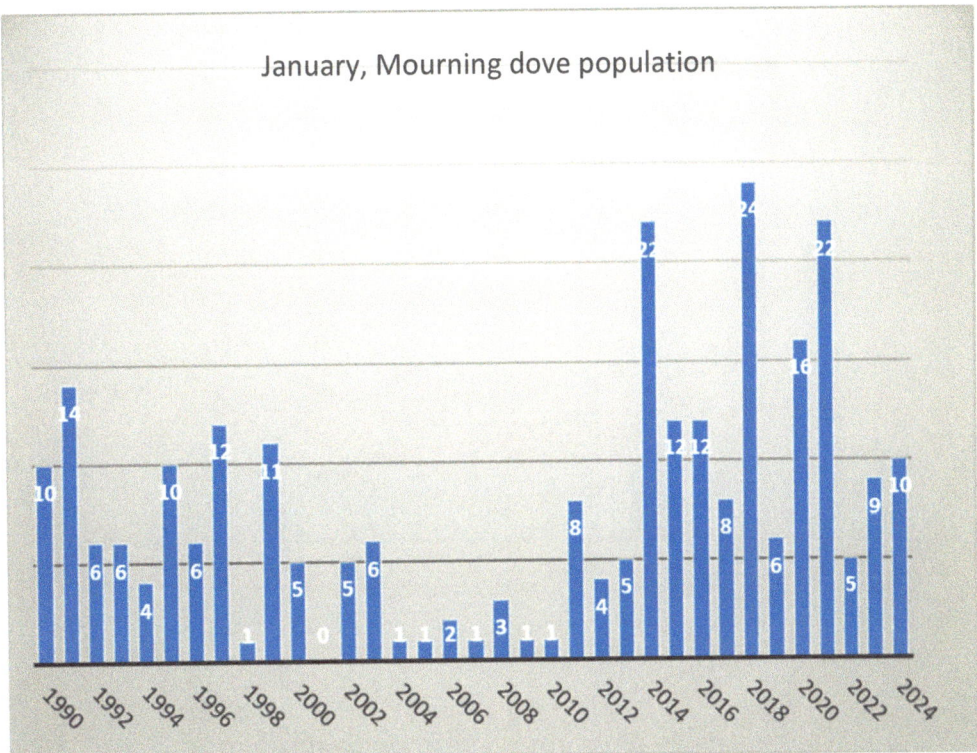

January mourning dove population

It was a good sign that the crow population was increasing but 2019 reversed the trend. The population went from eight to one, March 2020, back to 10.

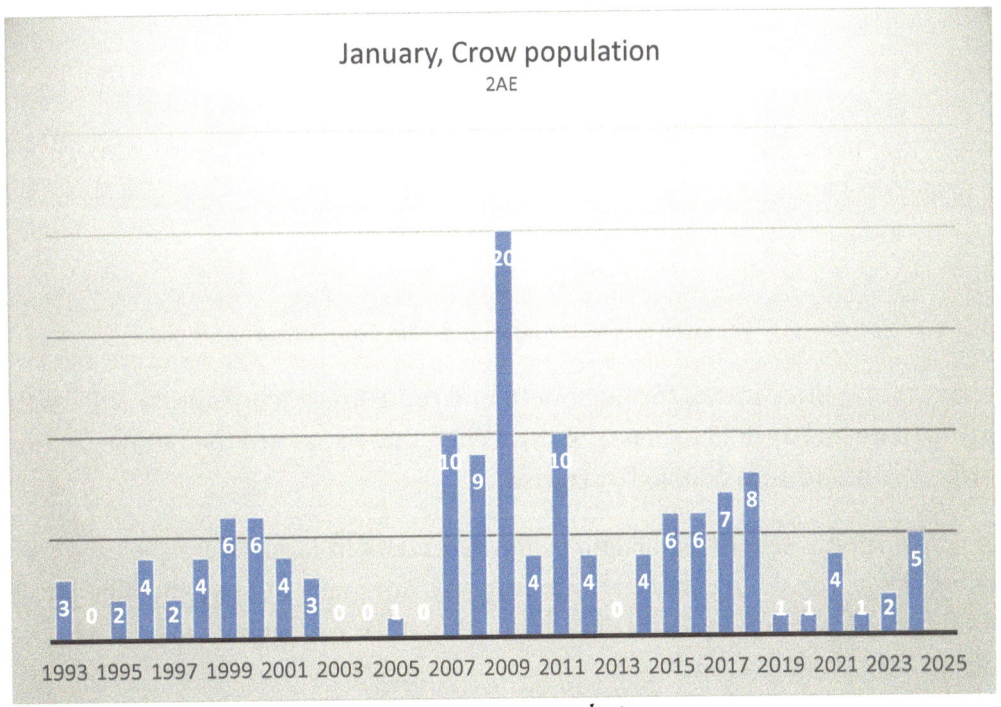

January crow population

JANUARY

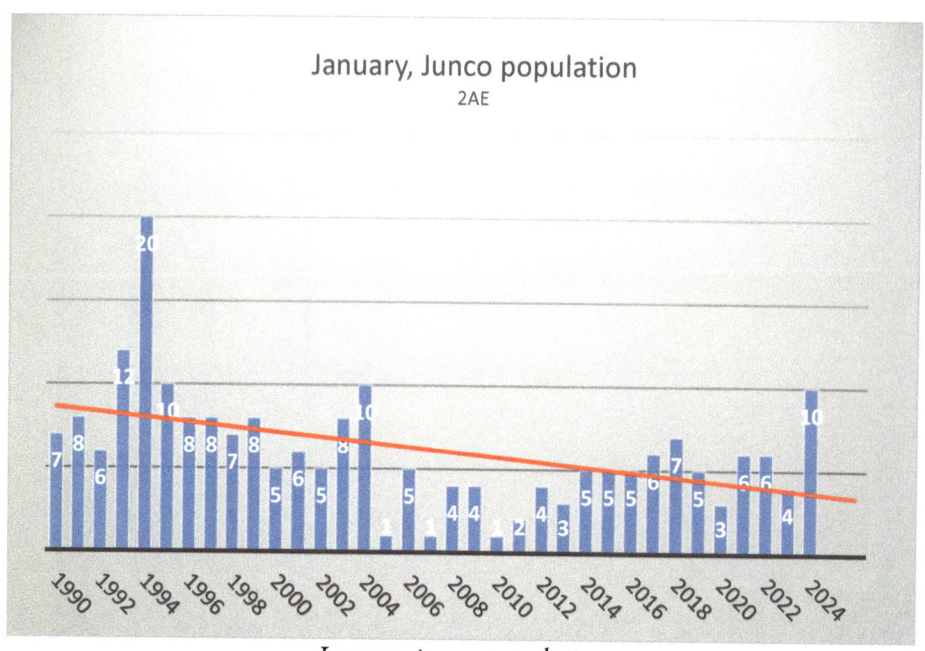
January junco population

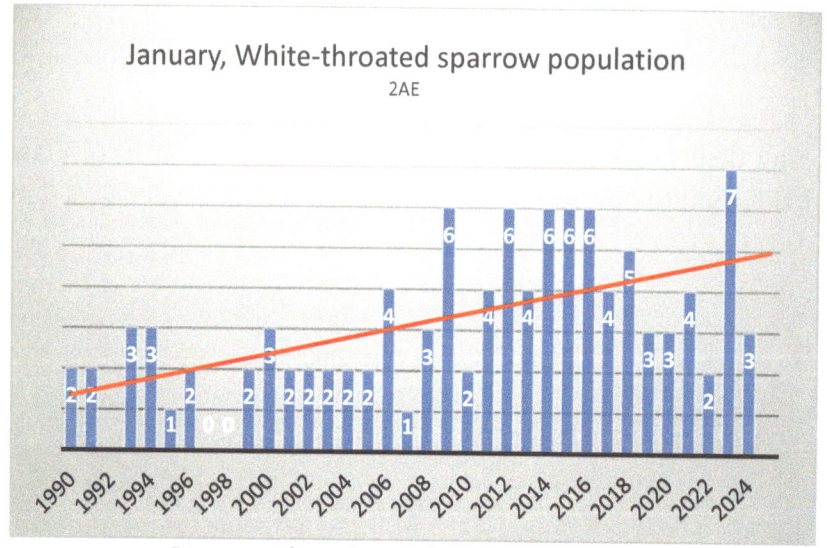
January white-throated sparrow population

The average rain is three inches for the month and the average temperature 30F for the period. In 2011 and 2016 we had over 25 inches of snow. When that happened the number of house finch, mourning dove, and cardinals doubled or tripled.

Fledglings: Three species that bring fledglings to the feeders in January are the cardinal, blue jay, and mourning dove. The young cardinals are easy to distinguish but not the blue jay. The male, female, and fledglings all look the same. The mourning dove young just look young. Doves bring fledglings from the first to the third week.

Migration: The black vulture was an unusual observation since they are normally in the south; maybe part of the climate change.

The one migration to see is the hundreds to a thousand of snow geese going north. Until 2013 this migration was seen during the morning of the first week, by 2017 it occurs during the fourth week. In 2018 they did not arrive from the north until late December. Migration depends on when the lakes freeze in the north.

The local flocks of overwintering Canada geese went from 200 to 50 and back like a yo-yo until 2010. None appeared in 2011 and 2012. Since then there has been a steady increase. In 2018 they were back to 100. In general, by 2018 the overall population seems to be increasing. By 2024, the effect of many large warehouse cropping up has the local down to 20 to 30. The large flocks are not seen.

Pecking Order: In the summer when there is plenty of natural food as well as the supplemental feed, the various species tolerate each other at the feeders. Winter is different, as the pecking order becomes very apparent especially during and after a snowstorm.

Tolerance, they continue to feed

The blue jays are the bullies expecting all the other species to leave as they come in screaming to land on the feeders. When they want to clear the feeders quickly, they mimic the cry of a sharpie, which is very effective.

Most times the mourning doves leave when the jays are present; one day a big bull dove stood its ground when the jay landed. With raised wings, it defied the jay as it approached. The bully left only to return with two other jays, then the bull dove retreated.

The jays are not the only species to apply the pecking order; a female house finch will chase a smaller junco. The junco takes revenge. It was above a house finch on the feeder with lots of extra safflower around. So, it dumped seed onto the finch, which yelled. Another time a junco landed on top of a house finch, again the finch yelled and flew away. Basically, a peaceful junco is aggressive at the feeders, especially during a snowstorm. A house sparrow dumped seed on a dove, the dove retaliated and chased the house sparrow from the feeder.

The larger white-throated sparrow will chase the Carolina wren, house finch, and junco from a feeder. It will then stay a while to feed. The male white-throat will chase a female white-throat from a feeder. White-throat knows that a song sparrow will chase them from a feeder. Two white-throats wait to the side until the song finished eating and leaves before going to the feeder. The cardinals will go after the white-throated sparrow, house sparrow, and house finch that are on the same side of a feeder. A teenage cardinal (bully) chases a white-throated from the feeder. The cardinal gives a very unfriendly yell (hiss) when chasing the other species. A female house sparrow will chase a female house finch. A male house finch will chase a white-throated sparrow and junco.

Pair of house finch

Congenial now and then

The red-bellied woodpecker chases a house sparrow from the split peanut feeder. The smaller chickadee will take seed away from a white-throated sparrow at the feeder.

When looking for a new nesting house, the house sparrow will chase a nuthatch from the same house selection. Yet the house sparrow will let a song sparrow sit next to it and feed during a storm.

A titmouse chases a white-throated sparrow off the raised feeder, then eats the seed on the feeder. A male junco will chase a female junco from the feeder. Junco and titmice will squawk at each other on the feeder.

Female downy at split peanuts

Junco going at a male house finch

Three pair of juncos, three pair of cardinals, a chickadee, and two pair of house finches, 16 altogether were on the feeder; that is pretty close quarters. When the different species get too close to each other a little scrap will develop. A male cardinal goes at it with a male white-throat.

Some other observations I have witnessed include when I put whole peanuts on the ground and feeder for the squirrels and jays, the jay can put a whole nut in its gullet and then put another in is beak and fly away. Carolina wrens will pull a lot of safflower from the bottom of a feeder before taking one seed. The birds and squirrels below the feeder love it. The small birds must be on extra alert after a snowstorm because a sharp-shinned hawk will come for them while they are feeding. If the sharpie misses, it will return in 10 minutes. It will repeat this until it gets a meal. When a song sparrow eats safflower, it crushes the seed in its beak and then swallows the crushed seed.

When it is warm enough for melting snow to form droplets at the end of rhododendron leaves, the house finch and chickadee will stretch up from the leaf below to sip from the droplet. A puddle of water sits on top of a rhododendron leaf, the bird leans down from above to get a drink. A chickadee with a sweet desire (birds do not have teeth so I cannot say sweet tooth) stretches up to get a drip from a sugar maple tree icicle.

In the fall turkey vultures used to migrate south and returned in May. Over the decade of the 2000s they started to return in late April, that progressed to March, and in 2009 they overwintered. On a cold, clear winter day of January, you may see their six-foot wingspan erratically soaring in the early afternoon.

All of January the juncos have a "chase," but this chase is low flying around the shrubs, up and over hedges. It is quite an acrobatic aerodynamic show. Since their body is slate gray, the show is accentuated when the shrubs and hedges are covered with snow. This "chase" may consist of three pair.

In 1970 the development was an open area with few trees, which was a good environment for mockingbirds. As the years passed and trees grew, the mockingbirds left the area. However, recently, for some reason, a mockingbird now returns the last week of December into the third week of January and then leaves for the rest of the year! It likes open space, so it stays in the front yard. When it snows, it may come to the backyard feeders.

Red nuthatch and chickadees take a safflower seed quickly then fly off five to 10 feet into a bush, open the seed, eat the meat, then fly back for another.

The titmouse stays on the feeder and opens the seed. White-throated sparrow comes and stays to eat maybe five to 10 seeds. It chases most other birds away.

Chickadee getting safflower

Junco

The junco comes to the feeder, eats three to four seeds and flies off, and then repeats the same trip, but not as fast as the titmouse. If there are other birds at the feeder it will stay and eat a few before leaving.

The song sparrow comes and likes to stay. It will chase smaller titmice and chickadees. It will try to stand up to the white-throated but the white usually wins.

White-throated sparrow

Cardinal ready to take off

When the Carolina wren comes to the feeder, it gets to where the seed comes out and just keeps throwing the seed without eating it. Now and then it stops to eat a seed but continues to throw the seed out. The cardinal is very leery; when it does come it takes a seed and flies off.

When the downy woodpecker comes for suet, it lands on the roof of the open bottom feeder. It dances back and forth across the top of the roof until it is satisfied it is safe to go underneath and get the suet. I used to put out peanut butter in the winter, but it would sit untouched and get moldy. However, when the parents are fledgling, I am replenishing the peanut butter continuously.

In winters before 2006, the male Carolina wren would arrive from the south; in 2010 a pair overwintered. In 2020 pairs are still overwintering. Another specie that sometimes overwinters is the Savannah sparrow.

A crow lands at the very top of a spruce and gives out loud caws, then waits for a returning caw. After a couple of returning caws, it flies toward that caw.

I usually do not see a bird and animal sit next to each other while feeding. A dove was sitting on a dogwood branch and four feet away was a squirrel eating a peanut. Late morning, hundreds of local Canada geese have risen from a nearby field of corn stubble. They land a half-hour later just to take off again; this is repeated for a couple of hours. After that they settled in the field. At 8:00 a.m., 24F, 200 geese fly over to get exercise; the sound of a jet plane taking off does not bother them. It is just an observation of how over the past 60 years wildlife has become acclimated to the sound of a jet engine or car, yet our voice puts them on alert.

I was at a nearby park that has a 10-acre plain that a thousand geese use to forage during the winter. Sitting on a park bench, I was observing the geese, when above me were six geese circling and honking. Suddenly a hundred geese rose straight up from the thousand. They rotated in a circle until they formed a traditional "V" formation and flew north. The six above honked again and a hundred more geese rose to form a "V" that went north. This procedure was repeated until all one thousand were going north. Then the six "air controllers" followed!

On the 20th of 2006, a brown creeper was in the yard. None have been observed since. On the 30th of 2006, a catbird was in the yard. Their normal arrival is April. In 2009 a teen cardinal has conjunctivitis.

On the 15th of 2010 a fox sparrow was running around; normal arrival is mid-March. In 2014, after a very cold period, I found a dead fox sparrow; it may have been frozen. Also in 2014, a female white-throated sparrow with no tail came to the feeder for four weeks. It seemed to fly okay!

Female white-throated sparrow with no tail

Pond: January is the coldest month of the year so there is ice on the pond. Lots of people ask, "What do I do with the fish in the winter?" I leave them in the pond which is three feet deep in the middle. The bottom foot is filled with muck so the frogs can hibernate. The ice can get six to eight inches thick. I can see the fish swimming slowly beneath the ice.

In the fall before the leaves drop, I put a net over the ponds. When leaves get in the pond, they will form methane gas under the ice and kill the fish. The excess leaves can cause a depletion of oxygen. I reduce the amount of feed to the fish because their metabolism slows down. I have had the same goldfish or generations of the originals for over 20 years (before the great blue heron).

The net is removed after most of the leaves fall. If left on over the winter, the heavy snow would rip it open. Now and then I see a cat or great blue heron on the ice looking down on the fish. So close, but so far away.

JANUARY

Great blue heron surveys the iced pond

During the last week, the punk of the cattail breaks up. Small birds and squirrels race to gather the soft punk for nesting material.

Flora

<u>Perennials</u>: When a warm spell of above 60F occurs, the daffodils emerge, above 45F, the wild onions would appear on the lawn. Snowdrops may be blooming in the fourth week.

So, ends January.

Index of February

February Overall	404
Weather	404
First Week	408
Second Week	411
Third Week	413
Bird Arrival Time at Feeders	415
Fourth Week	417
Pecking Order	422
Feeding of Split Peanuts	422
Flora	424
Review of Data	424
Birds	424
Table of the Gang	426
Population of Individual Species	427
So, ends February	430

February

Week number for the year of 48 weeks					
Relative Month	Month/Week	First	Second	Third	Fourth
Previous	January	41	42	43	44
Present	**February**	**45**	**46**	**47**	**48**
Following	March	1	2	3	4

The National Weather Service began February 7, 1870, with President Grant.
The amount of light on the 1st of the month is 10 hours, 10 minutes; on the 28th it is 11 hours, 15 minutes.
The average high temperature on the 1st is 36F, the low is 19F.
The average high temperature on the 28th is 42F and the low is 24F.
Extreme daytime high for the month was 76F on the 24th of 1985.
Extreme nighttime low for the month was -12F on the 9th in 1934, and -8F on the 24th of 2015.
The full moon in February is called the Snow Moon.

Overall Summary of February

February is like January in respect to temperature; it can be warm (50F) then plunge down to the teens with winds and feels like 0F. We seem to get a fair portion of ice storms and black ice, birds skid across the top of feeders and there is ice on the pond. The last week is when the birds start their repertoires early in the morning especially if it is clear and a crisp 11F. The Carolina wren is the loudest, followed by the cardinal, and even a pleasant trill from the blue jay.

Weather

Weather is a yo-yo. The mean temperatures trendline shows a six degree rise over 28 years. Stretching things a bit, it looks like the high temperature in Februarys occur in a four year cycle and the colds in a 10 year cycle. February 2020 was one of the warmest on record.

Snowiest Storms in the Lehigh Valley			
Month	Day	Year	Inches
January	23-24	2016	32
January	7-8	1996	26
February	11-12	1983	25.2
April	29	1874	24
February	16-17	2003	22
March	19-21	1958	20.3
February	10	2010	17.6

TRENDS DUE TO CLIMATE CHANGE

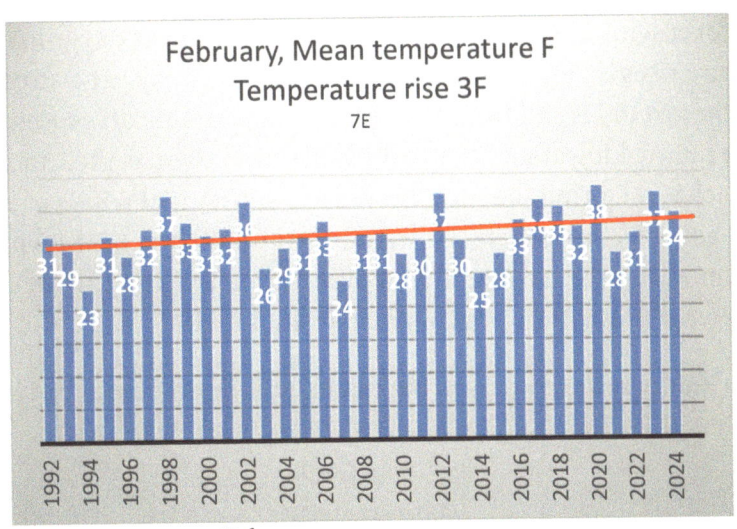

February mean temperature

February 2008 is the wettest on record breaking a 112-year record; 7.6 vs. 6.42 inches in 1896.

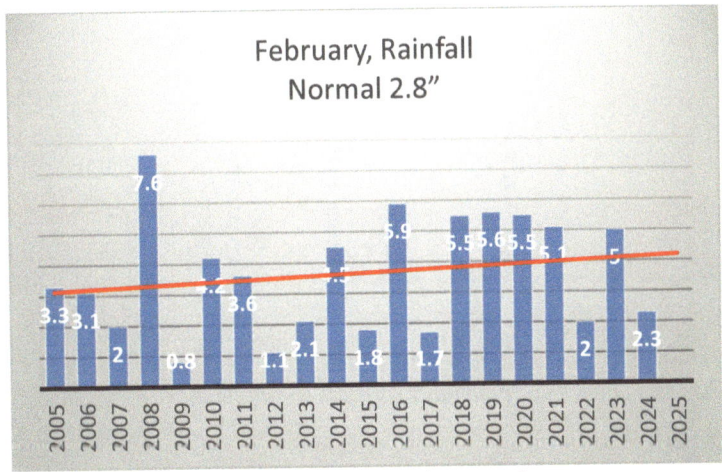

February rain in the month

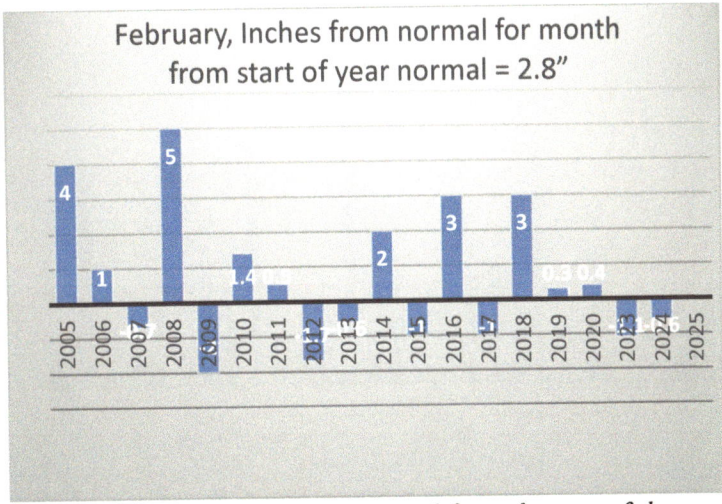

February inches of rain from normal from the start of the year

There is a snow cover with 20 to 30F temperature, then the next day it may rain that begins as ice. As the temperature rises to 50F a dense fog may form. We may get a strong wind followed by a cold front. Again, the ground is still frozen, so any rain that drains off causes flooding conditions. If it warms again, we may get fog at night or early morning, or worse, the runoff may freeze during the night causing black ice. During the month, we have substantial winds of 30 to 40 mph, which prevents birds from going to the feeders. Now and then thunder is heard with or without rain in the latter part of the month. When a seagull flies over it usually is indicative of a large storm to the east. This time of the year it usually is a snowstorm from the east.

Distinction: Lehigh County is the coldest in the nation at 8.7F on February 1, 2005.

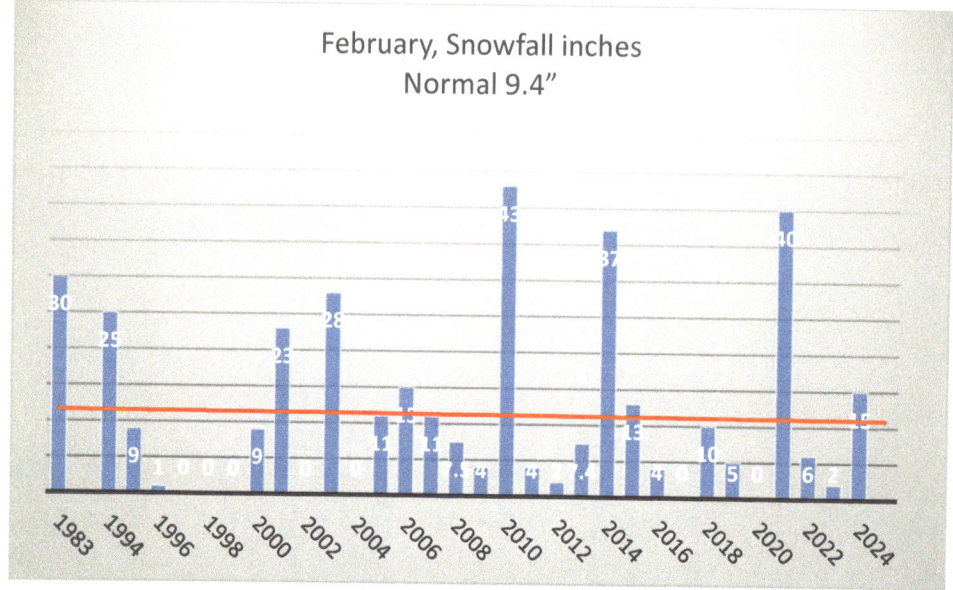

February annual snowfall

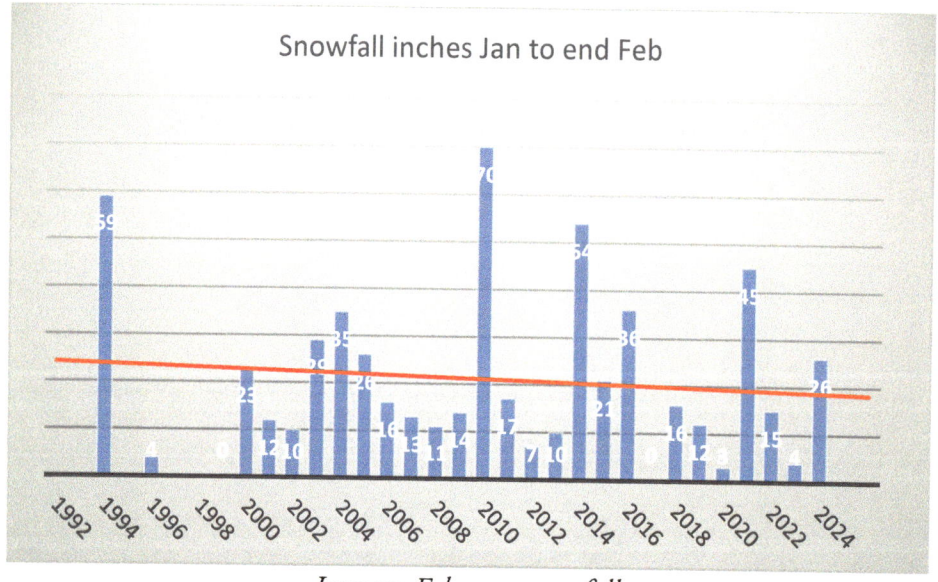

January-February snowfall

We can get 15 to 20 inches of snow. Cold winds make it feel like -20F, resulting in the monthly average of 22F instead of 42F. We have had wet Januarys and Februarys with no snow for the season. Other years there is no snow until the end of February, then the bad weather arrives as cold, wind, snow, and sleet. Worse is when freezing rain changes to ice, resulting in power outages, downed trees, and lots of accidents.

Some weather conditions produce dense fog, lightning, and snow thunder, or we get a bad nor'easter with winds to 50 mph, blinding snow squalls, followed by heavy snow. Other than a nor'easter, it can be a cold, windy month. Many years it is worse during the third week with 22 to 35 mph gusts.

On the 16th in 1903 the Lehigh Valley got 20 inches of snow; on the 17th of 2008 we have 17 inches. In 2008, even with all that snow, the average temperature was 31.3F versus the normal average of 29.9F. The second snowiest winter in Philadelphia was 2008, and the wettest in 112 years; 2014 is the snowiest winter since the 1990's. It is the third snowiest winter for the Lehigh Valley with a seasonal total of 67 inches. February 2015 was a very cold month. The average temperature was 9.7F below normal with the 24th having the second lowest temperature on record, -8F. Then again, February can also be warm, no snow, dry, and foggy. In 2017 it was warm, but 2020 was the warmest on record.

February 2017 – Record Warmth

February 2017 was the warmest on record until 2020; 2013 is the third warmest. Not only February, December and January all had days with temperatures over 77F. The snowdrop flowers were opened on the 8th with a high of 58F. On the 20th the silver maple was in full bud. On the 22nd, 61F in the morning, by mid-day the temperature rose to 74F. Little yellow flowers were out by the pond, the Easter rose was out, frogs were on the edge of the pond, a chipmunk was running around, skunk cabbage and rhubarb crowns were breaking through the soil, a raptor was soaring at over a thousand feet, a house sparrow took a dust bath, and the great blue heron was in the pond fishing. Then a cold front came through with a strong thunderstorm. By the next morning the temperature dropped 32 degrees to 42F. By then the daffodils were eight to 10 inches tall. The birds were confused, a mourning dove flew into a mockingbird that is the same color as the limestone wall. Skunks and chipmunks came out for fresh air. The skunk added a fragrance.

February 23rd – snowdrops

Bush cinquefoil

FEBRUARY

February 23, 2017 and 2020 – daffodils and Easter rose

With the extensive warm spell, perennials and trees bloomed two to five weeks ahead of expected blooming time and frogs were sunning themselves.

Blooming Time with Warm February vs. Expected Time			
Plant	February bloom date	Expected date	Weeks different
Silver maple	10th	March 20th	4
Easter rose	20th	March 20th	4
Purple crocus	24th	March 3rd	3
Daffodil up 6"	25th	March 7th	2
Snowdrops	15th	February 24th	2
Frogs out	24th	April 1st	5

First Week

Weather: The first week of the month could start off as 24F; dew point is 4F with a strong wind and it feels like -9F. Or it could be a warm winter, resulting in an unusual observation of a spider assembling its web! With a 50F day the birds will sing their spring songs. I may see one to 30 robins in the yard. It was unseasonably warm for the first week in 2007 and the grass looked greener.

Animals: Six years in a row the groundhog sees its shadow. Will we get six more weeks of winter? The squirrels eat the newly formed silver maple and swollen dogwood buds. February and March the squirrels are taking mouthfuls of leaves up the spruce trees.

A field mouse is eating hosta roots. In the morning there are fox and rabbit tracks at the base of the feeder. The rabbit could have become the fox's dinner. I smell skunk in the morning air; refreshing!

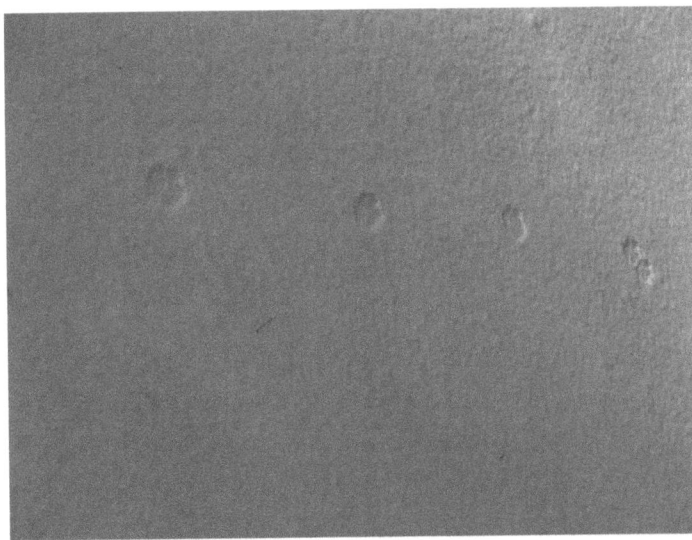
Fox tracks

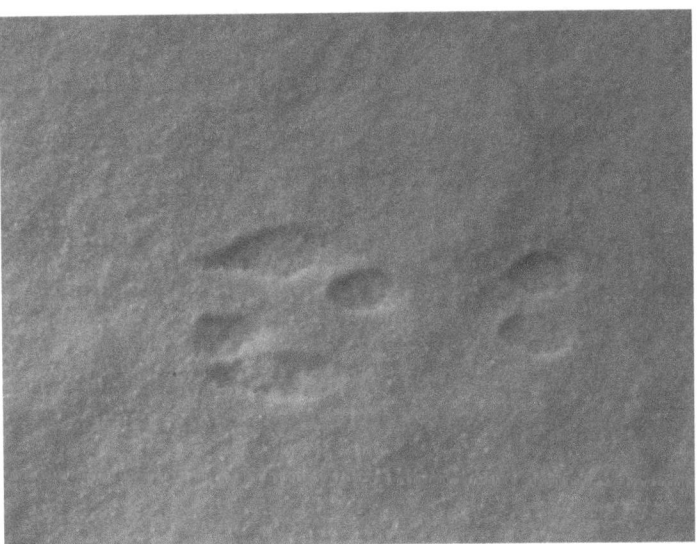
Rabbit track

<u>Birds</u>: On the 6th of 2010, 6:00 a.m., blizzard conditions, 25-35 mph gusts, 25F, received four inches snow last night. I can hear an eerie train whistle blowing through the snow six miles away in Emmaus. It is snowing so fast that I can only see the silhouettes of the birds on the feeder. There are lots of birds, but I cannot make out the species. Ten minutes later it let up enough to make out white-throated, cardinal, junco, house finch, and titmouse. Looking closer, the white-throats are chasing the junco, cardinal, and song sparrow from the feeder. The storm has completely let up by 11:30 a.m.; there are now a lot of birds. The population of the house finch has gone from four to eight to 10, white-throated from two to six, juncos from two to six to eight, cardinals from two to nine. A red-bellied woodpecker takes some split peanuts. A pair of blue jays and starlings have arrived. A hawk flies through and all the small birds scatter; the hawk flies on. Within five minutes all the small birds are back. A Savannah sparrow and three doves join the crowd. The brisk activity continues until dusk with a red sunset.

The next day the population is back to normal. During heavy snowstorms there is an absence of blue jays, chickadees, and titmice, they arrive after the storm. I observed this same trait during the blizzard of 2016. A bully white-throated sparrow initially chases a Carolina wren from the safflower feeder, but the smaller Carolina wren has a long beak and retaliates and chases the white-throated sparrow. In the afternoon of that clear, brisk day a pair of broad-winged hawks face the westerly winds to gain altitude then swing east to glide in a wide curve. They are higher than a passing hospital helicopter.

Ten starlings have arrived and have managed to take over the old owl house. They use their body group to keep warm. They dominate the feeders for the next three weeks before going further north. They try unsuccessfully to take over other house sparrow's houses.

The great blue heron flies over. I thought the blue was a migratory bird. Here it is February and it takes a daily trip from the swampy area in a nearby park somewhere north. In the evening I see it fly back. However, now and then it stops for a fish snack. It is amazing the way that six-foot wingspan glides through the large spruce to land on the edge of the pond. It can detect my slightest indoor movement and quickly retreats through the tall spruces.

A rascal of a junco is on the roof of the feeder and purposely dumps snow from the upper perch onto the head of a house finch on the lower platform. The finch shakes its head and continues to eat, not realizing who the culprit is. The junco flies away with satisfaction. During January I noted a junco dumped seed on a house finch. I wonder if it was the same junco.

There is a pecking order at the safflower feeder. The pretty cardinals are not very sociable to other species, no less their own kind. They hiss like a cat at the other birds. An adult male cardinal chases a white-throated sparrow from the feeder, then chases a young cardinal off the feeder, but not a female. He has his priorities right! He leaves, the white-throated returns, then the young cardinal sweeps in and chases the white-throated from the feeder. After the young cardinal leaves, the white-throated comes back and chases a junco who had just landed on the feeder. Meanwhile, the female cardinal continues eating. A pine siskin is here for a short visit.

A sharp-shinned hawk comes in for a fast strike on a dove. It appears to have missed the dove, then dove feathers come floating down to the ground. Another time a sharpie comes in low and snatches a starling that was sitting on top of the rhododendron. A blue jay does not call or caw but gives a clear human-like whistle. You would not know it was from a jay unless you saw the jay.

Blue jay

As evening approaches four house sparrows manage to squeeze into the deserted downy woodpecker's hole in an old stump. They are also actively trying to claim a birdhouse for breeding.

Twelve doves after a storm

When it gets very cold the birds do not drink the heated water. It is then that I see a chickadee and titmouse take a piece of snow, twirl it in their beaks and swallow. There is an ice storm, the birds have a hard time landing. When they try to land, they just keep on going and either bang into something or slide off a surface.

A backward "V" of Canada geese is going north, the apex of the "V" was to the rear of the formation! At least a thousand snow geese are up high going east at 9:00 a.m. The large "V's" come over every 10 minutes for 45 minutes.

When it gets cold like 14F, the birds come an hour later for breakfast.

Odd bed fellows; eight robins and eight starlings sit next to each other on the poplar tree. Three days later I again see robins and starlings sitting next to each other in a maple tree.

<u>Perennials</u>: If it is above freezing for an extended period the silver maple will form buds, with brute force the vegetation of the Stella-de-Oro and daffodils will start to poke through the hard soil. Snowdrop flowers are out behind the pond, daffodils are up three inches. Grape hyacinth leaves are greening up.

Second Week

<u>Weather</u>: The second week of 2015 was cold; the whole month was nine degrees below normal. The 13th was the calm before the storm; it was clear and windy and felt like -15F all day. A fox sparrow arrives on the 14th, it is starting to snow with 32 mph winds. A sharpie goes under the hemlock hedge to hunker down. The 15th, 6F, 32mph wind, the coldest day all winter; it feels like -13F. The gang of birds are at the feeders, a yellow-eyed Cooper's hawk flies by. February 16th, 6:30 a.m., -1F, now there is a red-tailed hawk in the spruce.

Animals: The squirrels are looking for cavities to make their nests. They give birth from the end of March to early April.

Birds: It is a clear 32F, the thousands of snow geese that I saw in January are not flying for exercise, and they are preparing to migrate north. Mid-morning, they rise from their winter location and flocks of thousands go northeast. I can see their white breasts accented against a robin's egg blue sky. Now and then there is a lone Canada goose or a flock of small birds flying adjacent to a "V" of snow geese. A flock of 60 to 100 local Canada geese are practicing for their trip north.

Canada geese

Flocks of starlings, robins, and grackles are going north; some stop and evade the feeders for an hour before continuing.

Good weather, 50F, the male titmouse, cardinals, and house sparrows are practicing their spring repertoire; usually they start singing in earnest on the equinox, March 21st. The temperature climbs to 56F and brings a flock of 30 robins to the yard! One joins the singing. One robin dips its head to get a drink from the pond and then brings its head up horizontally to swallow. A few robins might decide to take spring residence.

I put whole peanuts on the platform for the blue jays; they know the hand that feeds them. When they come in the morning and there are no peanuts, they start to yell from the maple tree until I come out to replenish the nuts. They land on the platform as I open the back door. I also put a few nuts on the ground under the feeder for the squirrels. The jays are shrewd, they take the nuts lying on the ground then fly up and eat the nuts on the feeder.

During a mild winter, a mockingbird may be here for a few days. It will not stay because it likes open space which my yard no longer offers.

A chickadee lands on the safflower feeder, it is immediately chased off by another chick which is chased off by another chick which is chased by another chick. The four chicks landed like an assembly line; none were there long enough to get a seed! One takes a mouthful of thistle to make its house fluffier.

On the 14th, woodpeckers are tapping their mating drumming onto a telephone pole; how romantic with a headache! Their skulls are made up of many bones that together provide a good shock absorber.

Perennials: Grape hyacinths leaves are greening up. Yellow crocuses are blooming.

Third Week

Weather: We can get a lot of heavy snow this week like five, 10 or 19 inches, day after day. Then wind the next day making it tough to snow blow because of the ice crystals blowing back on my face. Instead of snow or rain, there can be freezing rain any time of the month which is more dangerous than either the snow or rain. The ice brings down thick pine branches and power lines. Unattended parked cars on hills start sliding down the roads; people cannot navigate where they want to go.

We are expecting three to four inches of snow; on the 17th, 7:00 a.m., 12F, we got two inches of snow. The 18th, 2F, a hawk sits in the poplar tree, there are a few birds. Two squirrels are on the chase. Surprise; a house wren at the safflower! The 19th, 9F, another surprise, five pine siskins arrive; 20th, 6:30 a.m., 0F and with the wind feels like -25F. There is a dead chickadee on the feeder; there are 30 mourning doves on the ground. On the 21st at 7:00 a.m., 12F, snow predicted; house sparrows are chasing house finches from the feeder. Another surprise, a flicker is under the feeder. Squirrels are eating safflower which they normally don't do. It is snowing hard at 5:30 p.m., most birds have gone to roost. On the 22nd the temperature is finally up to 30F; five inches of snow last night. A sharpie hits the kitchen window as it misses a dove. At 12:00 noon the temperature is up to 44F and it is nice! The 24th and it is back down to -4F. The power goes out for 30 seconds at 4:40 a.m. On February 25th, 6:45 a.m., 14F, a Carolina wren is singing loudly. Blue jays are trilling. The average temperature for the month is 18.7F, the second coldest on record, 18F below normal. The last day of the month, 28th, 27F, gang is back. For the month we received 13 inches of snow, for the season 32 inches. A miserable 2015 month.

Animals: At the beginning of the month through the third week the squirrels are on the chase. When joined they roll over on the lawn or snow. They run too fast for me to take a picture. A chipmunk comes out of semi-hibernation.

The last week of the month in 2019 it is 30F with 30 mph winds, not nice outside. Groundhog Day was in the first week. I look out to see a groundhog at the top of the steps, it paused and went on; I'm glad.

Groundhog passing through

February 2019 – red fox

A couple days later a red fox was at the top of the steps. The fox has been around all winter. There are less rabbits and squirrels.

<u>Birds</u>: The following charts show for two different years that the birds arrive for breakfast a half hour after sunrise, a few come for lunch a half hour after noon, and most come for dinner an hour to half hour before sunset. A few come for breakfast a little before sunrise or just at sunrise. These schedules occur most of the year.

| Month February, Day 14-17, Year 2007 snow and cold Arrival hour and quantity ||||||||||||||||
| Twilight 6:00 a.m., Sunrise 6:50 a.m., Sunset 5:28 p.m. ||||||||||||||||
Hour a.m./p.m.	5	6	7	7:30	8	9	10	11	12	1	2	3	4	5	6
Blackbirds			15		11										
Blue jay				1							1				
Cardinal			1	3		1							1		
Chickadee				2	1						3			2	
Crow					3										
Goldfinch				1											
House finch				4	1					1					
Mourning dove				3	2									3	
Geese															
Junco			6		1				5	2				1	
Mockingbird				1											
Red-breasted nut				1	1					1	1				
White-breast nut										1					
House sparrow				4	4										
Song sparrow				1									1		
White-throated sparrow			1	1	1									1	
Titmouse				1	2		2			1					
Starling				4						3				1	
Downy wood															
Carolina wren					1	1									
19 species															
Squirrel				1		2				2					

TRENDS DUE TO CLIMATE CHANGE

Month February Day 14-16, Year 2009 Arrival hour and quantity																
Twilight 6:00 a.m., Sunrise 6:50 a.m., Sunset 5:28 p.m.																
Hour a.m./p.m.	5	6	7	7:30	8	9	10	11	12	1	2	3	4	5	6	7
Blue jay				1	2											
Cardinal			4	1	2	2							7			
Chickadee				2		2							1			
Crow					2	2										
Goldfinch						1										
House finch			2	4		2							4			
Mourning dove				2									1			
Geese						100										
Junco				4		1				4			4			
Pine siskin						1	1									
Red-breasted nut				1			1									
White-breasted nut																
Sharp-shinned hawk			2			1										
House sparrow				4	4											
Song sparrow				1	1	1							1			
White-throat sparrow			1	2	1	2							4			
Starling				2												
Titmouse			1	1		2										
Downy woodpecker					1								1			
Carolina wren				1		1				1	1					
18 species																

February 17, 2015 – 31 male cardinals in a friend's yard
Photo by Bob Goehler

On a warm day, the chickadees and juncos are doing acrobatics among the shrubs oblivious of the young Cooper's hawk cawing from the top of a tall evergreen, waiting for parents to bring breakfast. A broad-winged is being chased by 15 crows! On a beautiful clear morning, starlings are still trying to take over seven houses occupied by house sparrows. The cardinals and titmice are singing mightily; spring must be coming. Doves are hungry for safflower. A flock of 100 robins or blackbirds are going southwest.

Four tiny pine siskins pass through, they like millet and thistle seed. On a sunny afternoon there is a sharpie in the dogwood, it is looking down at a squirrel on a lower branch. The squirrel is too big a dinner so it sits and waits. Fifteen minutes later the sharpie snatches a small bird.

Sharpie looks over a squirrel as dinner   *Diner obtained, not the squirrel*

<u>Perennials</u>: Come April, I learned that not all the wheat seeds had been removed from the straw. What a crop of straw coming up among my perennials. No wonder the squirrels were happy.

The squirrels have been relishing the wheat seeds for weeks

With normal temperatures and rain the crocuses will be poking through. Even though it is 29F and feels like 17F with the wind, daffodils are popping through. I put up a new birdhouse and clean out others. Above normal temperatures by 8F encourages the yellow crocus to bloom. Easter rose has new leaves and the daffodils are up four to six inches. There may be swarms of bugs in the afternoon sun. I spread 10 wheelbarrows of compost on the perennial gardens.

Pond: There is still ice on the pond, but the fish can be seen slowly circling around. I can see frogs lying belly up on the bottom of the pond.

Trees: Spray dormant oil on the apples and blueberries. Used five ounces of dormant oil per gallon of water for the mixture.

Fourth Week

Animals: In 2012 we had a warm winter; maybe the squirrels did not want to cuddle up, on the other hand 2018 was very wet so maybe they just wanted to cuddle up. Who knows? Near the end of the month or beginning of March parents will arrive at the feeders with pups.

I don't know if it was a search for food but one year three clans of squirrels arrived. The clans from the north and west behaved themselves. The ones from the south caused problems at the feeders and were digging up the lawn. One clan arrived with four pups. The pups were very curious and ran and climbed all over. Some climbed up the dogwood to eat the new buds. Adult squirrels bring their pups to the feeders to teach them tricks of getting food and to scatter when they see me or a hawk. When there are whole peanuts, some squirrels take three at a time. If the pond is free of ice or if there is a strip of water along the edge of the ice, they will pause to get a drink. Now and then a squirrel will develop mange on its hind quarter. I have observed the same squirrel for three weeks; I don't know if the disease subsides. When the squirrels find a large deserted birdhouse, they will chew somewhere, sides or top, to get inside. Once a hole is made, they may get as many as eight inside to maximize body warmth. It was fun to watch the rising sunshine on the face of the house to waken the attendants and to then see all the bodies trying to pop out at the same time to get fresh air. To satisfy their sweet tooth, a squirrel will break off a three-inch long piece of a maple tree's icicle and chew it.

Squirrel eating ice

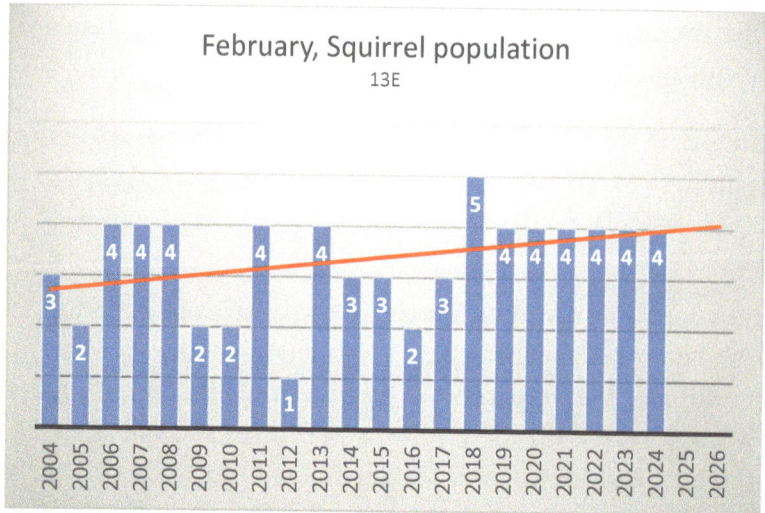
February squirrel population

It got exceptionally warm, in the 40's in 2016, the fourth week. Two chipmunks, a raccoon, a rabbit, and a skunk were observed. A groundhog was scampering on the hill. At night a skunk was under the feeder.

A rabbit is eating mountain pink (creeping phlox); they only eat the top, there is always enough crown left for the pink to recover.

On the 25th in 2008, 25F, there are two deer in the yard. In 2014 I had three. Nice surprise until they start trimming the shrubs. They stayed on for four days.

February 20, 2015 – red fox getting a drink at the pond

<u>Birds</u>: There were six male and six female white-throated sparrows feeding at the same time in the same vicinity, but they did not appear to be mated up. There were eight to 10 doves that were couples. A pair of broad-winged hawks is slowly soaring high with grace drifting to the east. A pair of turkey vultures are soaring with erratic aerodynamic motion.

Among a given species the pecking order by size is going on. Then between species the pecking order is in effect, such as a house finch chases a chickadee. Then a junco chases the house finch. The house finch chases the little red-breasted nuthatch. The male white-throated usually dominates the safflower feeder. It chases a male house finch, a song sparrow, and a chickadee; it chases the houses finches three times. However, when a smaller Carolina wren comes with its longer beak the white-throated leaves. A titmouse takes a piece of snow, twirls it in its beak and swallows. Dusk, five ducks circle the circle. The red-tailed hawk has reappeared and again is being driven off by five crows. A Carolina wren is at the feeder throwing safflower to the birds below.

Carolina wren shoveling safflower

The other end

February 26, 2010 and 2020 – robins

Chickadee getting thistle

At 6:50 a.m., 19F, the first grackle of the season arrives for a rest on the feeder. As it starts to take off, wham, a Cooper's hawk grabs it and flies off. All that migration effort to get here is gone in a second.

Grackle getting a split peanut

A cardinal with a bald head arrives, it looks like a miniature vulture. I observed the same thing last July. It is caused by mites.

Female cardinal with mites

Titmice bring a youngster to the feeders. One of the parents takes a bite of snow for its thirst. The spot on the white-throated is starting to change from a light gray to white. Pine siskin pass through.

TRENDS DUE TO CLIMATE CHANGE

Pine siskin

On the 28th when a Copper's hawk flies through, a downy woodpecker that is on the split peanut feeder goes flat and motionless. Other small birds scatter, leaving their scat on the kitchen window.

Downy flat against the feeder

Pair of downy

After the hawk leaves, a pair of downy arrive. Male woodpeckers are tapping the telephone poles and trees calling for a mate.

Snowing with 30 mph wind gusts, 25F, pecking order of the birds at the feeders is in full display. House finch chases a chickadee from the feeder. The small birds as a gang must be diligent as community protection from a hawk. Interesting that one moment the gang has its pecking order and the next there is community protection. "One for all, all for one." Titmouse, white-throated, chickadee all eat snow. Two hundred snow geese go north with a lone Canada goose honking and flying with them. A mild winter shows its effect, mating usually starts in the middle of March with bird species pairing up and males giving their spring songs. Six crows go west. Later in the day a turkey vulture is soaring with 31 mph wind gusts; its flight is very erratic. Red-bellied woodpecker start overwintering in 2010, in 2020 still overwintering.

FEBRUARY

At 6:45 a.m. cardinals have been practicing their mating call at this time for the past week. At 9:00 a.m. the song sparrow is also practicing. Sharpie comes for a dove and misses. Dark and snowing, 6:30 p.m., a flock of geese go over.

It is Leap Year, the 29th, for five minutes two female house sparrows are fighting each other. Other sparrows break up the fight. It must have been a dispute over a potential mate. A bull dove chases a female. Love (hormones) is in the air. The morning after an overnight four-inch snowstorm, there are 12 species of birds including 10 juncos and 20 starlings on the feeders.

	Pecking Order in February Not a very nice place, lots of bullying
	White-throated sparrow chases a Carolina wren from safflower feeder. Then the Carolina reverses and chases the white-throated from the feeder.
	An adult male cardinal chases a white-throated from the feeder, then chases a young cardinal off the feeder, but not the female!
	Young cardinal chases a white-throated from the feeder.
	Chickadee lands on the safflower feeder, it is immediately chased off by another chick, which is chased off by another chick, which is chased by another chick.
	Adult cardinal chases a white-throated from the feeder.
	Big white-throated chases a song sparrow off the raised feeder.
	Red-bellied woodpecker comes to the peanut flake tube but is chased off by a starling.
	White-throated is feeding, a junco is waiting for it to finish.
	Dove on the feeder raises its wings to chase off a blue jay.
	Fifteen crows chase a broad-winged from the area.
	House finch chases a chickadee from the feeder.

Odd to me, some birds are singing while it snows. Then a sharp-shin glides low over the yard, the small birds hush and scatter. The sharpie continues to glide through. When it is clear the birds start singing again. In the distance, two crows are chasing a red-tailed hawk.

Most bird like the split peanuts. I put them in a wire tube. It is interesting how the different species take the nuts from the feeder.

Bird position when feeding on the peanut flake tube	
Position on the tube	Bird
Horizontal (sideways)	Carolina wren
Vertical head up	Chickadee
Vertical head up	Downy woodpecker
Vertical head down	Goldfinch
Vertical head up	House sparrow
Vertical head down	Pine siskin
Vertical head up and semi-horizontal	Red-bellied woodpecker
Vertical head down	Red-breasted nuthatch
Vertical head up	Starling
Horizontal	Titmouse
Vertical head down, wings flapping	White-breasted nuthatch

Split peanut tube

    Come the third and fourth week I am looking for the migrating flocks of snow and Canada geese. The fourth week in 2017, 9:00 a.m., I witnessed thousands of both geese go north. What made it unusual is that there was a large raptor along-side a "V" of snow geese, not attacking but going along with the ride north.

    In the southside of Bethlehem, Pennsylvania, there is a huge rookery of crows. An hour before sunset they start to return to the rookery. At first there may be 10, then 100, then within 35 minutes there are 500. They just keep coming and cawing until the sun sets. This ritual is repeated every evening until spring. Quite a sight to see; not many residents take notice.

Insects: On a warm day boxelder bugs like to sun on the siding of the house.

Pond: In January I found a goldfish swimming upside down in the small pond. When I touched it, it would turn right side up and swim normal. A few hours later it was upside down again. This went on for three weeks. I asked someone who knew about fish, he said it had a bladder problem. He said feed the fish smashed frozen peas! I put the fish in a half-filled ten-gallon tote. It was still upside down. I added the peas, no taker. I added more peas. Finally, it ate some. After three days it was swimming normally. I kept feeding it peas for another week. Then put it back into the pond. It immediately rejoined the other fish. Everything was fine. Then one morning I looked into the pond. All the fish were gone. The great blue heron had a good meal.

Upside down fish

Flora

Perennials: The cold December and then warm January causes the daffodils to come up in February. Then with additional cold weather, growth stalls until March. Like February 2015 when the average temperature for the month is 10F below normal, snow covered the soil for most of the month; second lowest night temperature on record of -8F.

With normal temperatures and rain, the yellow crocuses and snowdrops will be blooming. Daffodils will be up two to eight inches. When it is warm during the first two weeks, the daffodils will be up six inches, the crocuses and snowdrops will be done.

Shrubs: A warm January will cause japonica's maroon and pink clusters to form. Brown winterkill is now starting to show on the shrubs.

Trees: A warm January will cause the maple to bud. Even though it was extremely cold in 2015, the silver maple buds started to swell by the 23rd. This is the month to prune the apple trees.

Review of Data

Birds: The yearly range of the number of species of birds for 1990 to 2018 has steadily increased from 13 to 27. Eight species only appeared once or twice during the period. They are the blackbird, catbird, duck, grackle, northern water thrush, pine siskin, purple finch, and seagull.

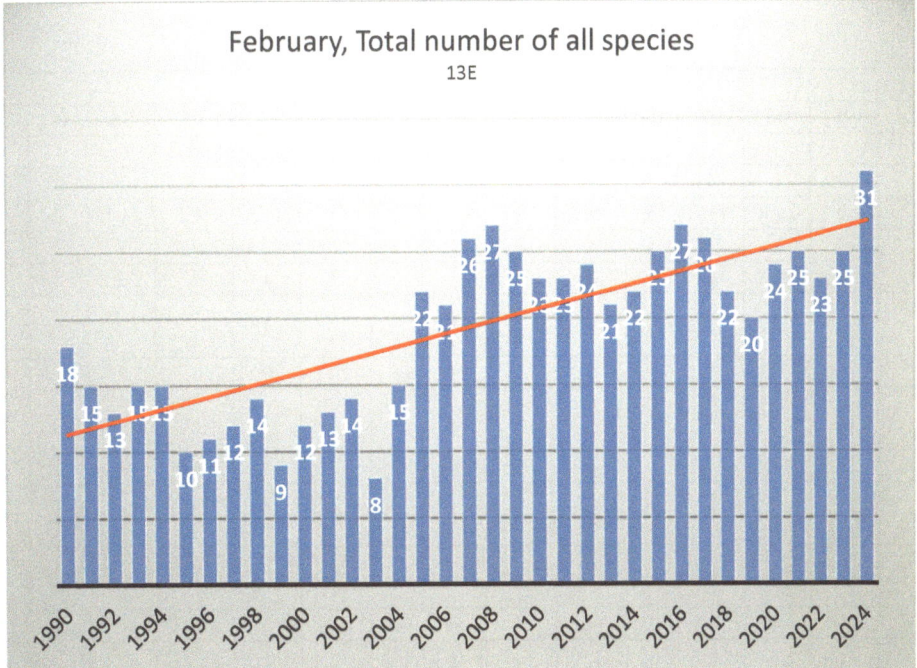

February number of all species

The population of all the species averages about 50. When a flock of 40 to 60 starlings pass through the quantity goes to 90 to 100. Other years the population may drop to the twenties.

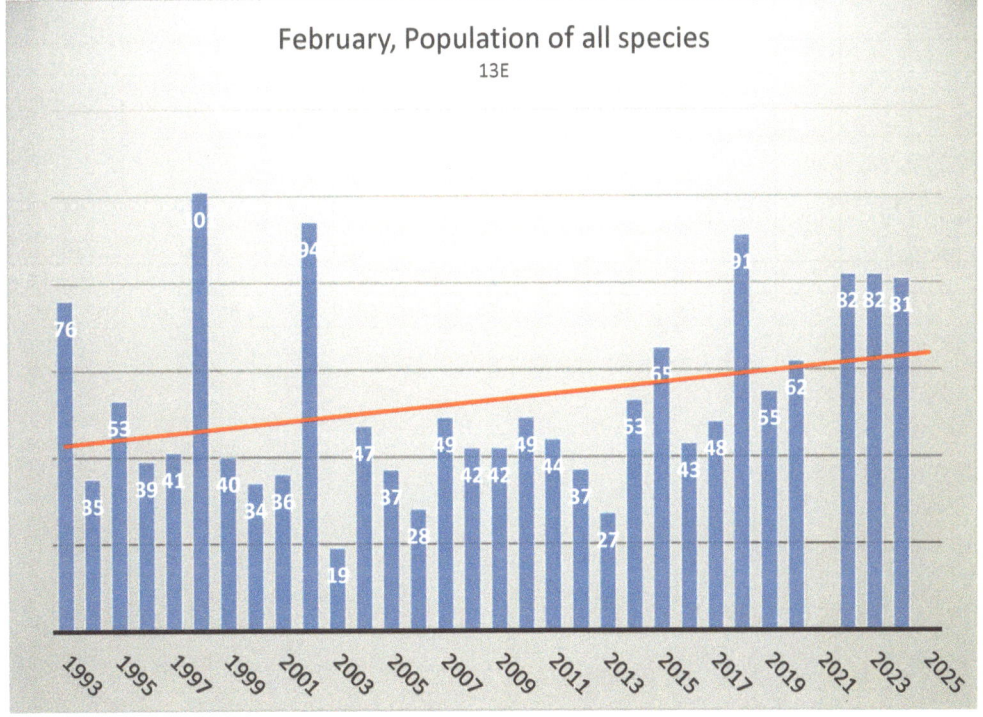

February population of all species

The number of the species in the gang has increased from six to 15 over the period, probably due to the environmental changes.

FEBRUARY

Table of the Gang

February Table of Gang																
Species/Quan.	05	06	07	08	09	10	11	12	13	14	15	16	17	18	19	20
Birds sing				1st		26th	11th				22nd				17	
Blue jay R, G	2	2	1	3	2	6	3	7	2	3	3	3	2	2	2	2
Cardinal R, G	1	1y,2	1y,2	3	1y,5	9	11	2	5	1y,5	1y,5	1y,5	4	5	5	4
Carolina wren	2	-	2	1	1	2	1	1	1	2	1	1	2	1	13/2	2
Chickadee R, G	3	5	1	4	3	2	1	2	2	1	3	2	2	1	2	2
Downy wood R, G	1	2	1	1	1	1	1	1	2	1	2	2	2	2	1	2
Goldfinch R, G	1	1	6	1	4	1	-	1	-	2	1	2	1	6	1	2
Hairy woodpecker R	1	-	1	1	1	-	-	1	-	1				1	1	1
House finch R, G	3	6	5	6	4	12	6	1	2	4	6	8	4	8	10	6
House sparrow G, R	2	2	6	6	8	2	1	2	2	2	2	4	2	4	3	4
House sparrow active						20th					20th		5th			
Junco R, G	4	6	6	10	4	8	11	1	2	3	6	7	5	2	3	5
Mourning dove R, G	3	8	3	2	4	2	4	12	2	18	23	12	9	8	6	10
Red bellied wood R			1	2		1	1			1	1	1	1	1	1	1
Red-breast nut R, G	2	1	1	1	2				1				2		1	
Savannah sparrow					1	2	1	2	1	0	1	1				
Song sparrow	1	2	2	1	1	1	-	1	1	1	1	1				
Titmouse R, G	2	2	3	3	3rd	1	2	1	2	2	2	2	2	2	1	1
Wood pecking	27th				14th				24th	3rd					17th	
White-breast nut R, G	1	2	1	1	1	0	1	0	1	2	2	2	1	1	2	1
White throat spar. R, G	2	2	3	3	4	6	6	4	2	8	6	4	4	4	5	4
Total Gang	15	14	15	15	16	14	13	13	14	14	15	15	15	14	14	14
Population	29	43	44	46	48	53	44	37	27	53	65	43	48	45	41	37

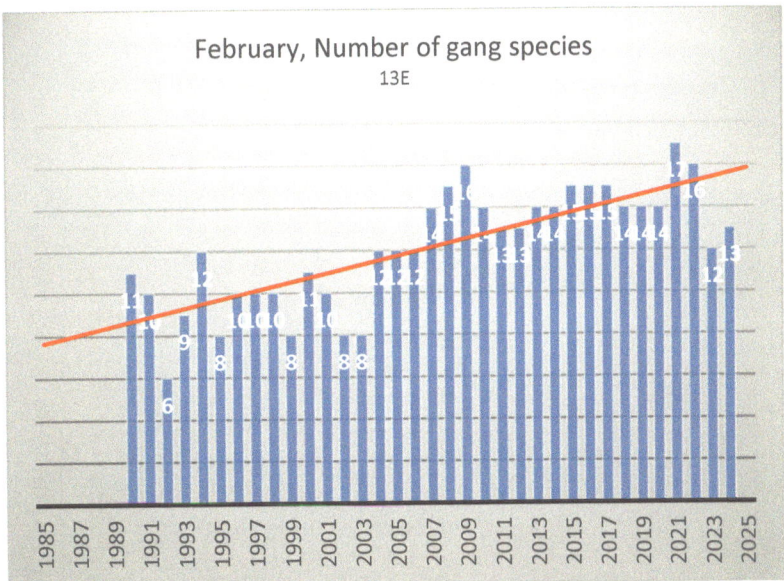
February number of gang species

Even though the gang's species has increased, the population is cyclic.

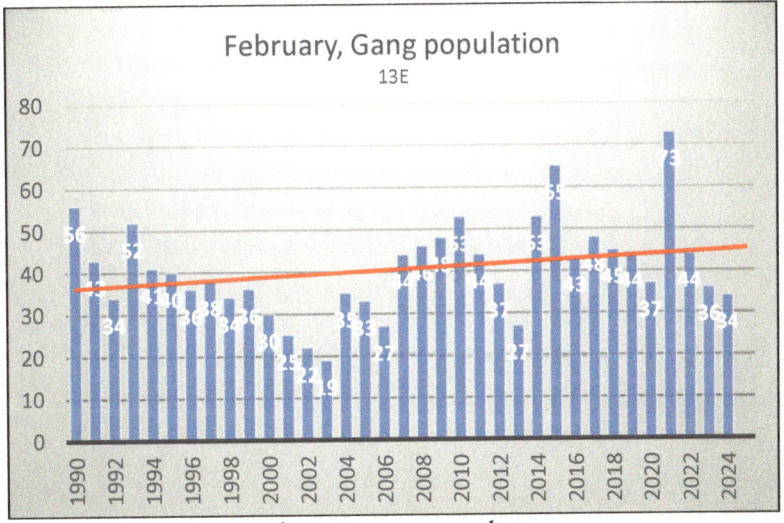
February, gang population

The population of many species is increasing. There were spikes in 2009 and 2015. The house sparrow and house finch show a sharp decrease but have been showing an increase since 2015. Surprisingly, even with the large warehouses and housing developments, the Canada goose population seems to be returning.

FEBRUARY

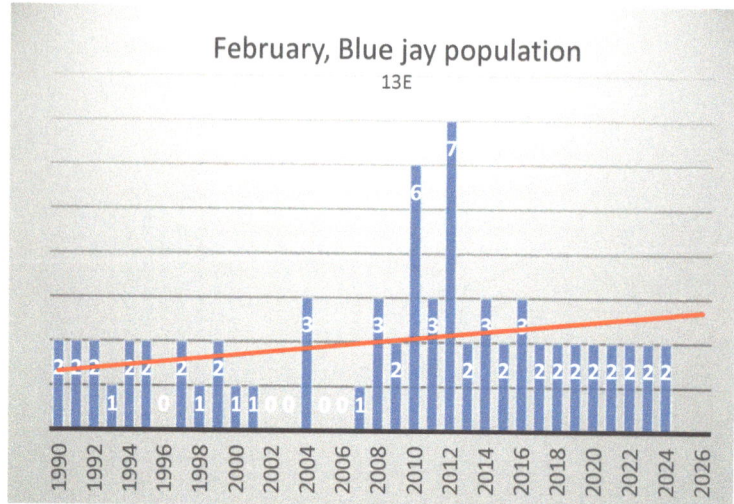

February blue jay population

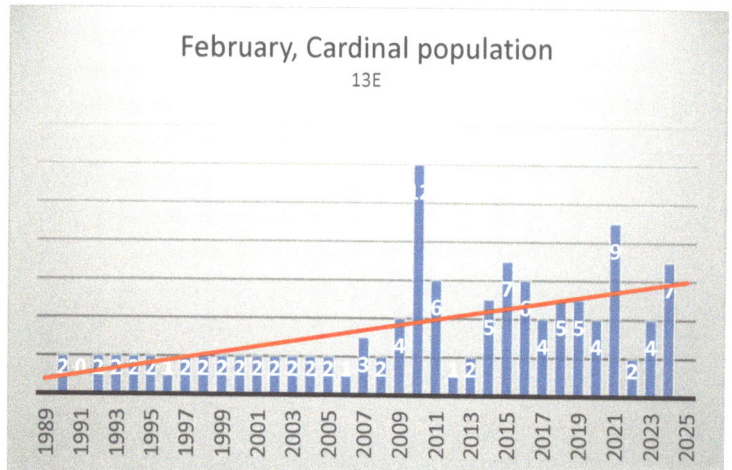

February cardinal population

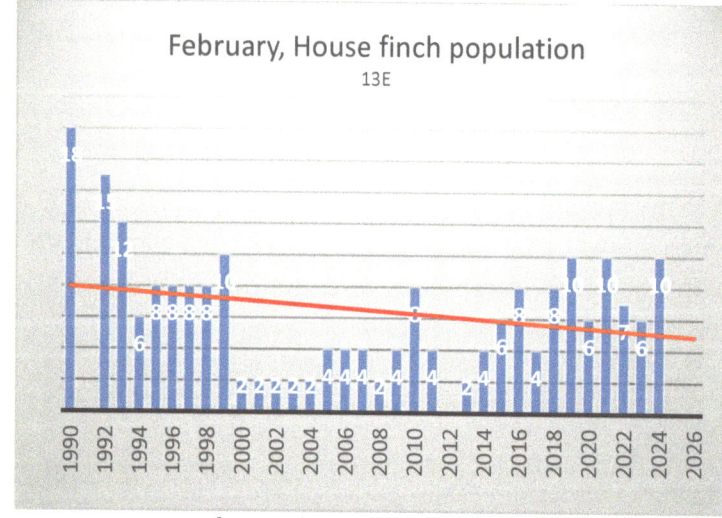

February house finch population

TRENDS DUE TO CLIMATE CHANGE

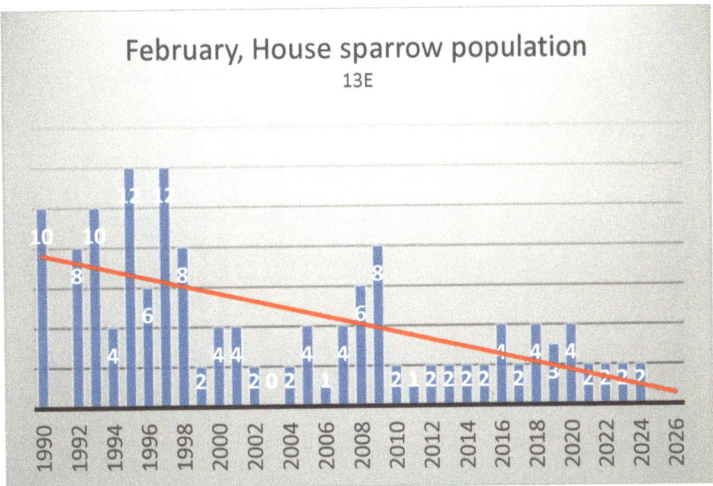

February house sparrow population

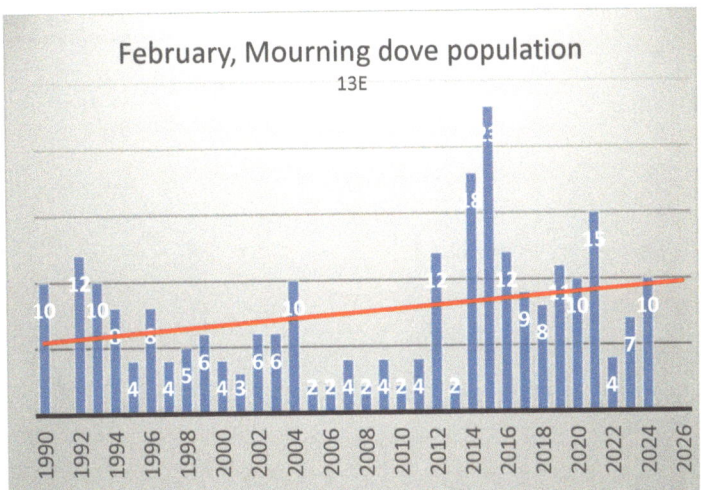

February mourning dove population

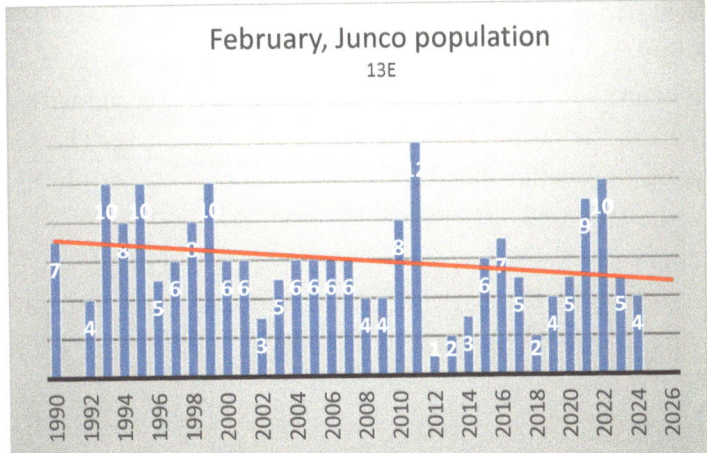

February junco population

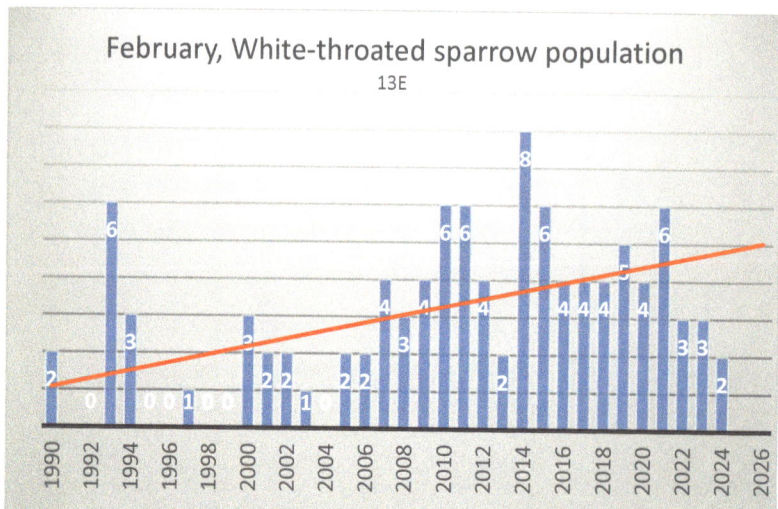
February white-throated sparrow population

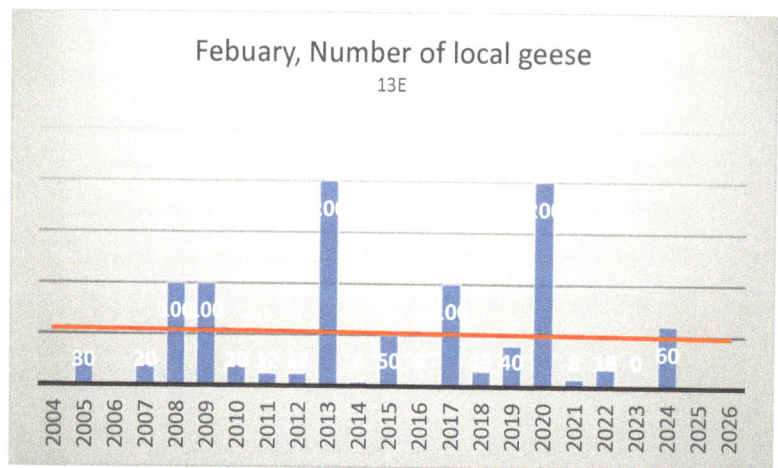
February Canada goose population

In 2010 and 2014 we had deep snows. The gang population increased. Then in 2015 we had high rain for the month and again the gang population increased to a high of 65. The most significant was the mourning doves. Not all the populations increased with a deep snow or heavy rain. Overall, for the period the number of total species is increasing. The number of gang species is staying at 14; the gang population has significantly increased, mainly due to mourning dove, cardinal, junco, and white-throated sparrow.

With a warmer January and February, snowdrop, crocus, and maple are blooming two to three weeks earlier than previous years. Rhubarb and skunk cabbage crowns are breaking through weeks earlier. Animals are coming out of semi-hibernation early.

So, ends February.

This is not the end of the cycle. The cycle shall never end. Each cycle is unto its own.

References

Laboratory of Ornithology, Cornell University, Ithaca, NY

PPL monthly average temperature readings

The Morning Call newspaper, Allentown, PA, daily temperature and precipitation publications

Birds and Blooms magazine, Milwaukee, WI

Birds of North America, Golden Press, New York, NY, 1966

North American Wildlife, Reader's Digest, Pleasantville, NY, 1982

Roy Hay and Patrick M. Syne, *Flowers and Plants for Home Gardens*, Crown Publishers, New York, NY, 1975

William Niering and Nancy Olmstead, National Audubon Society *Field Guide to Wildflowers*, Alfred A. Knopf, Inc., New York, NY, 1979

Lorus and Margery Milne, National Audubon Society *Field Guide to North American Insects and Spiders*, Alfred A. Knopf, Inc., New York, NY, 1979